[illegible] (Uncle) [illegible]

David [illegible]

[illegible] [illegible] [illegible]

Expanders for Oil and Gas Operations

About the Authors

Murari P. Singh, PhD, is the president of Safe Technical Solutions, Inc. (SAFETSOL). He has been involved in the design, development, and analysis of industrial turbomachinery for more than 30 years with Turbodyne Corporation, Dresser Industries, Dresser Rand Company, GE CONMEC, and most recently, GE Oil & Gas as consulting engineer. Dr. Singh has extensive knowledge and experience with fatigue and fracture mechanics, stress and vibration structures, reliability, life analysis, and probabilistic analysis. His practical application experience features a variety of rotating equipment, including CAES, warm gas and FCC expanders, steam turbines, and centrifugal compressors. He developed the widely used SAFE diagram for reliable evaluation of turbine blades. Most recently, Dr. Singh has been involved in developing and teaching applications of lifting strategy for many mechanical components. This includes subjects dealing with HCF, LCF, creep, and fracture mechanics. He has authored many technical papers, is coauthor of *Steam Turbines: Design, Application, and Rerating* and co-author of *Blade Design and Analysis.*

Michael J. Drosjack joined Shell Oil in 1975 as a member of the central Machinery Engineer Department in Houston, Texas. He retired from Shell in 2009 and since then has participated as a private machinery consultant in his company, Drosjack Consulting, LLC.

Through a variety of assignments over his Shell career, he was responsible for providing technical support for rotating and reciprocating machinery to Shell operations both Upstream and Downstream primarily in the United States but with international responsibilities. He was chair of the Shell US Machinery Technical Network for many years. He had assignments on projects involving specification, evaluation, installation, and start-up of machinery along with extensive field troubleshooting and failure analysis of machinery. An area of specialization was rotor dynamics. He was the FCC Hot Gas Expander subject matter expert for the Americas. Toward the end of his career with Shell, he spent considerable time on technology development projects, including evaluation and design of novel machinery for special applications.

Dr. Drosjack received his BS in mechanical engineering from Carnegie Mellon University. He received his MS and PhD degrees from The Ohio State University. Dr. Drosjack is a member of ASME and the Vibration Institute and is a registered Professional Engineer in the state of Texas. He has been a member of the Texas A&M Turbomachinery Symposium Advisory Committee since 1986 and currently holds emeritus status on the TAC. He was a cofounder of the Machinery Subcommittee of the Ethylene Producers Committee and vice chairman of the API 684 Task Force on Rotor Dynamics, having served on a number of task forces of the Subcommittee of Mechanical Equipment of API. He has authored numerous papers on turbomachinery issues.

David H. Linden is the president of D. H. Linden Associates, Inc. He has been involved in the design, development, manufacture, installation, and operation of turbomachinery for more than 40 years. Mr. Linden has worked for a number of original equipment manufacturer's including Conmec Inc., Dresser Rand Co., General Electric Co., Ingersoll Rand Co., and Westinghouse Electric Corporation. Mr. Linden is recognized as a renowned expert in the area of Hot Gas Expanders and industrial axial flow compressors. He specializes in the areas of turbomachinery design and reliability. He has authored or coauthored 14 technical papers for various technical forums including the ASME and Turbomachinery Symposiums. Mr. Linden is a member of ASME, ASTM, NACE, and is a contributor to the API 687 Rotor Repair subcommittee. Mr. Linden has a BSME from Rutgers University—College of Engineering.

Expanders for Oil and Gas Operations

Murari Singh

Michael J. Drosjack

David H. Linden

McGraw Hill Education

New York Chicago San Francisco
Athens London Madrid
Mexico City Milan New Delhi
Singapore Sydney Toronto

Cataloging-in-Publication Data is on file with the Library of Congress.

Expanders for Oil and Gas Operations

1 2 3 4 5 6 7 8 9 0 QVS/QVS 19 18 17 16 15 14

ISBN 978-0-07-179992-8
MHID 0-07-179992-3

The pages within this book were printed on acid-free paper.

Sponsoring Editor
Amanda Quinn

Acquisition Coordinator
Amy Stonebraker

Editorial Supervisor
Donna M. Martone

Project Manager
Shruti Chopra

Copy Editor
Alyson Platt

Proofreader
MPS Limited

Indexer
Sandi Schroeder

Production Supervisor
Pamela A. Pelton

Composition
MPS Limited

Art Director, Cover
Jeff Weeks

Contents

Preface

The widespread application of centrifugal gas expanders in Oil & Gas operations is only about 50 or 60 years old. The fundamental purpose of the expander is straightforward: Take a higher pressure gas stream, reduce this pressure for further processing, and recover energy through this expansion. There are numerous potential applications in oil and gas processes. The pressure reduction can be taken across a valve or orifice chamber with "throwing away" of the power. These applications become more exciting in the wave or energy efficiency–influencing industry.

Most of these services put significant strain on the machinery. And, particularly early in their development, these machines have experienced dramatic and repeated failures. However, as the design capabilities of the industry have grown, greatly aided by the more recent growth of computing availability and capabilities, performance has improved greatly.

This book has been written to discuss the application potential and the design requirements for gas expanders. The text is not written as an entry-level volume to develop awareness of the subject matter. Rather, the authors strive to provide sound, solid technical background for the more dedicated user, designer, and/or problem solver.

One of the critical factors influencing the design and operation of an expander is the process into which the machine is placed. These processes are discussed in a sufficient level of complexity that we believe is necessary to understand the demands being placed on the machine. Also included is enough information to form an idea of the value achieved by the applications.

The design is discussed in some detail separately for each of the classes of expander. Individual components are discussed as well as the tools needed to prepare and evaluate the final design. There is considerable technical detail including basics of advanced analytical tools and procedures.

Machine design technology has advanced greatly over the last 20 to 30 years, paralleling the growth in knowledge and advancements in computing capabilities. Use of finite element analysis for stress analysis is now common and relatively easy with personal computers. Computerized fluid dynamics is also a very standard tool for the evaluation of aerodynamics. Computerized machining capabilities make the production of complex shapes rather standard. Complex welding and forming technologies are applied. Where appropriate, application of these procedures and tools is discussed relative to the expander's design.

The principal purpose of this book is to open the door on these advanced tools and their application to the dedicated user of this machinery. Equipment demands are becoming more onerous. There is no indication from the industry that these demands will lessen. It is our hope that this text, along with its references, will permit the dedicated user to advance the technology and will enhance productivity and safety in the industry.

The authors would like to personally thank all of those individuals and companies whose contributions have been used in the preparation of this book. Coinciding with the analytical tools provided by the computer age has been an explosion in the availability of information at your fingertips. The fundamental sources of this readily accessible information can't be forgotten, as that is built on the backs of many other engineers and scientists who have made contributions to the technology over many years.

Acknowledgments

We would like to acknowledge those who preceded us in this technology and developed it to a point that would permit us to make careers following them.

Particular thanks go to:

Atlas Copco:
Joseph Lillard

Dresser-Rand:
Clint Jungerman

Elliott Group:
Art Titus
George Adda

GE Oil & Gas:
Leonardo Baldasarre
Duccio Manetti

MAN Turbo AG:
Kevin Kisor

CHAPTER 1

Introduction

Many oil and gas processes contain operating steps in which the gas is at temperatures and/or pressures higher than are needed in subsequent steps. Generally, the temperatures/pressures should be reduced to what is required. Thermodynamically, this reduction in pressure results in a Joule-Thomson effect. This is a thermodynamic effect that drops the gas temperature when it is throttled without the opportunity to reject the heat. In many processes, this is a desired result, particularly if there is a refrigeration step to the process or if the gas is saturated. The cooling will cause the heavier ends to drop.

At one time, this pressure-reduction phenomenon was accomplished across a valve, which was designed to get the desired cooling effect called the Joule-Thomson valve (J-T for short). However, there was a need to accomplish these phenomena in another manner. For one thing, the pressure drop, in effect, wasted some energy, and as operating units became larger, this loss became significant. When a gas has its pressure reduced, the gas velocity is increased. The energy that was lost to friction also introduced some damage into the J-T valves.

An alternative was desired and the solution that was developed was the turboexpander or expander turbine. Using this method, the gas pressure was dropped across what looked like a centrifugal compressor run backward. These devices were often coupled with a compressor, which injected the cooled gas into the refrigeration process. This method was widely adopted by natural gas production facilities, ethylene plants, and air separation plants. Gas liquefaction plants joined the parade along the way.

Around the same time, refineries began operating larger and higher-pressure catalytic crackers. With these devices, the pressure of the flue gas had to drop prior to entering the CO_2 boiler for incineration. There were several attempts to use an expander turbine to recover that energy and accomplish that pressure reduction without the need for a valve. However, that created another issue. That stream was quite hot and carried catalyst particles, which could cause severe erosion. Thus, catalytic cracker unit (CCU) hot gas expanders were introduced and, despite some initial problems, were improved and eventually became more widespread in the industry.

This book discusses gas expanders in the oil and gas industry. Separate chapters discuss each of the primary expander design types:

- CCU hot gas expanders
- Expanders for chemical applications (particularly nitric acid)
- Turboexpanders/cryogenic turboexpanders

The overall design and operating characteristics of the various expanders are distinct enough that this separation is required to permit a rational description and discussion of each.

For each class of expander, the following basic issues are addressed:

- Background and historical development
- Mechanical design features in detail
- Design and operating requirements and the manner in which they are met
- The processes in which they operate
- Problems that may have to be addressed and designed for

In presenting the material a considerable number of photos, sketches, and charts are used to illustrate the components and features of the machines, including their internal components and the manner in which they are assembled. In addition to offering detailed discussion of the individual machine types, this book presents more comprehensive, basic technologies that apply to these machines. These are presented in technical detail and depth, along with some examples of application that will be more limited in number and scope.

The basic technologies to be discussed include:

- An introduction to rotor dynamics with identification of features specific to these machines
- A discussion of vibration theory as related to the natural frequencies of the blades, vanes, and disks of these machines and the manner in which they should be evaluated
- Material strength and life estimation that addresses the stresses and strains and the manner in which they can or should be evaluated relative to the machines and their operation
- The introduction of probabilistic analysis into the evaluations and its use in life estimation

The desire is that this information should be utilized effectively by the reader. The application of this class of machine is growing in the numbers being employed in the industry, as well as in size, speed, and power capabilities. These factors will challenge the existing design and require design enhancements to meet future requirements.

The authors hope that the material presented herein will help the reader maintain and improve existing expanders and facilitate the user to challenge the designer and manufacturer effectively in the development of new machines.

CHAPTER 2

CCU Hot Gas Expanders

2.1 Historical Background

2.1.1 Introduction

Catalytic Cracker units (CCUs) are a part of most refineries. These process units take less valuable feeds and make more valuable products from them. They are one of the highest value-added processes in a refinery. The refining process and, in particular, catalytic cracking, will be discussed in much more detail in the next section of the book.

At the back end of the CCU is a CO boiler that consumes the CO produced in the catalyst regeneration process. Normally, it operates at atmospheric pressure. However, the gas stream coming to it is at 20+ pounds per square inch (psig) (depending on the specifics of the CCU design). This pressure must be reduced before entering the CO boiler. Prior to the application of Hot Gas Expanders, this pressure reduction was accomplished by using either a pressure-reducing valve or an orifice manifold.

Reducing the pressure in either manner "throws away" energy from the gas stream. In large units this can amount to considerable amounts of energy, potentially in excess of 25,000 recoverable horsepower (hp). In addition to the energy loss, the reduction in pressure produces very high gas velocities that can lead both to a very high level of noise as well as erosion in the valves or orifices. This is aggravated further as the gas stream also contains catalyst particles which are quite hard and abrasive.

The subject of how best to let down this energy was discussed for many years. The scales tipped in the direction of taking positive action as the Catalytic Crackers became larger and operated at higher pressures, thereby increasing the potential power recovery. The mitigating factor in all this was the risk associated with the potential reliability of an additional machine in the unit. Even in the largest units, the power recovery can't compete with the production losses associated with an unscheduled outage.

2.1.2 Birth of the CCU Hot Gas Expander

In an attempt to recover some of the energy that was thrown away, Elliott Company produced a CCU expander in 1950 that was installed in the Marathon refinery in Robinson, Illinois. The design was based on a turbocharger that essentially ran backward, that is, expanding the gas rather than compressing it. This machine operated for approximately 75 hr before erosion caused it to fail.

One of the roadblocks to CCU expander operation was the amount of catalyst in the gas stream exiting the regenerator. Figure 2.1 shows a schematic of a modern CCU. At the time the industry started to attempt to install expanders, the units did not have the Third Stage Separator (TSS), and the particle-laden flue gas exiting the process carried over into the expander. Cat cracker catalyst was fed into the riser, entering the reactor. This catalyst was then regenerated in a separate regenerator vessel. Despite further particle separation in the regenerator, a significant amount of catalyst was still carried over into the downstream equipment. This catalyst is extremely hard and very erosive, particularly at the high velocities that were encountered in an expander. Catalyst separation was performed in cyclones located in

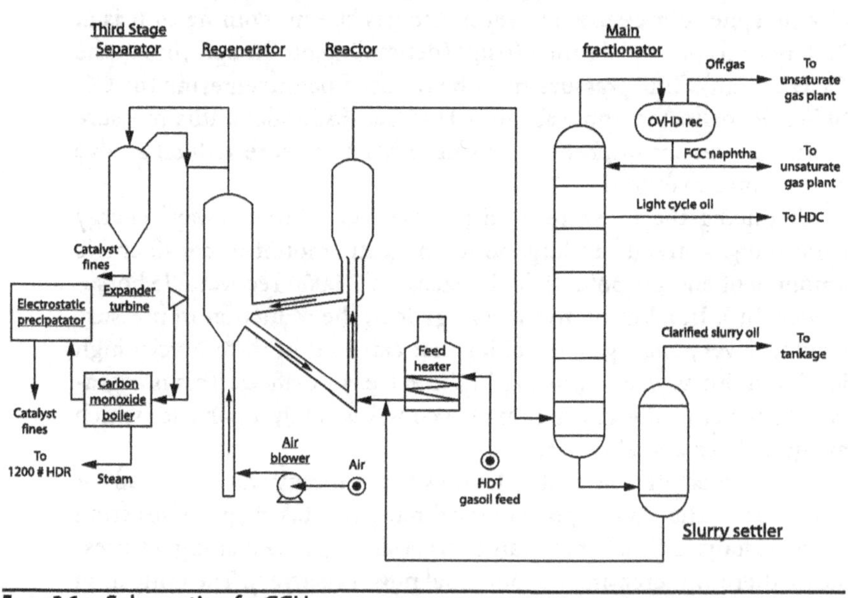

FIGURE 2.1 Schematic of a CCU.

the reactor and regenerator. Their performance permitted a significant amount of catalyst particles to pass through to the location in which the Hot Gas Expander would be installed.

A second issue in installing the initial expanders was that the materials being used for the expander blading and rotor were not very erosion resistant. These materials were those more commonly available and used in high-temperature turbines. In addition, there was limited design experience in dirty gas technology and in knowing how to accommodate the erosive nature of the gas stream. Also, in the 1950s, the computation fluid dynamics (CFD) and finite element analysis (FEA) tools were in their infancies and had not yet become design tools, or were not very widely available or used in commercial products or applications. As a result, the expanders had a very short run life.

Additional work proceeded and a breakthrough occurred in 1954 when the separation specialists and CCU process engineers in Shell Development Company designed a Third Stage Separator (TSS). This was an additional separation device upstream of the CO boiler, which was designed to remove a considerable amount of the catalyst that was being carried over. Figure 2.2 shows a drawing of this device. Considerable more discussion of separation and the Shell TSS is included in Chapter 3. A prototype using this device upstream of the expander was developed by Elliot Company and installed in the Shell Canada Montreal Refinery in 1954. The design of this machine was similar to a gas turbine power turbine stage (Geary, 1967).

After the prototype was operated with some success, a "full-size" expander was installed in the Shell Canada Oakville, Ontario, refinery in 1963. This machine produced 3,350 hp (2,500 kW) and operated at 7,800 rpm.

2.1.3 Early Expander History (Calandra, 1998)

These early designs and successes precipitated the early age of CCU expanders (1963–1974). A number of new designs were developed and installed in refineries throughout the world. The Ingersoll-Rand Company also enter the industry of producing CCU expanders. A partial list of the installations in this time frame includes the following:

- 1963 Ingersoll-Rand Shell installations
    - Deer Park, Texas
    - Norco, Louisiana
- 1966 Ingersoll-Rand installation at Shell, Martinez, California
- 1967 Elliott Company 22,500-hp TH installation at Exxon

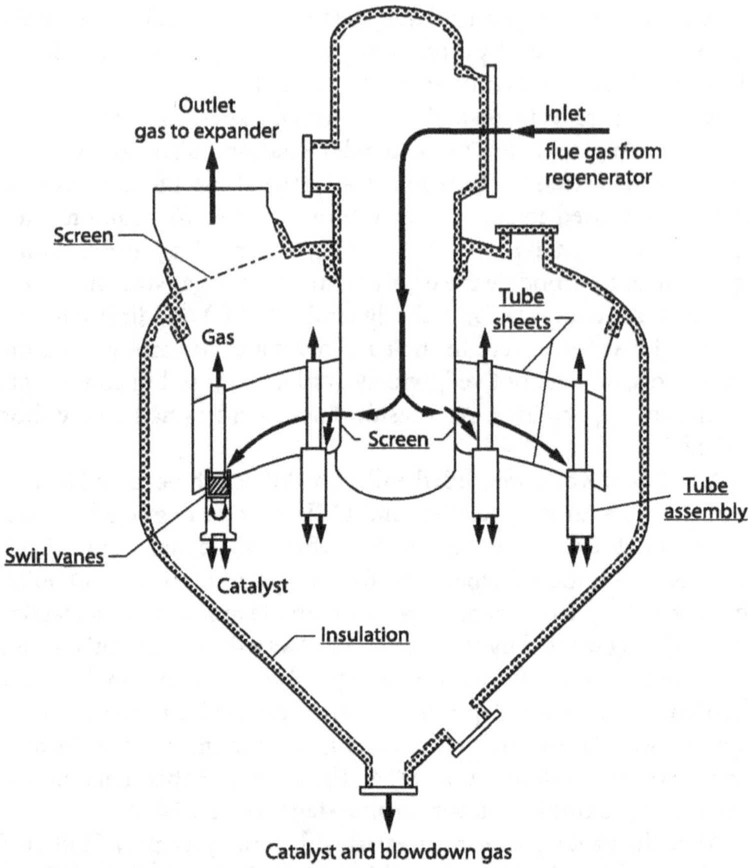

Figure 2.2 Shell Third Stage Separator.

- 1970 Ingersoll-Rand installation at Shell, Berre, France
- 1971 Elliott 22,500-hp installation at Gulf Alliance Refinery
- 1971–1979 Elliott focused on areas other than expanders
- 1971 Ingersoll-Rand installation at Gulf, Edmonton, Alberta, Canada
- 1973 Ingersoll-Rand installation at Sun, Toledo, Ohio
- 1974 Ingersoll-Rand installation at Texaco, El Dorado, Kansas

As the reader can see in this list, the floodgates opened on the application of the CCU Hot Gas Expanders. That is not to say that the operation of all of these machines was fully successful. During this time, there was a number of operational issues to deal with that included both the expander and the cat cracking units. As experience was being gained, design enhancements, particularly in the area of materials, were made. At the same time, remember that operation of CCUs was not yet well developed. In addition, control technology was limited. This was still the age of pneumatic controls and analog control systems. Computerized control systems were very limited in population and capabilities. Thus, numerous process upsets and unscheduled outages occurred. Outages put considerable strain on the expanders, and uninterrupted run lengths were not very long for either the CCU or the Hot Gas Expander.

2.1.4 Second Generation Expanders (Calandra, 1998)

A second generation of expanders was released from 1976 to 1985. Over this period, the installations were considered more standard and the machines were less experimental. During these years, as many as nine expanders per year were produced. These machines incorporated many design enhancements, including:

- Use of new metallurgy
  - 300 Series stainless-steel casings
  - Inconel X-750 and Waspaloy on blades
  - Waspaloy rotor discs
- Enhancements in design from operating experiences
  - Cooling
  - Seal designs
  - Bearing improvements

These enhancements permitted the capacities of the expanders to be expanded with:

- Higher temperatures from 1,100°F to 1,350°F (593°C to 732°C)
- Higher pressure ratios from 1.7 to 2.8
- Higher and lower flow capacities (larger and smaller machines)

CCUs were reviewed for potential power recovery. The payout from the energy recovered was compared to the cost of the installation of the expander and its peripherals. Many or even most new CCUs that passed the financial hurdle had expanders installed. A number of existing CC units were retrofitted with an expander. Almost all of these expanders were used to drive the CCU main air blower.

2.1.5 Expanders "Go to Sleep" (Calandra, 1998)

This burst of activity was followed by a lull of several years in which new expander installations slowed down. Lower energy-savings margins hurt installation returns. CCU run lengths were enhanced through improvements in catalysts, controls, and operating experience. These issues evolved into a series of operational reliability issues, including:

- Heavier erosion
- Heavier corrosion
- Deposition of catalyst causing blade-tip wear, tip cracking, and excessive rotor vibrations
- Flow-path plugging, particularly in Second Stage expanders
- Casing distress and distortion
- Flange leakage and stud failures

The combination of uncertainty over reliability based on production outages that had been experienced and the reduction in energy returns dramatically slowed the installation of new expanders.

A historical event in the evolution of expanders occurred in 1987. Dresser Clark and Ingersoll-Rand merged into a new company, forming Dresser-Rand. The expander operations were moved to the Olean, New York, manufacturing facilities. On the heels of this, a number of Ingersoll-Rand employees chose to break away and form a new company named CONMEC, Inc. They included several members of the Ingersoll-Rand expander design team. CONMEC initially pursued work with existing expanders. In 1988, CONMEC performed a major rebuild and performance rerate on the Valero, Corpus Christi expander. In 1995, CONMEC manufactured its first complete expander for installation in Dyfed, Wales. Since that time more than a dozen new CONMEC expanders have been manufactured. CONMEC was eventually purchased by GE Oil & Gas, which offers a complete line of CCU expanders.

2.1.6 Reliability Era for Expanders

From 1991 to the present day, a number of new expanders have been installed with a focus on operational reliability. By continuing to make operational enhancements, operators try to run longer between maintenance times with total reliability, that is, with no unscheduled downtime caused by the expander. This era was preceded by a rise in rotor blade failures that had generated concern:

- 1961–1990 <12 blade failures
- 1990–1998 20 blade failures
- 1996–1998 10 blade failures

Factors that had influenced the failures were an increase in operating temperatures, increased operating pressures, and aging of the installed base of equipment. Operating temperatures were increased from 1,100°F (593°C) to 1,355°F (735°C) as a result of process design upgrades and catalyst enhancements, availability of better metallurgies on all the CCU "hot end," and the fact that many of the machines were approaching or passing the nominal design life of 100,000 hr. The design margins of the machines were eroding as the initial designs were being pushed harder. With the coupling of all of these effects, the CCU expanders were experiencing more reliability problems.

However, there was an offset to these operational demands and age. Technical capabilities were improved with the advent of computer capabilities and design tools. Enhanced condition-monitoring capabilities provided significantly improved operating data and control information. These advances in the operation of the CCU drove the designers to make a concerted effort to design and redesign the machines. Major improvements were also made in the area of expander metallurgy, including the use of more advanced materials, and erosion- and corrosion-resistant coatings. Along with this, the industry experienced tremendous advancements in welding, forging, and other manufacturing processes.

The net result of these advances has permitted expander designers to improve expander design and redesign for the following:

- Increases in operating times
- Reduction in erosion
- Reduction in corrosion
- More forgiveness for upset conditions (off-design operation)
- More tolerance to external loads (e.g., piping loads)
- More operational structural margin

Although these efforts have been successful in enhancing design life and reliability, there is still a significant element of risk of downtime when operating an expander driving the CCU main air blower. To provide the operating reliability desired, many or most new installations use the expander to drive a generator with the potential of decoupling from the CCU operation. This is partly justified by the increase in the value of energy conservation which aids in project payout. It is also justified by the fact that an expander driving a generator may be taken offline without requiring the CCU to be shut down. This greatly reduces or eliminates the threat of production losses from outages that are mostly the costly result of limited reliability for the CCU expander.

Currently, CCU expanders are being applied virtually to all new CCU installations. A majority of these installations are installing expander generator sets. The expander population continues to grow at a rate of several new installations a year and at this writing is in excess of 130 units worldwide. The three major manufacturers of the CCU expander are Dresser Rand, Elliott Company, and General Electric Company.

2.2 Process Unit Descriptions and Applications

2.2.1 Introduction to Refining

The Hot Gas Expander is employed in the catalytic cracking process unit of most refineries that have motor gasoline (Mogas) as a major end product. They are employed as a means to reduce the gas pressure prior to the next step in processing. As a side benefit, they permit the recovery of energy that would otherwise be thrown away and lost. In some cases, this recovery will be in excess of the equivalent to 20,000 hp.

Many or even most refining processes in refineries present opportunities for energy recovery, either in pressure reduction, heat recovery, or both. Hot Gas Expanders and turboexpanders (to be discussed in Chapter 4) have considerable application in refinery processes. To help put the application in perspective, a brief and cursory review of the refining process is in order to help the reader understand this application. It may be best to begin with a look at a typical overview flowchart of a generic refinery. Figure 2.3 shows a simplified refinery flow scheme.

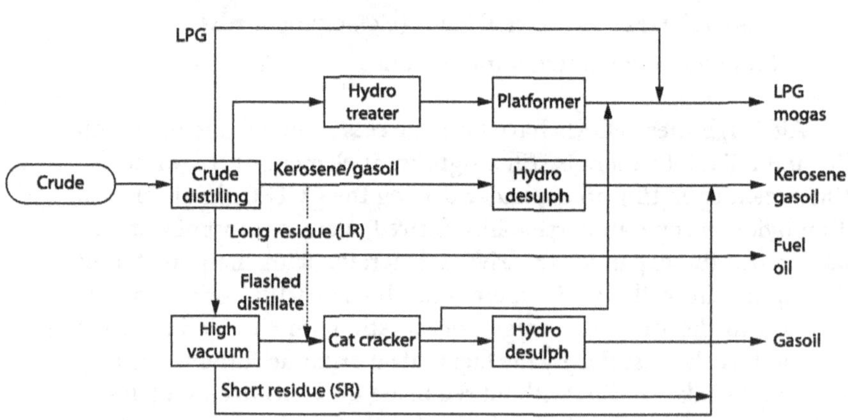

FIGURE 2.3 Simplified refinery flow scheme.

Crude oil enters the refinery. Crude oil is a generic name for a family of mixtures, which can contain a range of hydrocarbons, from gases in solution to heavy ends, as well as contaminants such as sulphur, water, heavy metals, and more. And, the crude viscosity (indicated as API gravity) may vary from components that will vaporize in the atmosphere to crude components that are near-solid with a temperature reduction. These variations are functions of the sources of the crude. And multiple crudes may be blended before entering the refinery. Some refineries may be configured for only light, sweet crude, whereas others may be configured for heavy, sour crude. Sweetness refers to the sulphur content; sour has higher sulphur content.

The crude composition is defined by both the availability of crude supply and the desired "mix" of products the particular refinery is charged with producing. This can change over time depending on margin changes in the products, availability and price of crude, and the operating capabilities of the individual refinery or refinery group. The study of supply and logistics for a refinery can be both very interesting and very frustrating.

The particular crude being used enters the refinery. After dewatering and first-pass desulphurization, it enters the refinery on the left side of Figure 2.3 and enters the Crude Distilling Unit, which performs the first separation. The crude is heated in the crude heaters, a process often termed "boiling the oil," and is separated into the various component fractions through a standard distillation tower. The lightest products leave at the top of the tower and the heaviest products are removed at lower levels of the tower. A crude distillation tower can be thought of as a very fancy moonshine still.

Liquefied petroleum gas (LPG) goes to product storage and delivery. Naphtha goes to the Hydrotreater and Platformer to produce gasoline and gasoline elements. Kerosene and gasoil go to the Hydro Desulphurfurization Unit to produce kerosene, gasoil, and long residue for fuel oil. The long residue can go straight to fuel oil or to the vacuum unit where it is split between flashed distillate and short residue. The flashed distillate goes to the Catalytic Cracker where it is split into gas oil and gasoline. Depending on feed composition, the CCU product may go through a Hydro Desulphurization unit before ending up as gasoil.

Lighter products such as LPG, gasoline, and naphtha are more valuable, and refinery process specialists will attempt to optimize the production of these products.

In principle, some of the refinery units are rather simple processes. For example, a Distilling Unit makes use of heat to boil the feed and tall towers with separation devices to separate the feed components. Lighter components go higher up the tower and are taken off there. However, the Catalytic Cracker and the hydro units are more complicated and make use of heat, catalyst, and separation to change the basic composition of the feed, making higher value products.

The Catalytic Cracker (also known as cat cracker, CCU, FCCU, RCCU, etc.) is one of the most expensive units to build and operate and has the highest value added of any major process unit in a refinery. The process lends itself to energy recovery but also adds severe financial penalties for any loss of reliability and availability.

One of the "takeaways" that should come from this overview of a refinery process is that there is considerable interaction between the individual processing units. An upset or shutdown on one will have an impact on other processing units both upstream and downstream of it. The "cascade" of impacts can be very dramatic. This is one of the reasons there is a considerable interest in maintaining equipment reliability in a refinery. A failure in any area will have a trickle-down effect on most or all of the other process areas.

2.2.2 Catalytic Cracker Unit

Most refineries have a cat cracker as a processing unit with gasoline as a primary product. Occasionally, a refinery may be too small to have a cat cracker and will ship the cat feed to another refinery. Currently, with the demand in diesel fuels, there may be refineries that take heavier feed and focus on heavier output products that do not require a CCU.

However, in a refinery that does have a CCU, it is most likely the most important unit in the refinery from a financial basis. CCUs may be one of the most capital-intensive process units to build and can generally return the best operating margins as long as the gasoline demand is reasonable. In most refineries and refining companies, news of a CCU outage travels fast.

Figure 2.4 shows the general flow diagram of a Catalytic Cracker. Hot gasoil enters the feed heater and is heated to around 525°C and about 2 bars. The feed then enters the reactor riser. It is mixed with catalyst transported by air. Depending on the age of the unit and the specifics of the process design, the feed is vaporized and atomized by steam, and a catalytic reaction occurs in the riser or in the reactor. Reacted feed and catalyst enter the reactor where catalyst is separated (typically by two-stage cyclone separators). Vaporized product and steam are separated and sent to the fractionator for further processing and separation into the finished product.

The catalyst can absorb some hydrocarbon on the surface during the operation. The heat in the unit will cause the hydrocarbon to "coke up" on the surface of the catalyst (meaning it forms a hard or thick carbon heavy coating). This coating would significantly reduce

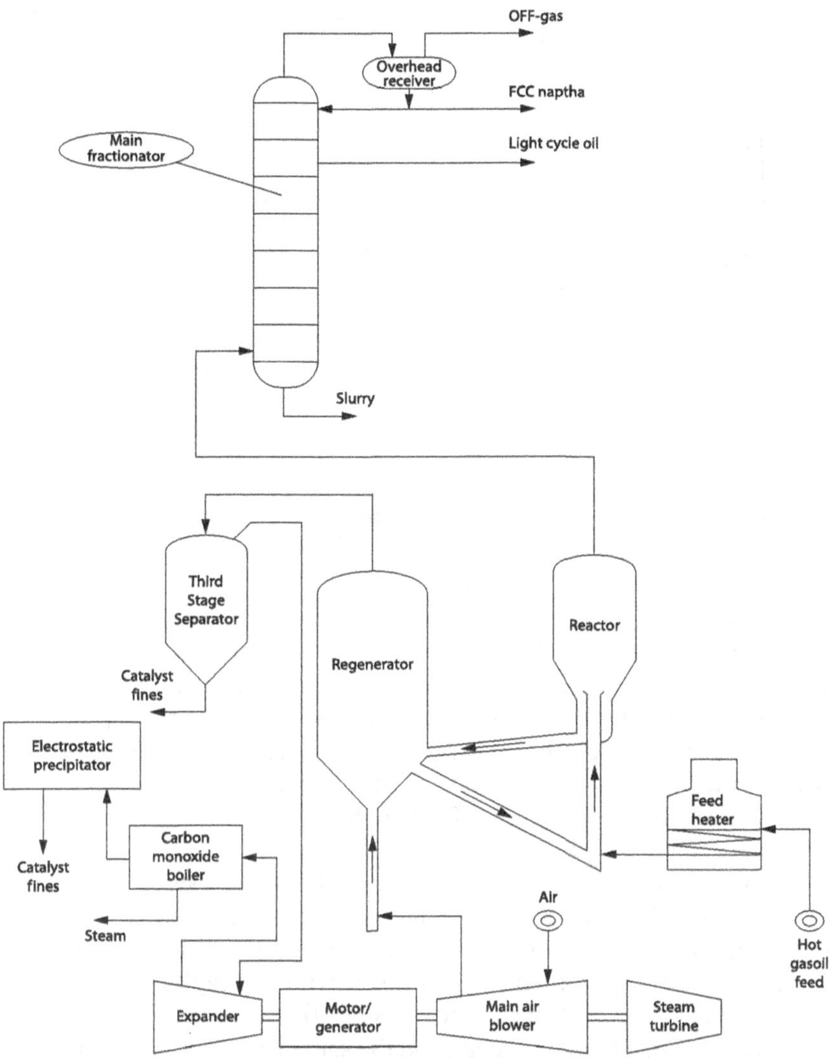

FIGURE 2.4 General flow diagram of a Catalytic Cracking process unit (Carbonetto et al, 2003).

the reactivity of the catalyst in additional contacts with feed. The reactivity is the ability of the catalyst to promote the catalytic reaction of the feed. A reduction in reactivity reduces the yield of the unit and, subsequentially, the money the unit produces. More catalyst could be introduced, but it is quite expensive. Additional detail on the catalyst is provided in Chapter 3.

Following separation of the catalyst from the liquid and gas streams, it enters the regenerator. The purpose of the regenerator is to

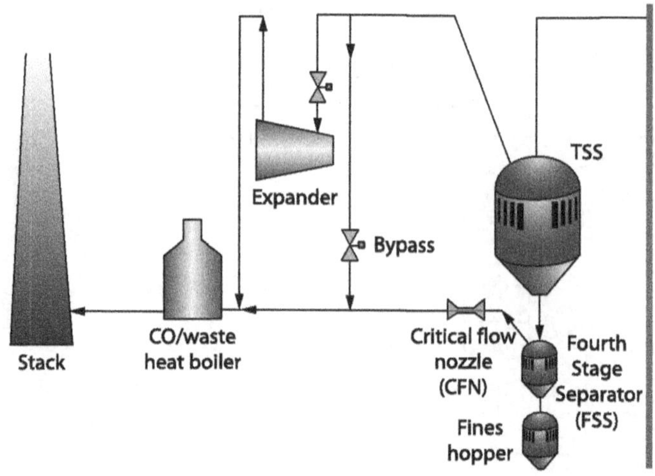

Figure 2.5 Sketch of the Third Stage Separator and flue-gas system.

remove the coke and any residual liquid hydrocarbon from the surface of the catalyst, and, thus, to "revive" the catalyst's reactivity for more cycles in the reactor. This coke and liquid is burned off in the regenerator and, with the aid of combustion, air is fed into the regenerator by the air blower. The catalyst is separated from the residual gas through the use of cyclone separators in the regenerator. Figure 2.5 shows a more detailed diagram of the configuration from the regenerator, through the expander, and to the CO boiler.

Separated catalyst is fed back into the catalyst feed system. The CO-rich gas that is produced in the regenerator is ultimately burned in the CO boiler. The gas that exits the regenerator will have a discharge pressure higher than the CO boiler which is typically operated at or very close to atmospheric pressure. The pressure of the flue gas leaving the regenerator depends on the cat cracker design but has been as high as 50+ psi in some units. It is the excess in energy from this exhaust stream that a Hot Gas Expander is configured to recover.

As noted previously, without a Hot Gas Expander, the reduction in pressure could be accomplished by either a specially designed control valve or orifice chamber. Even CCUs with Hot Gas Expanders will have a bypass valve to use in the event the Hot Gas Expander has to be taken out of operation.

The Hot Gas Expander's problem with the regenerator exhaust stream is that the separation system in the regenerator will permit a significant amount of catalyst and catalyst fines to pass downstream. To operate a Hot Gas Expander in the cat cracker and not have the catalyst in the regenerator exhaust erode it away, a TSS is installed.

A Fourth Stage Separator (FSS) can be included in the design to help recover catalyst fines.

There are multiple designs at present for a Third Stage Separator. We give a more detailed discussion of these in Chapter 3, in which we compare the separation technology employed. In essence, the TSS is configured to remove most of the catalyst particles above a given cutpoint (catalyst particle size). Depending on the separator, this cutpoint will be about 5 microns.

That does not mean, however, that the Hot Gas Expander will still not have issues with catalyst carryover. As discussed in Chapter 3 on Separation, most of the current design TSS will have a separation efficiency of 97% or better for 5 micron and larger particles. This corresponds to approximately 50 lbs/hr of 5 micron or greater catalyst in a 100,000 BPD CCU.

In addition, so-called super fines (<5 micron) cannot be separated by the TSS and have to be addressed in another manner. As shown in Figure 2.5 and 2.6, regenerator exhaust enters the TSS where the separation devices remove the catalyst and drop it into the fines hopper for reuse. In some units, a Fourth Stage Separator is also used to aid in catalyst recovery.

The Hot Gas Expander is located downstream of the TSS and drops the exhaust pressure to near atmospheric to match the operation of the CO boiler. If a Hot Gas Expander is not part of a CCU design, the

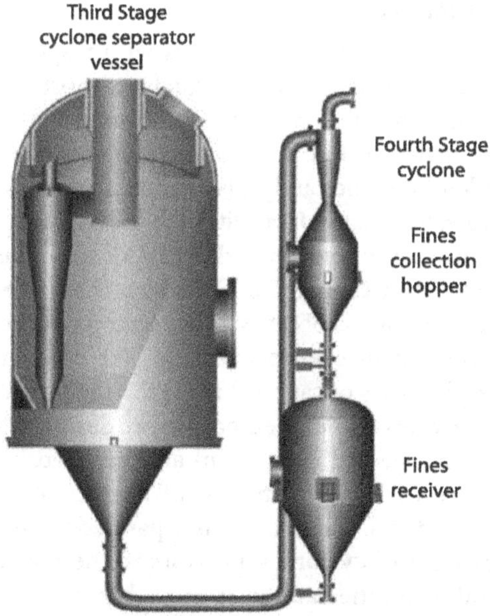

FIGURE 2.6 Fourth Stage Separation equipment.

pressure drop will be taken in a pressure drop device. A bypass valve is always included around the expander in case it must be taken out of service for one of the "cleaning" techniques that can be applied. It can also be used in the event of a catastrophic failure of the Hot Gas Expander.

Another feature shown in Figure 2.6 is the critical flow nozzle. This restriction is used to maintain the appropriate back pressure on the TSS. If the critical flow nozzle (CFN) is not in operating shape, the pressure balance in the separation system will be upset and may cause catalyst bypassing. Many of the critical flow nozzles incorporate a steam bleed to keep them clean and to maintain the operating condition.

Hot Gas Expanders have been installed with power recovery capability of upward of 50,000 hp. They can be directly coupled to the air blower as was the case in most of the early installations. However, the trend at this time is to use the Hot Gas Expander to drive an electric generator for energy recovery to decouple the cat cracker from the Hot Gas Expander. Extended run lengths on cat crackers are being executed. In those cases, catalyst life is the driving force on the outage intervals. A current "norm" is about five years; if the drive is to extend these runs through catalyst improvements, longer runs will challenge the Hot Gas Expander capabilities. This is a significant reason to move to the use of Hot Gas Expanders as generator drives, and this trend is likely to continue.

For considerably more discussion of the design and operation of the separations system, refer to Chapter 3, Dirty Gas and Particle Separation.

2.2.3 Cat Cracker Catalyst

Catalyst is one of the very biggest factors with respect to yield (light product produced versus feed). The catalysts most used are zeolite catalysts composed of silica, alimina, and caustic with a matrix having some other materials besides zeolite. It has a binder material and filler, typically kaoline. The catalyst is formed in spheres with the general look of white, light sand. In addition to the basic catalyst material, there may be other additives including a CO promoter, which aids combustion of CO to CO_2 and $DeSO_x$, which changes SO_2 to H_2S.

There is more discussion of catalysts and their properties in Chapter 3.

The development of catalysts with higher reactivity and better properties over the years of cat cracking is the major reason that older cat crackers have rates that exceed nameplate design rates.

The catalyst is injected with steam and hot feed into the reactor riser. It promotes the transformation of heavier feeds into lighter, more valuable products. The catalyst is separated from the liquid and gas streams by a set of cyclone separators in the reactor and is sent to the regenerator. As the catalyst is moved through the unit, some part of it is broken up into smaller catalyst particles, which are considerably less reactive. To maintain rates, "fines" are removed in the separation systems and fresh catalyst is added. The cost of catalyst is the third highest operating cost for a cat cracker exceeded only by the cost of feed and fuel.

The presence of catalyst downstream of the regenerator was the cause of the challenge in early Hot Gas Expanders to extended run length. Both larger and smaller particles have deleterious effects on the Hot Gas Expander and will be discussed in more detail later in the book.

2.3 Expander Applications

2.3.1 Air-Blower Drivers: Power Recovery Train Arrangements

Fluid catalytic cracking (FCC) expanders were first applied in the late 1950s. These early expanders were used to drive the FCC main air blower. It was relatively easy to control these trains as the early expanders did not produce more power than that needed by the air compressor. To make up for any deficiency, an auxiliary driver was used in the power recovery train (PRT). Steam turbines were the auxiliary drive of choice as the standard steam turbine controls could be used to control the air-blower output and power recovery train speed. Because many of the initial expander installations had retrofitted the expander into the existing steam turbine–driven compressor train, the turbine and its controls were already in place. The expander power was used to supplement the steam turbine power. As expander power outputs increased, the steam turbine power could be reduced, thereby saving

FIGURE 2.7 Early expander retrofit to an existing air-blower train.

the steam that would have been required for the steam turbine to drive the compressor. Figure 2.7 shows an early expander retrofit to a steam turbine–driven centrifugal compressor. Figure 2.8 illustrates a flow schematic of the initial power recovery train.

In order to maximize efficiency, newer power recovery trains replaced the centrifugal compressor with an axial compressor as can

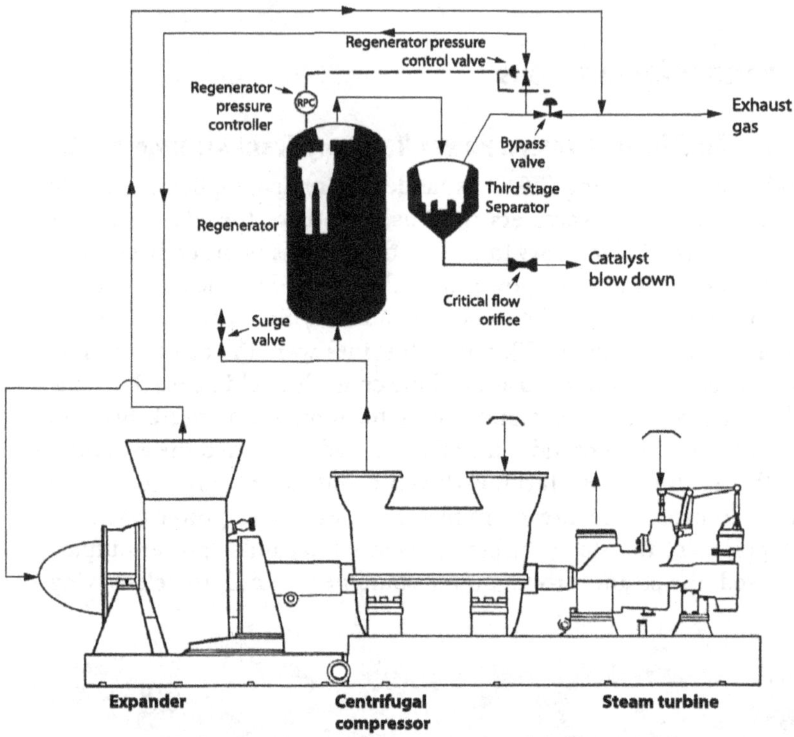

Figure 2.8 Expander Train Arrangement 1: Expander/centrifugal compressor/ steam turbine.

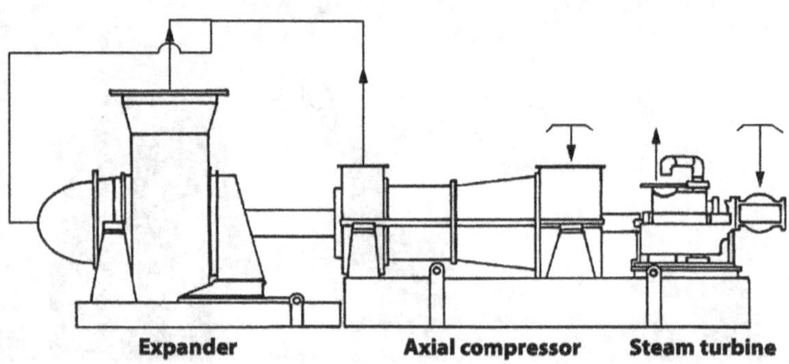

Figure 2.9 Expander Train Arrangement 2: Expander/axial compressor/ steam turbine.

be seen in Figure 2.9. In both compressor train arrangements, the expander power does not exceed the compressor power and control can be via the steam turbine speed controller. It is important to make sure the steam turbine is not unloaded to a point where the steam turbine governor can no longer control the train speed. The process flow is controlled by variable speed operation and in the axial compressor it can be variable speed, variable stator geometry, or a combination of both.

Power Recovery Train Arrangement 3

As the process operating conditions increased and the expander technology developed, the expander power output could exceed the main air-blower power requirements. Either this excess power was throttled back and wasted or it was recovered in a four-body power recovery train such as that shown in Figure 2.10. As shown in the figure, the power recovery train includes a steam turbine, compressor, motor/generator, and expander. This train arrangement could be applied to existing air-blower trains or was the preferred arrangement for new larger catalytic cracking units. This four-body train arrangement can be applied to both centrifugal and axial compressor applications.

An induction motor/generator is used in this train arrangement because it can either supplement power (if the expander does not produce enough power), it can generate power if the expander produces excess power, and it can be used as the train speed controller. Because the start-up capabilities of induction motors are limited from zero speed, a smaller steam turbine is used in the train to bring the train up to a speed initially where the motor could be put online without excessive electrical demand. This is usually in the range of 80% to 90% of the full operating speed of the train. With both the motor and steam turbine providing power, the process could be brought to full operation. It should be remembered that the expander is a power consumer until the process is brought up online. As the expander produces more and more power, the steam turbine and/or motor power requirements are reduced. At full power, the expander might produce more power than the compressor

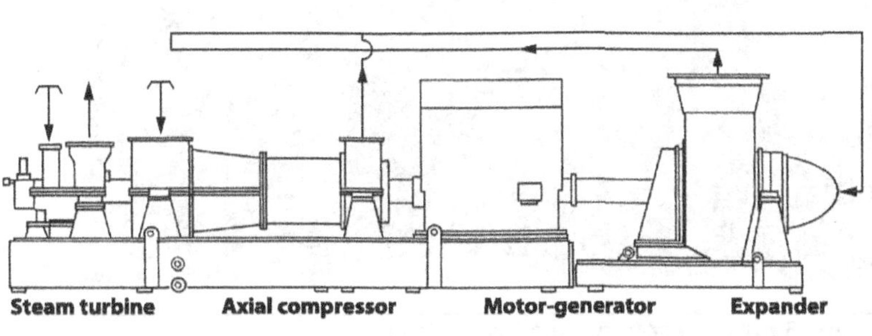

Figure 2.10 Expander Train Arrangement 3: Steam turbine/compressor/motor/expander.

demands, and the steam turbine input would be zero. The excess power would drive the train speed above the synchronous speed, and the motor would begin to export power in a generating mode.

The use of the induction motor/generator in the power recovery train was the first application of electrical power generation in catalytic cracking units. Approximately 75% of existing power recovery trains use induction motor/generators. Figure 2.11 shows a typical large-size FCC power recovery expander train using a steam turbine, compressor, induction motor/generator, and Hot Gas Expander.

Advantages of the induction motor/generator include:

- Its ability to be used as the start-up driver.
- The ability to use the induction machine as a motor or a generator.
- Low initial and installation costs.
- Simplified controls compared to a synchronous motor or generator.
- A cost-effective and relatively efficient means for converting excess expander power into electrical energy.

Some of the disadvantages of the induction motor/generator include:

- Lower efficiency when compared to a synchronous generator.
- Limited power capabilities (generally less than 10 megawatts [MW]).

Figure 2.11 Type 3 power recovery train installation.

- Operation with a negative power factor.
- The need for a gear box since the most common designs are four and six poles that operate at 1,800/1,500 rpm for 60-cycle operation or 1,200/1,000 rpm for 50-cycle operation.

The sizing of the steam turbine is generally based on the start-up characteristics of the induction motor/generator and the electrical grid capabilities of the refineries.

The largest CCU power recovery trains are in expander train arrangement 3 and use axial compressors with a power rating of more than 65,000 hp (48,500 kW). The expanders are capable of producing more than 70,000 hp (52,200 kW). In recent years, CCUs that have gotten even larger and multiple mid-sized power recovery trains of this arrangement have been utilized.

Power Recovery Train Arrangement 4

Power recovery train arrangement 3 requires all components of the train to operate at synchronous speed (3,600 rpm for 60-cycle applications). When the train operates on a 50-cycle electrical system, the arrangement is modified to a five-body train (arrangement 4), which is shown in Figure 2.12.

The gear box is required because without it the expander and axial compressor efficiencies would be unacceptable and would require larger equipment at a lower synchronous speed of 3,000 rpm. Lower speed motor/generators (1,000 to 1,500 rpm) are also more economical and are readily available.

Power Recovery Train Arrangement 5

As power costs increased over time, it became more economical to apply expanders to smaller CCUs (between 25,000 and 50,000 BPD). Because steam availability is often a problem in small refineries prior to starting the CCU and it is more economical to upgrade the

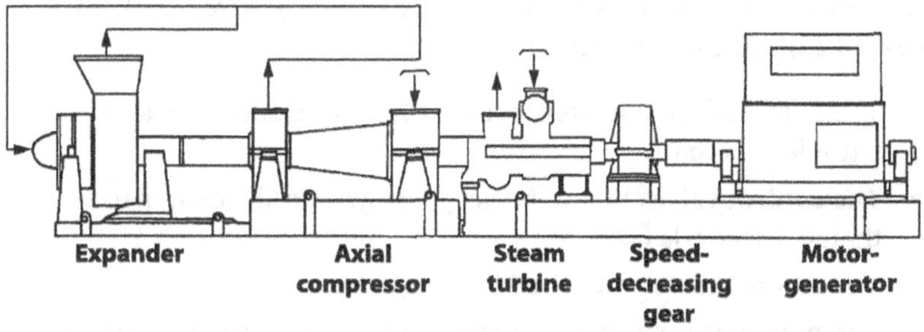

Figure 2.12 Expander Train Arrangement 4: Expander/compressor/steam turbine/gear/motor.

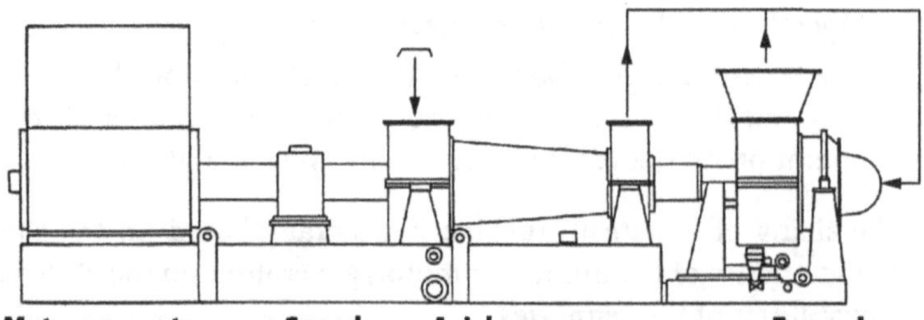

FIGURE 2.13 Expander Train Arrangement 5: Motor/gear/compressor/expander.

electrical system, power recovery trains were developed without a steam turbine. This can be done with an induction motor/generator as the compressor is smaller and has a lower start-up power requirement. Figure 2.13 shows a four-body train that includes the motor/generator for start-up and power supplementing, a speed-decreasing gear, axial compressor, and expander. The motor needs to be sized to bring the train from zero speed to full operating speed. This typically takes less than one minute. A gear box is required since the air blower and expander operate at higher rotational speeds that are not coincident with the electrical grid frequency. Expander and compressor rotational speeds can be as high as 7,200 rpm.

2.3.2 Expander Generators

Many CCU operators have not been able to take advantage of the economic benefits of a power recovery expander owing to space limitations, incompatability of the expander to the existing compression equipment, and/or the high capital equipment and installation costs. These limitations led to the development of the stand-alone power recovery train. The stand-alone power recovery train, however, had a number of technical issues of its own that needed to be resolved for a successful application, including:

- The potential for a power recovery train overspeed on an electrical loss of load.
- The control valves had to be very fast, with full closures in less than 0.5 seconds (s).
- The control valves must be 100% reliable.
- The potential for process operational upsets existed—particularly rapid pressure swings on a power recovery train trip.
- The control system was much more complicated compared to

a conventional power recovery train. The controls needed to provide both process control and electrical generator control, including very fine speed control for synchronizaiton, load control, and overspeed protection on a loss of load.

Most importantly, the controls needed to be designed so that changes in the generator loading did not affect the process normal operating conditions. The process needs must always override the control needs of the electrical system. After all, the intention of the system is to refine oil and not to produce electrical power.

The power recovery train control task is particularly difficult because the process and expander operate in a high-temperature, corrosive, dirty environment with high flue-gas flows controlled by large process butterfly valves. Expander inlet valves range from 36 to 84 inches (in) (90 to 213 cm) in diameter and with discharge valves as large as 120 in (305 cm) in diameter. The process and expander control valves in a conventional power recovery train application typically take 3 to 20 s to travel full stroke. In the expander/generator applications, the full-stroke closure time must be less than 0.5 s. To meet the high-speed demands of the application, it has been found that the FCC's main digital control system (DCS) is inadequate. The DCS instrument sample times and control responses are too slow. A stand-alone dedicated programmable logic controller (PLC)–based control system must be used to assure that the control needs are met.

New technology developed in the 1980s allowed the use of stand-alone power recovery expanders to drive electrical generators. The stand-alone expander/generator application was found to have a number of advantages that include the following:

- The expander/generator set can be installed remotely from the FCC unit or main air blower.
- The expander/generator can be easily added to existing FCC installations.
- No changes need to be made to an existing air-blower train to make it a drive-through train. The existing equipment does not have to have a double-ended shaft or be sized to accommodate the drive-through power.
- The expander/generator sets are compact and occupy relatively small plot plans.
- There is no need for the expander/generator train to match the main air-blower operating conditions such as speed. Expanders tend to operate at higher speeds than the air blower.

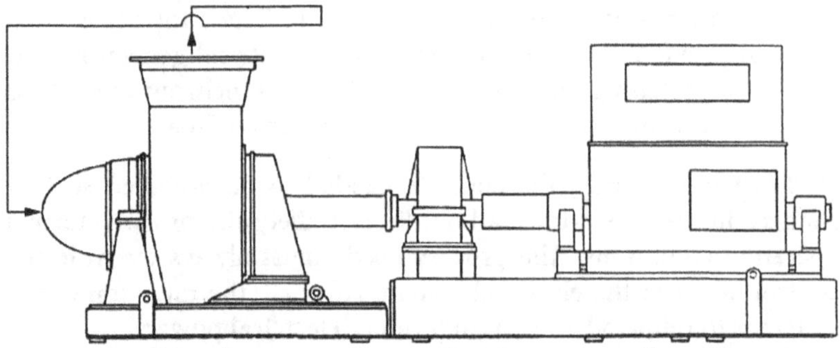

FIGURE 2.14 Expander Train Arrangement 6: Expander/gear/generator.

- The power recovery train and auxiliary equipment can be installed while the FCC unit is in operation and tied into the process at a scheduled outage or turnaround time.
- The power recovery train can be taken on- or offline at any time without affecting the FCC unit operation.
- The expander/generator train generally operates with higher efficiency because of the optimization of the machinery.
- The installation costs are low.

Though initially slow to be accepted, the stand-alone expander/generator set applications have proven themselves to be both very economical and reliable. Since the year 2000, 75% of all new power recovery trains are being built as stand-alone expander/generator sets.

The most common PRT configuration for an expander/generator set is an expander driving a synchronous generator through a speed reduction gear as can be seen in Figure 2.14.

Expander

The design requirements of the expander used in generator applications is slightly different from those applied to a conventional power recovery train. The expander operating speed is generally selected based on high expander efficiency and rotor structural safety margins (the expanders are sometimes slightly slowed down to increase the structural margins).

Two major ways the expander driving a generator differs from the compressor drive expander is that the generator drive applications are more cyclic in operation and the start-up/heat-up rate is much faster. Conventional PRT start-up with the FCC unit that has a slow heat-up of generally less than 200°F/hr (93°C/hr). The generator drive applications start up with full temperature flue gas. The heat-up rate needs to be controlled in the generator application. Heat-up rates are generally held to less than 400°F/hr (204°C/hr) by mixing cooling steam into a small flow of flue gas. The more cyclical operation and faster heat-up rates will reduce the life of the expander components.

In addition, the expander and other rotors used in the power recovery train need to be evaluated at transient and nonuniform torque levels that may originate from the generator or electrical system.

Generator

Standard synchronous generators versus induction generators are used in the expander/generator applications. The synchronous generators are more efficient and allow positive VARs to be put into the electrical grid. This can be very important as an FCC expander/generator set can produce a significant amount of power in comparison to the refinery's electrical system. To date, the expander/generator sets have produced between 10% and 40% of the entire refinery's electrical requirements. FCC expander/generators have ranged in size from 13,000 to 50,000 hp (10 to 37 MW). Generator designs can be 4 pole (1,800 rpm for 60 cycles, 1,500 rpm for 50-cycle applications) and 6 pole (1,200 rpm for 60 cycles, 1,500 rpm for 50-cycle applications). Generator efficiencies are typically greater than 97.5% at a 0.85 power factor.

The most commonly used generator design is a horizontal TEWC (totally enclosed with air/water coolers) design equipped with a

Figure 2.15 Typical synchronous generator with exciter housing removed.

brushless DC exciter, depicted in Figure 2.15. Integrally forged, one-piece solid salient rotor construction is the general standard. Class F insulation ratings are the most common.

Gears and Couplings

Because the optimum expander operating speed is much higher than the generator speed, speed-reducing gears are included in the PRT. Designers have used standard parallel shaft gear designs. The gear box ratings are greater than the maximum power of the expander because of the generator application. Special gear box design requirements are required to accommodate a generator short circuit or out-of-phase synchronization. Depending on the generator design, this can increase the normal torque ratings by three to seven times the normal torque. As does the gear box, the PRT couplings must be designed for the transient and overload conditions of the generator. In some installations where the short-circuit torques are very high, the gear-to-generator coupling is designed with an overload provision that prevents the high torques from being transmitted into the gear or expander rotors.

The primary components that make up an expander/generator power recovery train are shown numbered in Figure 2.16. They are as follows:

1. Third-stage particle separator
2. Expander inlet control and trip valves

Figure 2.16 Typical expander/generator installation: Key components.

3. Hot Gas Expander
4. Speed reduction parallel-shaft gear box
5. Synchronous generator
6. Power control room
7. PRT oil console
8. Expander cooling and seal system
9. Expander abrasive cooling system
10. Expander bypass valve (out of sight in the figure)

The rotating equipment used in the power recovery train is generally of a standard design. The rotating machining can be sized and designed for optimum efficiency and reliability.

Control Valves

The key to successfully operating the expander/generator set is the expander and process control valves. The valves are built to withstand the severe environment of the fluid catalytic cracking unit (FCCU). The gas is at high temperature, erosive and corrosive in nature. Continuous operating temperatures as high as 1,400°F (760°C) are common, and transient short-term operation as high as 1,600°F (870°C) is common. The valves operate above the critical pressure ratio and differential pressures as high as 40 psia (2.81 kg/cm^2). The resulting supersonic gas velocities can be very erosive. Erosion is minimized by overlaying the erosion-prone areas of the disc and valve body with hard Stellite 6. Butterfly valves are cost-effective and are used because of their simplicity, rugged design, reliability, and near-linear control characteristics. Figure 2.17 shows typical valve flow characteristics for a 66-in (26-cm) expander inlet valve. The control range is usually between 20 and 70° open.

Most expander/generator installations include four major valves: expander inlet control valve, expander inlet trip valve, bypass valve, and expander discharge isolation valve.

Expander Inlet Control Valve The expander inlet control valve has a number of functions that include process pressure control, expander/generator speed control, expander heating, expander/generator load control, and emergency shutdown. (See Figure 2.18.)

Expander Inlet Trip Valve Whereas the primary purpose of the expander inlet trip valve is emergency trip/shutdown, it can be used in some applications to assist in PRT speed control and generator synchronization.

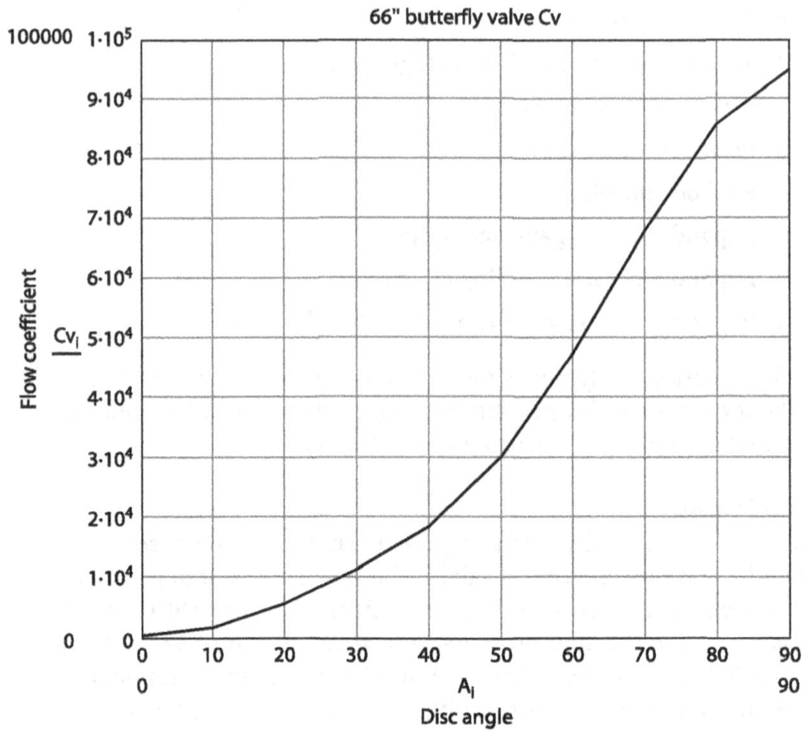

Figure 2.17 Typical FCC butterfly valve flow characteristics.

Two expander inlet valves are used to ensure a positive gas shutoff to the expander, to minimize flue-gas leakage, and to provide shutdown system redundancy. Over time, valve leakage tends to increase owing to valve body and disc distortion, catalyst depositions, and erosion. To assure minimal leakage, a smaller-sized bleed valve is often installed between the expander inlet control and trip valve, giving a typical double block and bleed arrangement.

The expander bypass valve is used for FCC unit start-up and process pressure control. The bypass valve is sized to pass 100% of the process flue gas. The process is often started up through the bypass valve and, after stable process operation is achieved, the expander/generator set is brought online. In addition to the start-up function, the bypass valve maybe operated in a partially opened position to control the process pressure should the flue-gas flow exceed the expander flow capability or if the expander power is being limited.

Bypass Valve The bypass valve has some of the most difficult operational requirements. It must operate continually under full system pressure drop, have very fine sensitivity for process control, and have high-speed operational characteristics. High pressure drop operation can be problematic with high noise levels, vibration, lower control sensitivity, and increased erosive wear. To reduce these concerns, the bypass valve pressure drop is generally reduced downstream by the use of orifice plates, an orifice chamber, or an addition valve.

The most demanding operation for the bypass valve occurs on a generator loss of load. When the PRT trips offline, the inlet valve goes fully closed in a fraction of a second. During this very short time period, the bypass valve must open and take control of the process pressure. A properly designed and tuned system can do this with virtually no effect on the process operation. A poor operating system, however, can cause major process pressure swings and, in the worst case, complete close off or bottle up the regenerator.

Expander Discharge Valve The expander discharge valve is used strictly as an open or closed isolation valve. The critical operational requirement is the valve must be a tight shut-off valve that will allow the expander to be isolated. These valves are physically very large (as large as 120 in [305 cm] in diameter) and even a small leakage rate translates to a large amount of high-temperature flue gas. Critical to the system design is the prevention of the discharge valve to go closed during expander operation. Over-pressurization of the expander discharge casing and exhaust ducting will occur if the discharge valve is inadvertently closed. Because the discharge valve does not perform control functions, electric motor-driven actuators are typically used.

Although a majority of the expander discharge valves are butterfly valves, several expander installations have used standard gate or double-disc gate valves. The gate valve operating experience has been poor owing to a greater susceptibility for catalyst to pack into the moving parts and prevent proper operation.

Other Valves Some of the early expander/generator installations used additional control valves. It is not uncommon to see a second expander bypass valve. The small bypass valve is used on split range control to provide very fine process pressure control. Experience has shown that today's hydraulic power units can provide very fine control of even the larger valves and meet the process pressure control requirements.

Another valve that was used in some early expander/generator applications was a speed control bypass valve that bypassed gas from the inlet to the exhaust of the expander. This valve was used to provide the fine speed control for generator synchronization and to limit PRT overspeed during an emergency shutdown. Similar to the small bypass, experience has shown that the expander speed control valve was not required. In addition, field experience has shown this valve to be susceptible to freeze-up when it remains idol for long periods of time.

FIGURE 2.18 Typical expander inlet control valve.

Control Valve Design Features

The control valve design is critical in assuring proper long-term operation of the valves. Figure 2.19 shows a cross-sectional drawing of a CCU control valve. Long-term operating experience has highlighted potential problem areas and the designs have evolved over time to become very reliable. Some of the more significant operational

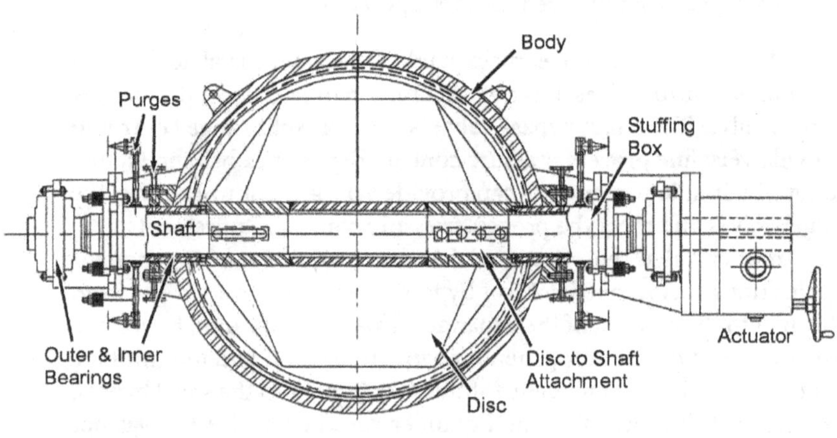

FIGURE 2.19 Control valve cross section.

problems have included freezing in last position, sticking, closed-valve leakage, body and disc cracking, and seat and shaft erosion.

Valve Body The valve bodies are the pressure containing part of the assembly and are designed in accordance with the ASME pressure vessel code. In addition to containing the high temperature and pressure flue gas, the valve bodies need to be rigid enough to accommodate piping loads, thermal growths, or strains, and maintain the critical valve and bearing alignment.

Like the expander inlet duct, the expander inlet control and trip valves are of a hot wall design. The hot wall valve bodies are normally fabricated from a 300 Series Steel such as ASTM SA-240 Type 304H. Most expander bypass valves and discharge valves are of the cold wall design (refractory lined on the inside of the duct). As can be seen in Figure 2.20, the cold wall valve bodies are fabricated from a high-quality carbon-steel boilerplate such as ASTM SA-516-70 and lined with 4-in (100-mm) thick refractory. Integral to the valve body are the valve bearing and packing box housings.

Figure 2.20 Welded in refractory-lined cold wall valve.

Hot wall valves have the valve seat and disc-shadowed areas overlaid with hard Stellite to reduce erosion caused by the particle-laden flue gas. Cold wall valves use specially designed seats that accommodate the differential thermal expansion between the seat and the body. The seats are also overlaid with Stellite for erosion resistance. The valve seat areas of both designs are precision machined for metal-to-metal sealing and reduced leakage.

The valve bodies can be manufactured either in a "welded-in" construction or with flanges. The welded-in construction has a lower initial cost and eliminates the potential for flange leakage. The flanged designs have a higher initial cost and can experience flange leakages. The obvious advantage is that they can be readily removed during a unit turnaround for maintenance.

Valve Disc The valve disc is designed to give an optimum compromise between maintaining a smooth gas-flow field and the mechanical design requirements. The disc is designed to accommodate both the high temperatures and the pressures of the flue gas.

Both the hot and cold wall valve discs are commonly fabricated from 300 Series Steel such as 304H. (See Figure 2.21.)

The process of mounting the disc on the shaft is critical as there are large differential expansions to accommodate and high torque and pressure loads. Disc mounting is generally accomplished by pinning

FIGURE 2.21 Valve disc being machined.

the disc on the drive end and allowing expansion on the drive via axial splines or keys. In a cold condition, the disc is often offset in the valve body to accommodate the in-service thermal growths.

Shaft/Stem The valve shaft is subject to high pressure and thrust loadings as well as high operating temperatures. To accommodate these loadings, the valve shafts are manufactured from forged Inconel materials such as Inconel X-750 or Inconel 718 per ASTM B637. The shaft is fixed with a thrust bearing on the actuator side and thermally grows toward the opposite of the valve often referred to as the free or idle end.

Bearings In most valve designs, the shaft is supported with two inboard and two outboard bearings. The inboard bearings shall be located as close to the disc as possible to reduce shaft deflections. The inboard bearings are metal-to-metal bearings and made from high-temperature materials such as Nitronic 60 or solid Stellite. The outboard bearings are high capacity roller bearings. A thrust bearing is generally used on the actuator side of the valve.

Shaft Seals and Stuffing Box Shaft coatings, overlays, or solid Stellite closure blocks are often used to prevent shaft erosion. Closure blocks are most effective when they are spring-loaded against the disc. (See Figure 2.22.) Many closure blocks use internal purging to prevent ingress of flue gas and catalyst into the bearing and seal areas.

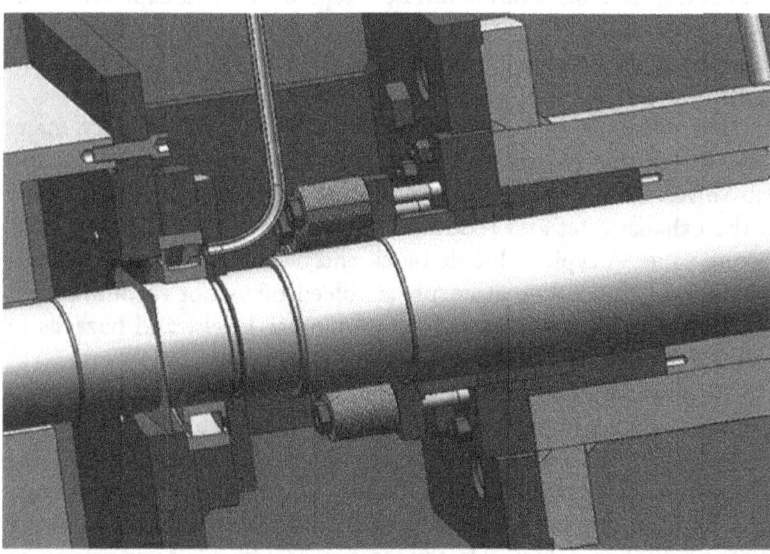

FIGURE 2.22 Typical CCU butterfly valve stuffing box.

Spring-loaded stuffing boxes are commonly used to seal the shaft end. The stuffing boxes typically are packed with braided graphite packing material. The stuffing box region often has a second injection port that can be used to inject sealant if the stuffing box should leak.

Testing To assure proper operation in the field, all valves should be thoroughly inspected throughout the manufacturing and assembly periods. In addition, shop testing is commonly conducted. Shop testing includes hydrostatic testing of the body, which is a hot functional test that confirms the valve operation at full operating temperatures, and a leakage test to assure positive sealing at operating temperature.

Leakage One of the more problematic issues with the FCC control valves is valve leakage. Excessive leakage is a serious problem because it will prevent the isolation of the Hot Gas Expander and, most importantly, can prevent the shutdown of the power recovery train. A number of PRTs have continued to rotate after both the expander inlet and trip valves were in their fully closed positions. This poses a major safety problem because major machinery damage can occur should there be a need to stop the train or in an emergency condition such as the loss of lubrication oil. Leakage is best minimized by scrutinizing the valve design. The various valve manufacturers have various design features that include the seat designs, precision machined seats, eccentrically positioned discs, and more.

New valves can be manufactured to have guaranteed leakage rates of less than 0.5% of the full flow capability. Field experience has shown that valve leakage rates of 2% or more are common and should be expected. Leakage rates can exceed 10% if the valve is worn or has significant catalyst deposits.

In addition to the valve design and manufacturing quality, many users install flue-gas bleed lines between the expander control and trip valves. This allows leakage from the upstream valve to be bled to the exhaust duct and reduces the pressure differential across the second valve. A typical double block and bleed arrangement is commonly used. Intervalve atmospheric bleed off is not recommended because of environment consequences, noise levels, and hazards to personnel and local equipment.

To allow safe entry into the expander or to allow expander disassembly while the process is in service, process blinds need to be installed. The original system design should include inlet and discharge blinds along with the installation hardware, tooling, and structure. Blinding provisions need to be included on both the expander inlet and discharge ducts. Heat protection and fresh air provisions are generally required by personnel during the blinding operations.

Valve Actuators and Hydraulic Power Units

The demanding control requirements for the FCC expander/generator set application are met by the special valve actuators and hydraulic power units (HPUs). These actuator systems are state of the art and have very demanding service requirements. They must be capable of precise high-speed and high-sensitivity operation in a very demanding service. Figure 2.23 shows two valve actuators mounted on an expander inlet control and trip valve assembly. Generally, individual HPUs are installed for each control valve (see Figure 2.24).

Actuator The valve actuator is the key element that physically makes the valve move and perform its required functions. These electrohydraulic actuators are mounted directly to the valve via the mounting bracket. Included in the actuator assembly are the precision-honed hydraulic cylinder, stainless-steel piston, actuator-to-shaft coupling, position sensors, mechanical position indicator, limit switches, and mechanical positioning provisions. The linkages and mounting hardware of the actuator are quite robust and are designed to accommodate a minimum of four times the maximum actuator stall torque. The actuator includes two methods of positioning the valve on loss of

Figure 2.23 Inlet control and trip valve assembly with mounted actuator assemblies.

FIGURE 2.24 Typical hydraulic power unit.

power or control signal. There is a locally manually operated hydraulic control provision and a hand wheel provision.

The actuator systems operate at high pressure and can deliver hundreds of thousands of pounds of force.

Hydraulic Power Unit The hydraulic power unit is the brains of the actuator system. The HPU takes the input from the process and generator set controllers and translates the requirements to the actuator. In addition, the HPU monitors the condition of the actuator and provides feedback and emergency operations via its own self-contained PLC.

As can be seen in Figure 2.24, the HPUs are located in close proximity of the valve and actuator assembly. They are located remotely because of the high ambient temperatures around the valve and for convenient access for operators and maintenance personnel.

Some of the critical design features of the HPU include the following:

- Designed for area Hazard Classification.
- Designed to be weather proof.
- Are completely self-contained including a dedicated PLC for emergency operations and system monitoring.
- Capable of control sensitivities of better than 0.1°.
- Contains dual-pumping hydraulic systems with each pump connected to its own single filter manifold with isolation valve, gauges, block and bleed valves, and accumulators, as well as other components.
- Built with fully redundant hydraulic pump assemblies that accept power from different power sources. Pump drivers can be electric, steam, or air.
- Contains zero-leakage internal seals that are used to assure no loss of hydraulic fluid and minimum valve movement in the event of a fail in position mode.
- Functions at high-speed operation. Full closure occurs in less than 1 s.
- Has adjustable end-of-stroke cushioning.
- Hydraulic accumulators sized to allow a minimum of two complete opening and closures of the valve in the manual operating mode.
- Uses dual-coil servo valves.
- Designed for dual- or triple-system redundancy with full online testing.
- Is self-monitoring and will produce local and remote alarms for failure or abnormal conditions. System health monitoring includes systems pressures, temperatures, filter differential pressure, positioning errors, and signal failures.
- Uses stainless filters.
- Contains stainless-steel piping and tubing for all fixed piping.
- Provides online testing capability.
- Can provide an online function test to assure the valve is not sticking. A waveform generator can be used to provide protection from valve sticking via small cycle movements that will not affect the process variables.

Expander/Generator Controls

Expander Control Panel The brains of the expander/generator set is the control system. Multiple control panels are required to house the process controls, expander controls, and generator controls. Control is accomplished by redundant programmable logic controllers. The control panels are usually housed remotely from the power recovery train in either a stand-alone power control room (PCR) or in a central refinery control room. Figure 2.25 shows typical expander-generator control panels.

The expander control panel includes the following features:

- Operator HMI
- Process controllers—dual PLCs
- Start/stop switches
- HPU controllers
- Overspeed protection system
- Expander auxiliary system controls
- Annunciator panel
- PRT supervisory instrumentation including radial vibration, axial position, bearing temperatures
- Data recorder
- Digital transfer switch

Figure 2.25 Power recovery train and generator control panels.

The power recovery train start-and-stop sequences, expander heat-up sequences, PRT speed control for synchronization, expander loading control, condition monitoring, overspeed protection, auxiliary equipment (lube oil and seal) control, and PRT monitoring are all accomplished in the controllers.

Many expander/generator installations continue to maintain existing digital control systems in the main control room. These systems have been found to be inadequate for controlling the PRT because the input sample and controller response times are too long. The DCS can be used to operate the process without the PRT and is often used to start up the process. Once the process is at a stable capacity and steady operation, control is turned over to the expander controllers and the PRT is started. This can be accomplished seamlessly with the use of a transfer switch that is built in-to the expander control panel.

Generator and Switch Gear Controls

The generator controls used for the expander/generator set are standard synchronous generator controls; these are protection and instrument packages that facilitate control, protection, and supervisory instrumentation.

Control Control features include the following:

- Automatic and manual generator synchronizer
- Voltage regulation—automatic, manual, power factor control or field current regulation
- Digital excitation control system
- Frequency control
- Breaker controller
- Load control
- Power system stabilizer

Protection Protection features include the following:

- Phase under or over voltage
- Under and over frequency
- Reverse power or antimotoring
- Over and loss of excitation
- Negative phase sequence or unbalanced current
- Voltage phase reversal
- Stator thermal protection

- Accidental/inadvertent energization
- Breaker failure
- Transformer fuse failures
- Stator ground fault
- Current directional—ground
- Trip coil supervision

Supervisory instrumentation Supervisory instrumentation features include the following:

- Winding temperature detection
- Cooling circuit detectors

Dynamic Simulation

One of the key analytical tools used in the design of any power recovery system is *dynamic simulation* (see Figure 2.26). Dynamic simulation involves developing analytical models of the individual processes and control systems and integrating them in a system model that can be used to simulate and study process phenomena and equipment behavior that are typically characterized by dynamics rather than steady state. This allows the system to be emulated for design, development, optimization, and validation of the system dynamics well

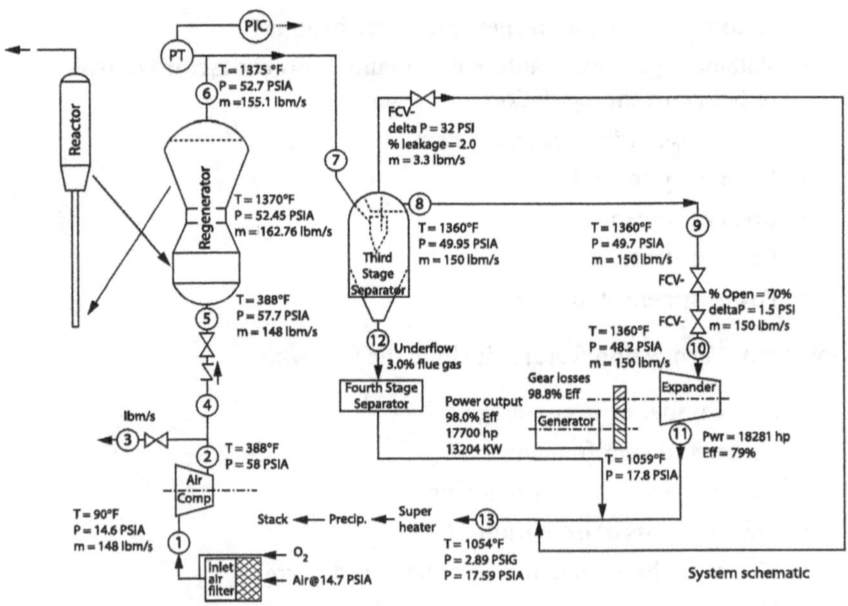

Figure 2.26 Dynamic analysis system model.

before the equipment is actually built and installed. Typical applications of dynamic simulation are the design and development of a process system, testing of control configurations and tuning parameters, the development of operating procedures, operator training, system safety analysis, and advanced process control. Some of the benefits evolving from the simulation include improved process and machinery operations, understanding process and machinery transient operations, reduced capital and operating costs, safer operations, and improved performance.

As can be inferred from examining Figure 2.26, a significant amount of data and information is required before performing a detail dynamic simulation study. Typical input data includes:

- Process vessel volumes
- Process vessel conditions (pressures, temperatures, flows, gas compositions, etc.) throughout the system
- Process control details
- Other system volumes including ducting and other vessels such as separation equipment
- Machinery operating characteristics (main air blower, blower driver[s], expander, generator, etc.). The operating characteristics include performance maps, flow characteristics, speed torque, among other characteristics
- Controller characteristics, actions, sampling rates, sensitivity, and tuning parameters
- Machinery mechanical characteristics such as weights, inertias, and more
- Mechanical process control characteristics including valve flow characteristics

The simulation studies allow the designers and operators to evaluate the process and machinery operating characteristics at any time throughout the simulation period. For example, the PRT overspeed on a generator loss of load is of high interest. Figure 2.27 shows the PRT speed versus time during a generator load loss. The following responses can be noted for a very short time period of only 2 s:

- At time zero, the load is lost and the generator breaker opens.
- The PRT begins to accelerate and reaches its trip speed in 0.50 s.
- The expander overspeed trip system initiates a PRT emergency shutdown.

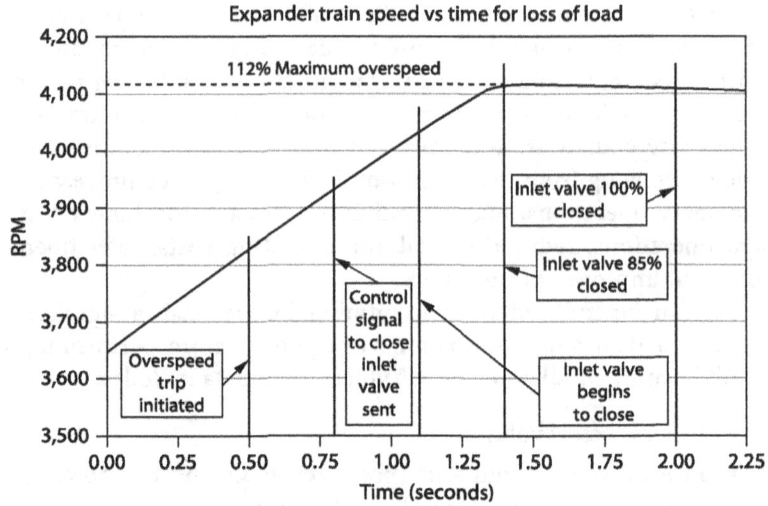

FIGURE 2.27 Typical dynamic analysis output with control actions.

- A short time delay (0.3 s) is experienced as the signals are processed by the controllers and a signal is sent to the valve controller.
- Another short time delay (0.25 s) is experienced as the controls initiate the valve actuator system and begin to physically close the inlet valves.
- The inlet valves go from 100% (full) open to 15% open in 0.3 s.
- Throughout this period, it can be seen that the PRT continually accelerates until the valve is close to being fully closed. The maximum overspeed of 112% is reached before the train begins to slow down.
- Finally after 2 s, the inlet valves are fully closed and the PRT continues to slow down.

Similar process detail can be studied with a system response analysis. One of the strengths of a dynamic simulation is being able to perform various "what-if scenarios." Typical scenarios that can be analyzed in the application of an expander/generator set include the following:

- Generator loss of load
- Evaluation of normal start-up and shutdown modes
- Generator synchronization
- Emergency shutdown
- Coupling failures

- Process upsets such as after burning, depressurization, or over pressurization
- Optimization of control system parameters such as instrument sampling time, control sampling times, controller action times, mechanical action times, among others.
- Loss of power to a particular function or device
- Loss of control signal
- Control or mechanical component failures

Control system development and tuning is another feature that is very helpful in optimizing the system responses. Figures 2.28 and 2.29 show the effects of including bypass valve predictive biasing on the critical regenerator top pressure. By commanding the bypass valve to move to a predicted position (versus waiting for the pressure control to note a deviation), the regenerator top pressure spike can be reduced from more than 2 psi to less than 0.25 psi.

2.3.3 Overview of Design Objectives and Challenges

The design objectives of a CCU Hot Gas Expander are limited and can be difficult to attain. The primary objectives may be simply stated as:

- Do not, in any circumstances, reduce or impair the operation of the CCU. This means that there should be no shutdowns scheduled solely to maintain or repair the Hot Gas Expander.

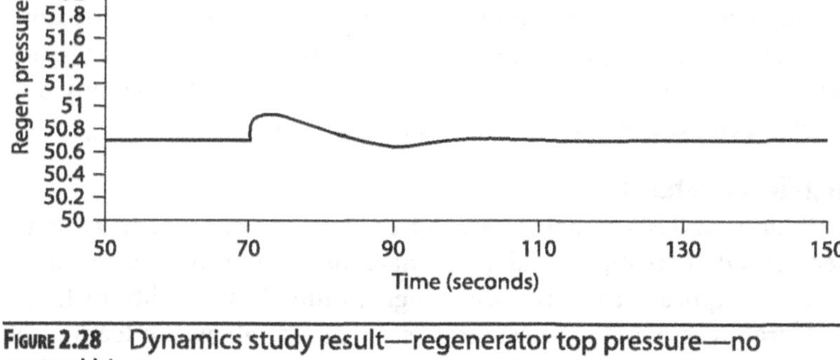

Figure 2.28 Dynamics study result—regenerator top pressure—no control bias.

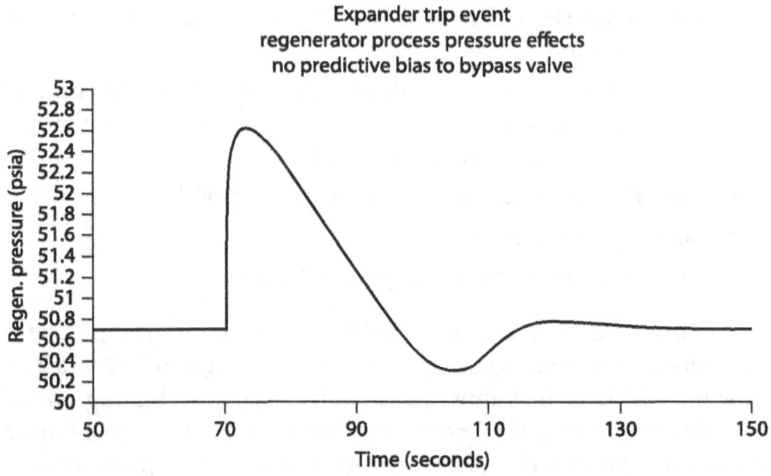

FIGURE 2.29 Dynamic study result—regenerator top pressure—with control bias.

In addition, the Hot Gas Expander must reach the design goals that in no way impair the ability of the CCU to reach the maximum rates desired by the unit operations plan.

- Do not require any maintenance that will require the CCU to be shut down at a time less than that planned or required for other CCU maintenance activities (e.g., catalyst management or boiler inspections).
- Do not become the critical path item on a CCU maintenance turnaround or shutdown.
- Operate at the maximum capacity required to meet the operating demands of the CCU.

The early application of CCU Hot Gas Expanders did not necessarily meet these objectives. The design of the expander faced a number of very challenging requirement that have limited its operating run life and successful performance. Perhaps one of the major issues is that there have been a considerable number of challenges, any one of which could make it miss its design objectives.

High Temperatures

By their basic design, CCU Hot Gas Expanders are exposed to gas temperatures that are high for all turbomachinery other than gas turbines or aero-engines. In the earliest designs, limited availability of high-temperature materials caused concern with the allowable loading of discs and blades and led to limited operating runs as well the need to inject cooling steam to lower temperatures at critical junctures. As time has gone by, better materials have become available and the technology has taken very large steps in two areas to help mitigate this concern:

1. There have been major advances in the development of high-temperature, high-load capability alloys that greatly increase the design margins, which in turn have reduced the potential for failures due to elevated temperatures.
2. The control technology that has been applied to CCUs has grown dramatically in its capabilities and its accuracy. Current CCUs are able to manage temperatures much better, and the variances from design points have been reduced considerably. These developments have reduced the concern for significant temperature variations. However, these machines still may experience the highest gas temperatures of any turbomachinery outside of a gas turbine.

One of the fallouts of the temperature issues in the past was the presence of emergency shutdowns due to the loss of temperature control in the operation of the CCUs. A review of the operating foreman's diary on one particular unit showed that in the early days of operation of a CCU (stated in the late 1960s), there was an emergency outage at least once per month. Most of these were the result of control issues. Current CCUs may run four or more years without experiencing an emergency shutdown. Although the high-temperature operation and excursions remain a challenge, they are much less of a problem today.

High Power Density

Another design challenge for any rotating machine is high power density, that is, the amount of power generated by one stage or wheel of the expander. In the case of CCU Hot Gas Expanders, the demand on a single wheel commonly exceeds 20,000 hp and may be as high as 70,000 hp. This may be the highest demand per wheel of any piece of turbomachinery except for the largest land-based gas turbines. The single-stage design is required because of the difficulty in operating multistage expanders in the catalyst-laden environment. CCU expanders of two to four stages have been installed, but unfortunately most have had poor operating success.

The presence of the high power density translates into high design loadings which must be managed carefully. This has translated into particular care being required for blade and disc root design, material selection, and heat treating of the materials of construction, calculation accuracy of stress levels and flow capabilities, and exactness of fabrication to minimize error in the design.

Current performance levels of the machines would indicate that the designs have been able to meet most of the challenges of CCU Hot Gas Expander operation. However, if considerable vigilance is not maintained, these machines can fail dramatically.

2.3.4 Early Design Features

Some design features that were employed in the early applications of CCU expanders still exist. A number of these have shown themselves to be the cause of potential failure. Because many of them have been in place longer than those charged with operating and maintaining the machines, failures have occurred.

Steam Cooling and Quench Hardware

Many CCU expanders are equipped with cooling steam that, in theory, bathes the disc in the area where the blades are attached. This is the portion of disc that has the highest relative loading. In the early days of the CCU expanders, the materials used had much lower stress margins between the design points and material capabilities. As the material capabilities were improved, the need for the cooling steam became less dramatic. As a result, less attention was paid to the cooling steam. A common problem related to the cooling steam is the quality of the steam. In the early applications, it was typically a low-pressured steam header. And, since the flow was low, and low-pressure steam headers aren't given the greatest amount of attention, there were numerous incidents in which water and/or wet steam was injected onto hot components of the expander.

2.3.5 Expander Design Definitions

Design temperature. The inlet temperature for which the expander is designed and guaranteed to perform. The inlet temperature is used to establish component operating temperatures around which the designs of the components are based. Thus, any reference to life or time is referred to the design temperature.

Maximum continuous inlet temperature (MCIT). The inlet temperature of the expander that is the highest inlet temperature under which the expander can operate continuously. Note that continuous operation at this temperature will reduce the design life of many of the expander components. Operation above the MCIT can damage the expander.

Rated/trip temperature. The maximum temperature to which the expander can be subjected; above which it should be taken out of service immediately.

Maximum exhaust temperature. The maximum temperature at which the expander exhaust system can be operated on a continuous basis. The maximum exhaust temperature often limits the expander inlet temperature at reduced flow and pressure operation.

Operating temperature. The actual operating temperature of the component when the expander is operating at the design temperature and pressure.

Design pressure. The inlet pressure for which the performance of the unit is guaranteed.

Maximum pressure. The highest allowable pressure under which the expander can be operated. The maximum inlet pressure is used in the design basis of the expander components. Expanders have both a maximum inlet and discharge pressure.

Rated pressure. The maximum pressure at which the expander is designed to operate.

Design life. The guaranteed minimum life of the hardware. Generally equal to the 100,000 hr of continuous operation under normal loading conditions.

Normal operating life. The calculated life in hours for continuous operation under normal load conditions. The normal operating life must equal or exceed the design life.

Effective operating life. The calculated life in hours for operation under combined steady-state normal and limited load conditions. The effective life may exceed the design life if the expander has been lightly loaded over time or is less than the design life if the expander has been operated under conditions that exceed the design conditions.

2.3.6 Design Overview

A majority of all CCU expanders are of the single-stage overhung designs similar to that shown in Figure 2.30. Other designs used are between bearing and multistage designs, but those units have generally had more operational problems, and only a few currently remain in service. The CCU expander is made up of three major assemblies: the inlet casing assembly, the exhaust casing assembly, and the rotor, bearing, and seal assembly.

The basic design is very similar to the power turbine "end" of a gas turbine and has been redesigned multiple times to meet the specific demands and function of the CCU Hot Gas Expander. In this section of the book, we describe the individual design components and present a detailed discussion of the construction and design methodology.

Figure 2.30 Overview of a CCU Hot Gas Expander. (courtesy of Dresser-Rand Co.)

Characteristic of the CCU expander is the large horizontal inlet duct connection and vertical exhaust or discharge duct connection. This geometry allows the power recovery train to be mounted near grade. The power recovery train is most commonly mounted on single monolithic solid-concrete foundation. Inasmuch as a number of power recovery trains have been mounted on steel structures, the steel structures need to be designed carefully to prevent structural dynamic vibrations. CCU expanders are mounted on steel baseplates that may or may not be an integral part of the remaining power recovery train baseplate.

The expander assembly is made up of several major subassemblies including the inlet casing assembly, the exhaust casing assembly, and the rotor, bearing and seal assembly (see Figure 2.31).

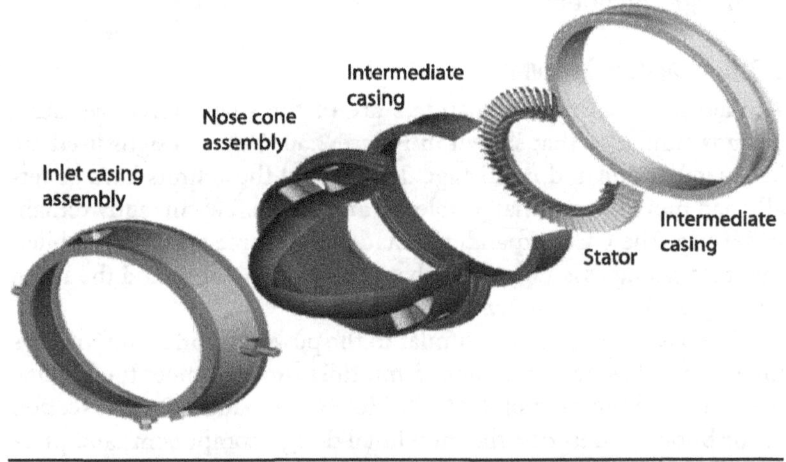

Figure 2.31 Expander inlet casing assembly. (courtesy of the Dresser-Rand Co.)

Inlet Casing Assembly The first major sub assembly is the expander inlet casing assembly. As can be seen in Figure 2.31, the inlet casing assembly is made up of the following components:

Inlet casing. The inlet casing is the main structural support and pressure containing section of the expander. The expander inlet duct bolts directly to the inlet casing. The casing is directly supported by trunnion pins mounted on the baseplate pedestals.

Nose cone. The nose cone is the flue-gas flow guide that directs the gas into the expander flow path. If is often the structural support for the stator vanes and shroud ring.

Shroud ring. The shroud ring forms the outer flow path over the stator and rotor blades. It mounts to the rear of the nose cone.

Stator vanes. The stator vanes accelerate and direct the flue gas into the rotor blades. They are mounted to the nose cone in the design shown in Figure 2.31.

Intermediate casing. The intermediate casing is a structural member that joins the inlet casing to the exhaust casing. Removal allows access to the shroud ring fasteners and removal of the nose cone from the inlet casing.

Exhaust Casing Assembly The second major subassembly is the exhaust casing assembly made up of the discharge casing and exhaust diffuser. The diffuser can be bolted into the exhaust casing or an integral part of the exhaust casing.

Rotor, Bearing, and Seal Assembly The third major subassembly (see Figure 2.32) includes the following:

Rotor assembly. The rotor assembly includes the rotor blades, disc, shaft, tie bolt(s), seal sleeve, and coupling hub. The thrust collars are usually integral to the shaft.

Bearing support and housing. The bearing support is the main structural support for the rotor and houses the expander bearings. The bearing support is bolted directly to the expander baseplate.

Inner exhaust casing. The inner exhaust casing forms the inner flow path as the flue gas exits the rotor blades and is directed into the exhaust casing. It is supported by both the outer exhaust casing and bearing support.

Bearings and oil seals. The expander journal, thrust bearings, and oil seals are mounted in the bearing housing. They operate with critical tight clearances to the rotor.

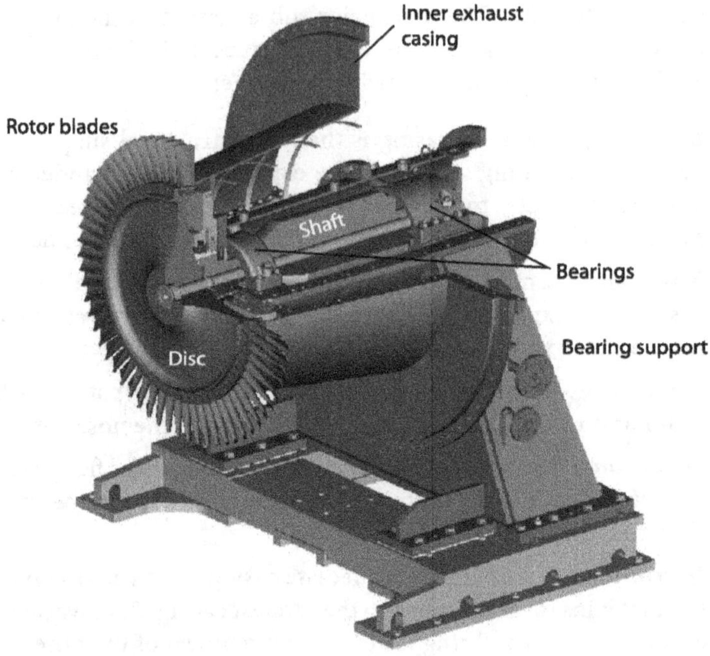

FIGURE 2.32 Rotor, bearing, and seal assembly. (courtesy of the Dresser-Rand Co.)

Seal housing. The flue-gas seals are mounted into a seal housing that is also mounted to the bearing support.

In the aforementioned design, the expander rotor is removed by pulling the rotor, bearing, and seal assembly out of the casings in the direction of the driven equipment. This allows the rotor to be removed without disassembling the expander outer casings. In some manufacturer's designs, the stationary flow-path components can also be removed from the outer casings without removing the inlet piping or disassembling the outer casing. Two disadvantages of this design are that it requires a very long coupling that can affect the expander's rotor dynamics characteristics, and the inner exhaust casing is usually horizontally split, which is undesirable for hot casings. More detailed descriptions and design details follow for all of the major components of the expander.

Several expander manufacturers offer designs that are similar to that described above. Another major manufacturer offers a second design of the CCU expander. The alternate design shares many of the same parts and components as those previously described above and can be seen in Figures 2.33 and 2.34.

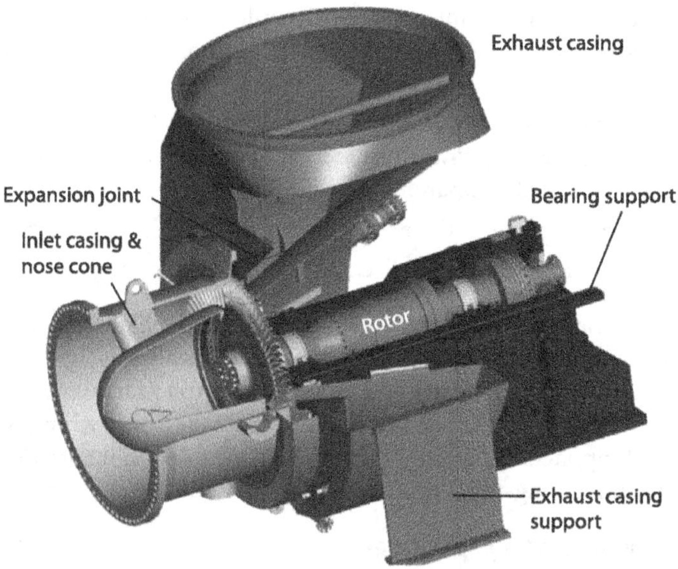

FIGURE 2.33 Overview of a CCU Hot Gas Expander. (courtesy of the Elliott Co.)

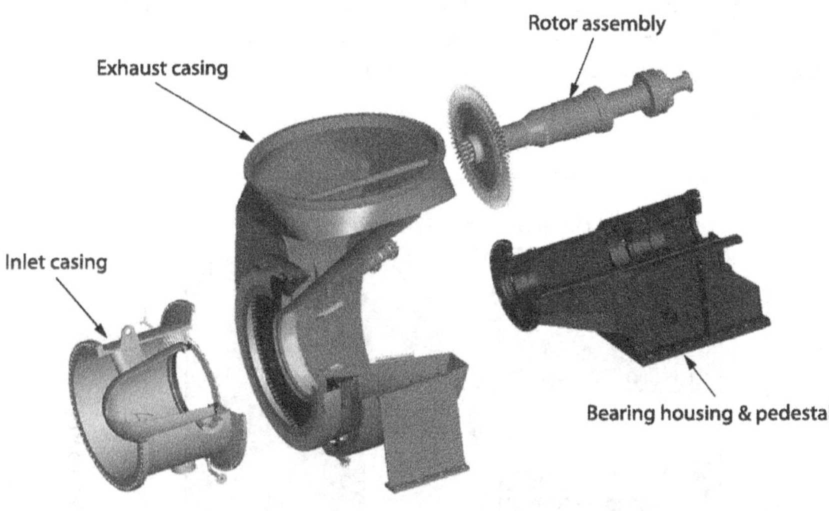

FIGURE 2.34 Alternate CCU Hot Gas Expander. (courtesy of the Elliott Co.)

The second design offers a number of notable features that represent differences in design philosophy and include:

- A single-piece inlet casing and nose cone assembly.
- A single-piece inner and outer exhaust casing assembly.

- An expansion joint between the inlet and exhaust casings to allow fixing the exhaust casing in position and separate supports.
- The rotor is removed by lifting and pulling through the outer exhaust casing toward the inlet ducting. This requires the removal of the expander inlet casing.
- Close coupling of the expander rotor to the driven equipment.
- Stator vanes mounted directly into the inlet casing.

As can be seen, each of the two major designs has its own set of advantages and disadvantages. The debate as to which design is better has carried on for more than 50 years. Each expander user needs to assess its individual needs and priorities and determine which works best for its application. It can certainly be concluded that both designs have provided long-term reliable service in many installations.

CCU Expander Frame Capabilities

Catalytic Cracking units come in a wide variety of designs and capacities. CCU expanders have been applied to CCU units as small as 35,000 barrels per day (BPD) to as large as 180,000 BPD. To cover this wide range of operating conditions, each of the expander manufacturers has designed a number of expander "frame" sizes. Figure 2.35 shows the frame coverage for one manufacturer. Four expander frame

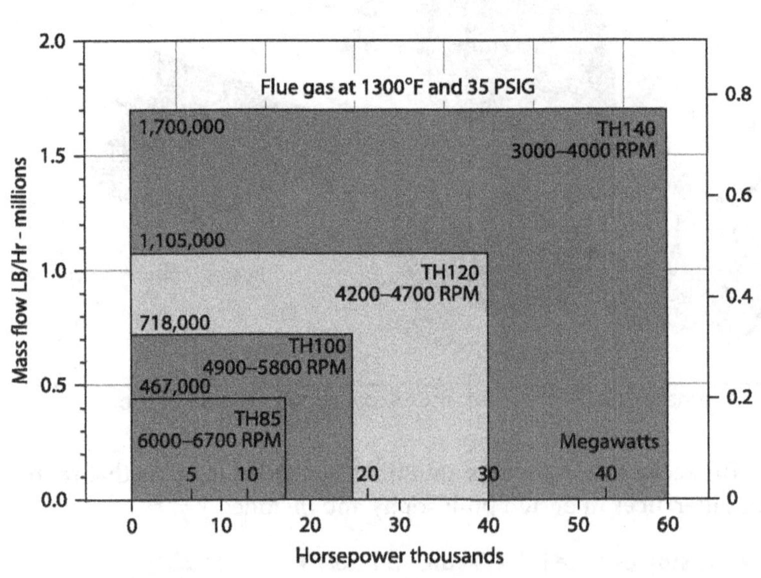

Figure 2.35 CCU expander frame capabilities. (courtesy of the Elliott Co.)

sizes are used to cover flue-gas flows from 250,000 to over 1,700,000 lb/hr (113,000 to 771,000 kg/hr). The frame sizes may also be built as a two-stage unit for the higher expansion ratios. The power capabilities and operating speeds vary with each of the frame sizes and are illustrated in Figure 2.35.

2.3.7 Expander Flow-Path Components

Nose Cone

The expander nose cone is the first component of the Hot Gas Expander to act on the inlet gas and is a major part of the expander flow-path design. The nose cone's primary function is to accelerate the catalyst-laden flue gas into the stator blading uniformly. A uniform controlled acceleration helps to minimize flow-path friction and pressure losses, prevents flow separations, and very importantly distributes the erosive catalyst uniformly in the flow path. A poorly designed nose cone can actually concentrate the catalyst and cause accelerated local wear on the stator and rotor blades. Design enhancements and improvements in fabrication technology have permitted the manufacture of nose cones that are much more aerodynamically shaped than was present in earlier designs.

Nose Cone Bullet The bullet section of the nose cone is the inner flow path into the expander and controls the flue-gas acceleration. The bullet is subject to a significant pressure drop. Along the length of the bullet, the flue gas is accelerated from a typical inlet velocity of 150 ft/s (46 m/s) to 400 ft/s (122 m/s). The nose cone often includes various expander rotor cooling systems that are used to control the operating temperatures of the expander rotor. Many expander nose cones are insulated on the inside of the bullet to control the operating temperatures. The nose cone bullet is often a structural support member for the stator vanes, so dimensional stability is critical.

Nose Cone Struts The expander nose cone struts are used to support the nose cone "bullet." The struts are attached to the bullet outer or flow-path surface and the inside of the outer flow guide.

The outer flow-path guide can be either an integral part of the inlet casing or a separate flow guide referred to as the nose cone outer liner. Figure 2.36 shows the two different nose cone designs. The struts are attached to the inlet casing in the left photograph, while the right design shows the struts attached to an outer liner that is not part of the inlet casing. The outer liner design is a double-wall design used to reduce the stresses and temperatures in the inlet casing.

Larger and/or wider inlet struts tend to deflect the catalyst in the flue gas and shield the downstream stators from the catalyst. Increased

FIGURE 2.36 Two CCU expander nose cone designs.

erosive wear has been noted in the stators that are on either side of the inlet strut. Designs in which the nose cone assembly is a separate part from the inlet casing require the nose cone to be supported and thermally guided to allow differential expansion between the two parts.

Nose Cone Structural Design Considerations

The expander nose cone assembly is both geometrically complex and subjected to significant pressure and temperature loadings. The nose cone bullet is subjected to the pressure drop that occurs across the stator vanes. This typically represents 50% to 60% of the expander's total pressure drop. The pressure drop causes a high axial thrust loading on the nose cone bullet. This bullet thrust load is transferred to the outer casings via the inlet struts as can be seen in Figure 2.37. Locally high stresses and deflections can develop in the regions where the struts join to the nose cone bullet or outer casing. Although the nose cone pressure loadings can be significant, thermal stresses are generally the life-limiting component of the nose cone parts.

High-thermal stresses result from high axial and radial temperature gradients, high-transient temperature gradients, misdirected cooling flow, liquid carryover, and process upsets. The high local stresses in combination with the high inlet temperature often result in nose cone strut cracking. Typical strut distress is discussed in Chapter 5, Common Operational Problems and Failure Case Studies.

Expander nose cones are often the main structural support for the stator and shroud rings. Dimensional stability is critical to maintaining stator to rotor and rotor to shroud clearances. The nose cone

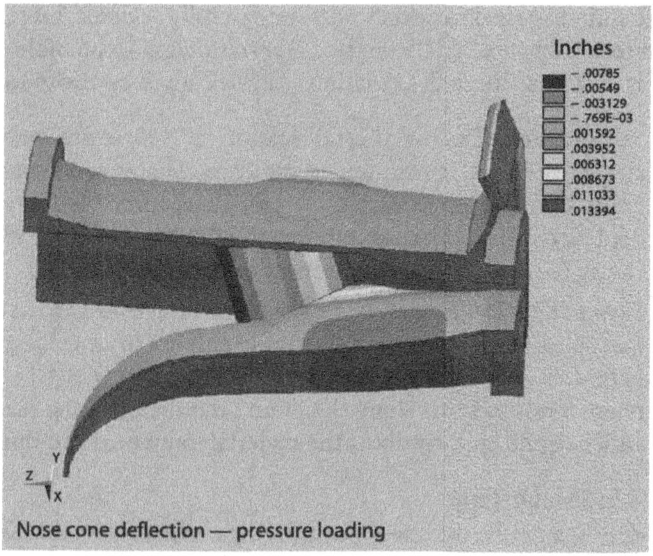

FIGURE 2.37 Expander nose cone deflections.

is subjected to the full inlet temperature of the expander. These temperatures are sufficient to cause high-thermal displacements, stress, distortion, and creep damage. A number of expanders have experienced nose cone to rotor rubs that have resulted in major expander rotor failures.

Nose Cone Materials of Construction

CCU expander nose cones are typically fabricated from 300 series stainless-steel plate, formed hemispherical heads, and forgings. Typical materials of construction include ASTM A-240/A-182 Types 304, 304H, 321, and 347. Although the 300 series stainless-steel cones provide good oxidation and sulfidation resistance, they can be susceptible to aqueous corrosion during a unit shutdown. Corrosion from the formation of polytheonic acid is not uncommon.

The highest-operating-temperature CCU expander nose cones are fabricated from Incoloy 800H conforming to UNS-N08810. The Incoloy 800H material provides enhanced material capabilities and in-service aging resistance. H grade materials are often used to take advantage of the higher-temperature allowable stresses of the high-carbon versions of the alloy. The optimum material properties are achieved if the fabrications are solution heat treated and precision machined.

The stainless-steel materials must be carefully selected, fabricated, and periodically checked throughout service owing to possible material degradations. These degradations can be caused by the following:

- *Sensitization*. During the fabrication of 300 series stainless-steel nose cones, chromium carbide precipitates can form at the grain boundaries of the material. These precipitates locally deplete chromium around the grain boundaries, which significantly reduces the corrosion resistance of the material.
- *Sigma phase embrittlement*. Sigma phase is an iron-carbon phase that is formed when the material is subject to operating temperatures between 932 and 1,652°F (500 and 900°C). Sigma phase tends to form along the grain boundaries of the material, which significantly reduces the material toughness and ductility.

Flow Path: Shroud Rings

Expanders generally use two shroud rings. An outer shroud ring forms the outer flow path and often covers both the stator vanes and rotor blades. The outer shroud ring can incorporate the stator vanes and is then referred to as an integral stator/shroud (see Figure 2.38). The outer shroud ring can be supported either off the nose cone on the upstream side of the shroud or off the exhaust casing on the downstream side. The second shroud ring is the inner shroud that forms the inner flow-path downstream of the nose cone. The inner

Figure 2.38 Expander cross section showing typical inner and outer shroud rings.

shroud ring is used only in the area of the stator vanes and may incorporate a rotor seal.

The flue-gas flows in the vicinity of the shroud rings are at high velocity and a strong erosion potential can exist. In addition, catalyst is thrown out against the outer shroud ring and has a tendency to erode the shroud or cause deposits in the area over the rotor blades. These deposits can develop sufficient thicknesses to cause blade-tip rubbing. The shroud design can significantly affect the erosion and deposition potentials.

Shroud Structural Design Considerations

Although the shroud rings generally have low differential pressure stresses, the thermal gradients along the axial length can cause significant thermal stress, displacement, and distortion. The shroud ring shape is critical in both guiding the flow as well as in maintaining the stator vane and rotor blade radial clearances (see Figure 2.39).

Shroud Materials of Construction

Shroud rings are generally manufactured from forged rings or fabricated from plate. Shrouds are generally fabricated from 300 series stainless steel such as Type 304, 321, or 347. The shroud rings are stress relieved and precision machined. Because the shroud rings can be subject to catalyst erosion, the flow-path side of the ring is often coated with an erosion-resistant tungsten or chromium carbide coating. The fabrications are usually solution heat treated and precision machined.

FIGURE 2.39 Rotor blades installed into an outer shroud ring.

Flow-Path Stator Vanes

Expander stator vanes are used to accelerate the flue gas into the rotor blades. Flue gas enters into the stator vanes at a velocity of 400 ft/s (122 m/s) and is accelerated to close to the sonic velocity of the gas or approximately 1,800 ft/s (550 m/s). At the same time, the gas is turned from the axial direction into a tangential direction at an angle that is compatible with the downstream rotor blades.

The aerodynamic shape of the stator vane has typically been of the free vortex design. To improve the overall expander efficiency and reduce stator erosion, many newer expanders have been manufactured with compound lean or forced vortex stator vane designs.

Stator vanes need to be designed to accommodate the high-pressure differential and aerodynamic thrust loadings under which they operate. The vanes often operate at a critical pressure ration that can produce sonic and supersonic velocities that result in areas of overexpansion and flow separation. The vanes can experience erosive wear and consequentially are designed with high structural margins.

The vanes are also subjected to the full expander inlet temperatures that are sufficient to cause creep and in-service material aging and degradation. The thermal displacements and distortions need to be minimized to maintain the stator structural margins and stator to shroud and rotor clearances. Stator damage can result if the expander is subject to rapid temperature changes. The thinner trailing edges of the stator vanes respond very quickly to temperature changes while the leading edges and attachment regions do not.

Stator vanes are subject to significant aerodynamic and pressure fluctuations and vane-to-vane variations. The nonuniform flows can excite the vane natural frequencies and have caused vane failures in some expanders. The vane natural frequencies need to be evaluated in the initial design phase and tuned as required to prevent fatigue failures. The passing frequency of the rotor blade is the strongest excitation to the stator vanes. The integral stator-shroud vane assemblies have a distinct advantage over the cantilevered designs in that the natural frequencies are much higher. (See Figure 2.40.)

Although the stator vane nominal stresses are low, vane failures can occur from excessive erosion, improper vane tuning, or high-transient thermal stresses. Rapid temperature changes from process upsets or the ingestion of a liquid are causes of high-transient thermal stress. The thermal mass of the vane varies as well as the vane to the shroud rings. With a rapid temperature change, the strain on the vane will be very high in a material that has relatively low ductility. The result can be major cracking of the vane or vane attachment

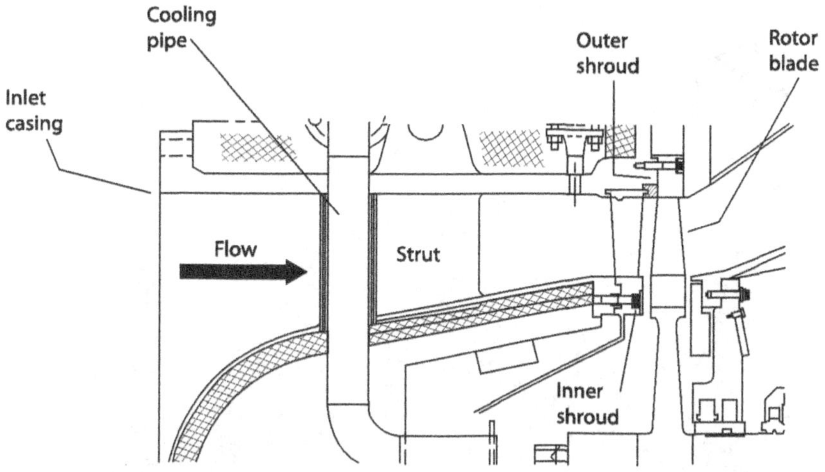

Figure 2.40 Integral stator – shroud assembly.

to the shroud rings. An example of stator vane cracking is given in Chapter 5.

Stator Vane Design Considerations

Expander stator vanes can be mounted in a variety of ways depending on the individual manufacturer. The vanes can be attached either individually or as an assembly to the nose cone, inlet casing, or shroud ring. Because the stator vanes take at least 50% of the pressure drop across the expander, the attachment must be rigid and the loads must be transmitted into the outer casing supports.

Two types of stator vane constructions have been used in the CCU expander:

- Vanes can be cantilevered (from either the inside or outside diameters). (See Figure 2.41.) Cantilevered vanes require clearance between the vanes and the shroud ring. This clearance represents a performance loss and can contribute to localized erosion of the shroud ring and rotor blades. The cantilevered vane design was the design employed in the original CCU expander and has been used less in the more recent designs. (See Figure 2.42.)
- Vanes constructed with an inner and outer shroud ring are referred to as an integral stator/shroud design. Most new expanders are constructed with the integral stator design and many of the original cantilevered installations have been retrofitted with integral stator/shrouds.

FIGURE 2.41 Cantilevered stator vane design.

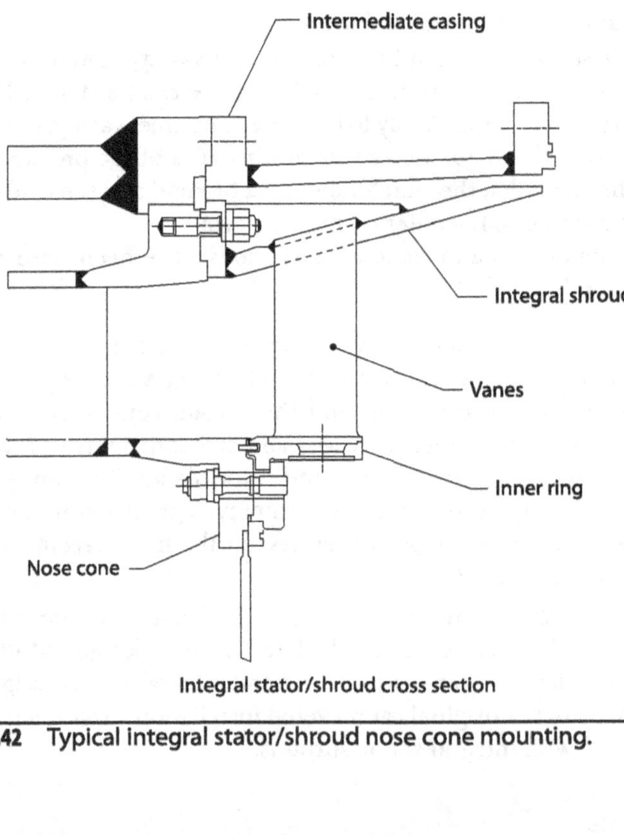

FIGURE 2.42 Typical integral stator/shroud nose cone mounting.

Two stator vane mounting methods have been commonly used in the CCU expander:

- Individual vanes are inserted into a stator housing tangential groove. These vanes are carefully spaced and retained in position with locating pins.
- The welded ring assembly, which can be seen in Figures 2.41 and 2.42, involves mounting the vanes into a solid ring via welding the vane or a vane extension to the shroud ring. The assembly is bolted into place.

Stator Vane Materials of Construction

Stator vanes have been manufactured for a variety of materials including nickel-based alloys Waspaloy or IN-738, or from cobalt-based alloys Stellite 31 or X40. All of these materials provide high-temperature strength and creep resistance, good high-temperature ductility, as well as high erosion resistance. Additional stator vane erosion resistance is achieved by coating the pressure or concave side of the airfoil and the leasing edges of the vane with an erosion-resistant coating. Tungsten or chromium carbide coatings have been commonly used. Experience has shown that the erosion life of stator vanes is almost twice as much as that of the rotor blades. The stator aerodynamics and material selections contribute to this increased life.

Exhaust Diffuser

An integral component of the expander flow path is the exhaust diffuser. The exhaust diffuser is located directly on the downstream side of the rotor blades and mounted in the exhaust casing, as can be seen in Figure 2.43. The function of the diffuser is to efficiently decelerate the flue gas exiting the rotor blading. Gas typically will enter the exhaust diffuser at a velocity of 1,000 ft/s (305 m/s) and exit at 600 ft/s (183 mm/s). Increased expander power and efficiency can be achieved with an optimized diffuser design. An efficient diffuser converts the kinetic energy of the flue gas into static pressure. This conversion effectively reduces the back pressure on the expander blading and increased recoverable energy is available. A properly designed exhaust diffuser can effectively increase the expander power by +1.5%.

Exhaust Diffuser Structural Design Considerations

Exhaust diffusers operate with low stresses owing to only a small pressure drop between the inside and outside. The most significant stresses are the result of temperature gradients in the region where the diffuser attaches to the exhaust casing. This is particularly true during expander transient operation such as unit start-up, rapid load changes, or unit afterburns.

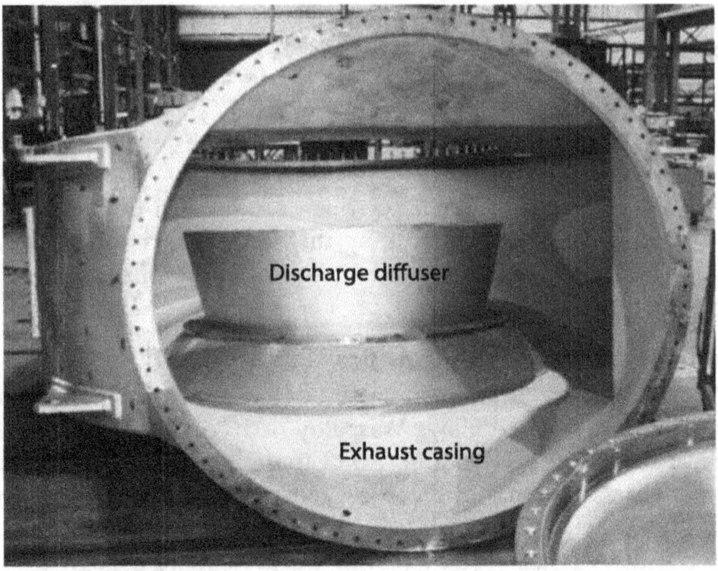

FIGURE 2.43 Expander exhaust diffuser mounted in exhaust casing.

The main attachment flange and its mated flange on the exhaust casing are fairly thick and massive and respond slowly to rapid temperature changes in the expander.

The fasteners used in the bolted-in design are very susceptible to these high-transient thermal stresses. The stresses have been reduced by adding stiffening gussets to the flanges, using tapered or weld neck design flanges, and using Belleville washers on the fasteners.

Exhaust diffusers can either be bolted in and removable or integrally welded to the expander exhaust casing.

Although exhaust diffuser failures are not common, most of the failures have been the result of flow-induced vibrations. A typical diffuser failure example can be found in Chapter 5. In the extreme, large sections or even the complete exhaust diffuser has separated from the exhaust casing. A common fix to exhaust casing vibration issues is to add circumferential stiffening ribs or a discharge end stiffener (see Figure 2.44).

Exhaust Diffuser Materials of Construction

Expander exhaust diffuser casings are typically fabricated from 300 series stainless-steel plate and forged rings. Typical construction materials include ASTM A-240/A-182 Types 304, 304H, 321, and 347. The fabrications typically receive a stabilizing stress relief after fabrication.

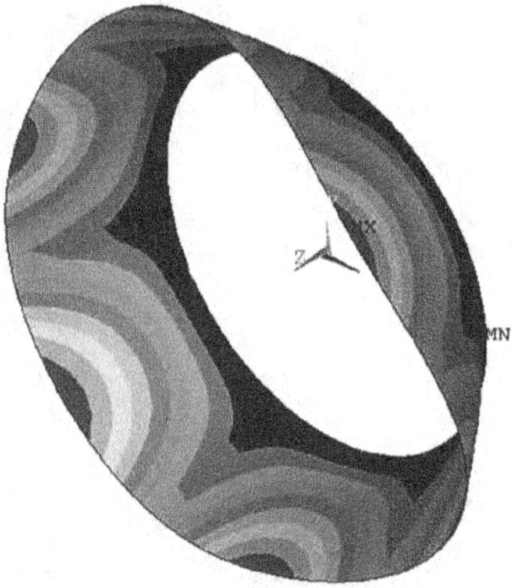

Figure 2.44 Vibration analysis of exhaust diffuser.

2.3.8 Expander Outer Casings

Inlet Casing

The expander inlet casing is the main structural support and pressure-containing wall of the Hot Gas Expander . Two inlet casing designs can commonly be found for the CCU expander. The first design (shown in Figure 2.45) uses the inlet casing as the outer flue-gas flow path. In many ways it is an extension of the inlet ducting. The single-piece casing has fewer parts and is a less complicated design. The single-piece inlet casing design sees higher operating temperatures because it operates at the full inlet temperature of the flue gas. This casing generally responds faster to temperature changes into the expander.

The second inlet casing design is shielded from the flue-gas main flow via a nose cone outer liner, as can be seen in Figure 2.46. The shielded design operates at a temperature that is less than the flue-gas inlet temperature. This is particularly true when the nose cone outer liner is insulated to reduce the heat transfer between the two casings. Lower operating temperatures generally result in lower operating stresses and higher materials capabilities.

Both inlet casing designs need to accommodate the 300°F (167°C) temperature drop that occurs across the axial length of the expander. In the single-piece inlet casing design, the temperature gradient is pushed further back in the casings toward the exhaust casing. In the

FIGURE 2.45 Single piece inlet casing design. (courtesy of the Elliott Co.)

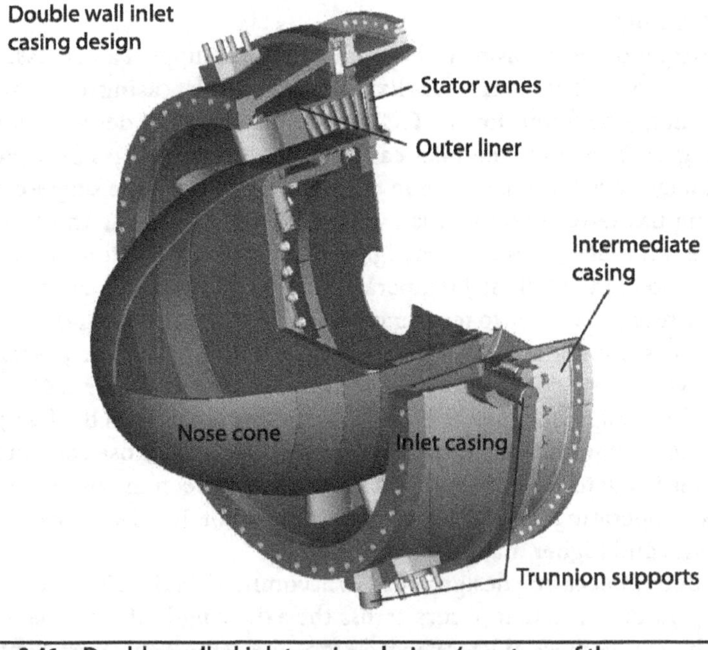

FIGURE 2.46 Double-walled inlet casing design. (courtesy of the Dresser-Rand Co.)

double-wall casing design, the temperature gradient occurs mainly in the inlet/intermediate casings.

Inlet Casing Design Considerations

Inlet casings must be designed to accommodate the maximum-rated inlet pressure and temperatures to which the expander may be exposed. Whereas the inlet casing sees the highest operating pressures, the casing hoop stresses that result from the internal pressure are quite low and are generally a small part (less than 20%) of the overall casing stress. The highest stresses are the thermal stresses that result from radial and axial temperature gradients in the casing. Additional stresses result from various piping and aerodynamic loads. The casing design is generally controlled by thermal stresses, piping loads, and the limiting of casing deflections.

Expander inlet casings do experience significant radial and axial thermal growth. Any restraint on this thermal growth can result in large increases in casing stresses and nonuniform deflections and distortions. Since the inlet casing often is the main support system for the expander flow-path components, any distortion or nonuniform deflection can result in adverse changes to the rotor clearances. The thermal growth of the casing is controlled by supporting the casing on trunnion pins or guide keys. The pins or keys allow the casings to grow freely, while maintaining the casing alignment to the rotor.

Expander inlet casings are attached directly to the expander inlet ducting. This ducting is physically large in size and is constructed as a hot wall design. The weight of the duct and thermal movements are high and often result in significant loadings being applied to the expander inlet flange. These loadings can easily cause excessive force and moments to be imposed in the inlet casing and can lead to high casing stresses and distortions that affect the rotor to casing operating clearances.

Horizontally split casings are not typically used in Hot Gas Expanders owing to casing distortions that often result from nonuniform thickness casings. In addition, the flange heights tend to be minimized and are often insulated to reduce radial thermal temperature gradients and stresses in the casings.

Expander outer casing stresses are multidimensional and complex. The use of FEA has proven to be the best analytical tool for the design of the casings.

Typical structural design criteria for expander casings are as follows:

- The maximum nominal stresses developed in the casing during normal operation should be less than the yield strength of the material at temperature.

- The maximum bending fiber stresses developed in the casing during normal operation should be less than 1½ times the yield strength of the material at temperature.
- Casing stresses are evaluated against creep strength, considering bending and pure tension/compression. The design must meet the following on a local point basis:

$$0.75\ \sigma_B + \sigma_{P/A} \leq 0.5\%,\ 100{,}000\ \text{hr creep strength}$$

σ_B—bending stress
$\sigma_{P/A}$—tension or compressive stress

The inlet casing upstream flange needs to be compatible with the inlet duct flange designs. These flanges are gasketed joints generally designed in accordance with the ASME Pressure Vessel Code. Flat-faced, raised-faced, slip-on, and seal-lip weld neck designs are commonly used for the inlet duct-to-casing flange. Even though the inlet casing-to-duct flange is gasketed, gaskets are not typically used in the flanges between the expander casings. Metal-to-metal joint designs are common, although high-temperature joint compounds are typically used. Some manufacturers include the use of an Inconel packing in the flange faces to ensure minimal leakage.

Rotor Part Containment

In more recent years, machinery manufacturers and operators have become more concerned about part containment from a major rotor failure. Obviously a lack of containment represents a major safety hazard within the refinery. The expander inlet and intermediate casings are often the components that need to contain any separated rotor parts or pieces. The basis for containment design is to ensure the energy absorption capability of the casing exceeds the kinetic energy of the released part(s).

Most controversial in the design of the casings is what type of failure the casing needs to be designed for. The expander rotor components are large and massive and generate a significant amount of energy upon separation. Design requirements vary from the separation of one to six rotor blades, including the attachment region mass. It is not a typical practice to design for the separation of a portion or the entire disc. Fortunately, there have been very few reported failures in which the expander casing(s) have failed to contain separated rotor parts. In the few cases where failed rotor components did exit the casings, it involved a major rotor disc failure. Occasionally, rotor blade or disc failures have penetrated the outer casings sufficiently to cause the pressure boundary to be breached and allowed significant flue-gas leakage into the surrounding area.

Because of the high energy levels of the rotor, rotor failures are quite devastating with both major expander internal and external damage. Bearing and seal failures often result from a rotor blade or disc failure. High rotor vibrations can lead to the failure of other components including bearings, supervisory instrumentation, and lubrication oil supply and drain lines.

Inlet Casing Materials of Construction

CCU expander outer casings are typically fabricated from 300 series stainless-steel plate and forged rings. Typical construction materials include ASTM A-240/A-182 Types 304, 304H, 321, and 347. Even though the 300 series stainless-steels casings provide good oxidation and resistance to sulfidation, they can be susceptible to shut down aqueous corrosion such as polytheonic acid. The CCU expander outer casings designed for the highest operating temperatures have been fabricated from Incoloy 800H conforming to UNS-N08810. The Incoloy 800H material provides increased material capabilities and in-service aging resistance. H grade materials are often used to take advantage of the higher-temperature allowable stresses of the high-carbon versions of the alloy. The optimum material properties are achieved if the fabrications are solution heat treated and precision machined.

The stainless-steel materials must be carefully selected, fabricated, and periodically checked after service owing to material degradations that can occur from the following:

- *Sensitization.* During the fabrication of 300 series stainless steel, chromium carbide precipitates can form at the grain boundaries of the material. These precipitates locally deplete chromium around the grain boundaries and significantly reduce corrosion resistance of the material.
- *Sigma phase embrittlement.* Sigma phase is an iron-carbon phase that is formed when the material is subject to operating temperatures between 932 and 1,652°F (500 and 900°C). Sigma phase preferentially forms along the grain boundaries of the material. Significant reductions in the material toughness and ductility result from the sigma phase formation.

Intermediate Casing

Some expander designs include an additional casing between the inlet and exhaust casings referred to as the intermediate casing. The intermediate casing serves several purposes depending on the inlet casing design. The primary purpose is to allow installation and removal of the nose cone and flow-path shroud and/or stator assembly in many designs. Two-stage expanders also require intermediate casings for installation and removal of the second-stage flow-path components.

Many intermediate casings include borescope provisions that allow flow-path inspections during unit shutdowns. A spring-loaded plug that seats in the shroud ring is removed and a borescope can be inserted into the region between the stator vanes and the rotor blades.

Intermediate Casing Structural Design Considerations

The intermediate casing is designed with the same techniques as the inlet casing. Because the intermediate casing generally operates at the expander exhaust pressure, the pressure stresses are generally insignificant. The intermediate casing is generally subjected to lower temperatures and temperature gradients than the inlet casing. Overall, the result is greatly lower operating stresses.

Intermediate Casing Materials of Construction

Intermediate casings are typically fabricated from 300 series stainless-steel plate and forged rings. Occasionally, the intermediate casing has been manufactured from a casting. Typical materials of construction include ASTM A-240/A-182 Types 304, 304H, 321, and 347. The lower operating temperatures reduce the amount of in-service aging.

Exhaust Casing

The expander exhaust casing is a flow guide that receives the flue gas exiting the expander diffuser and turns the flow from an axial direction to a radial direction. Depending on the manufacturer, the casing made be a single-piece fabrication or multipiece construction. With the rotor removal from the rear design, the exhaust casing is split into the outer and inner exhaust casings. The inner exhaust casing is horizontally split.

The exhaust casing assembly usually includes the exhaust diffuser, inner exhaust casing/flow guide, casing supports, guide keys, viewports, and drain connections. (See Figure 2.47.)

One of the unique aspects of the CCU expander is that the expander casing is not rigid and is supported in multiple locations. The main casing support is on the inlet casing assembly. The inlet casing assembly is supported on inlet struts that are part of the main baseplate. The rotor, bearing, and seal assembly are supported from the bearing support. The two support points are maintained and aligned to each other via the expander baseplate. Under normal operation, the expander inlet casings grow toward the bearing support and the bearing support grows toward the inlet casing. This thermal growth needs to be accommodated or else excessive stress and casing misalignment is likely to occur. The differential thermal growth is

FIGURE 2.47 Expander exhaust casing.

accommodated in the expander exhaust casing assembly. Two major designs are used in CCU expanders to accommodate the differential thermal growth between the inlet casing assembly and the bearing support. The first design uses a flexible exhaust casing that deflects in the axial direction during normal operation. This design concept is commonly referred to as "oil canning" of the front and back walls of the exhaust casing.

The second design uses an expansion joint in the casing assembly to accommodate the axial movement, as can be seen in Figure 2.48. The expansion joint absorbs the axial movement in a compression mode.

The expander exhaust pressure is determined by the pressure losses in the downstream flue-gas ducting and pollution control equipment such as flue-gas scrubbers, electrostatic precipitators, waste heat boilers, and exhaust stack. These losses can be as little as 0.5 psig (0.035 kg/cm^{2}g) to as high as 5.0 psig (0.35 kg/cm^{2}g).

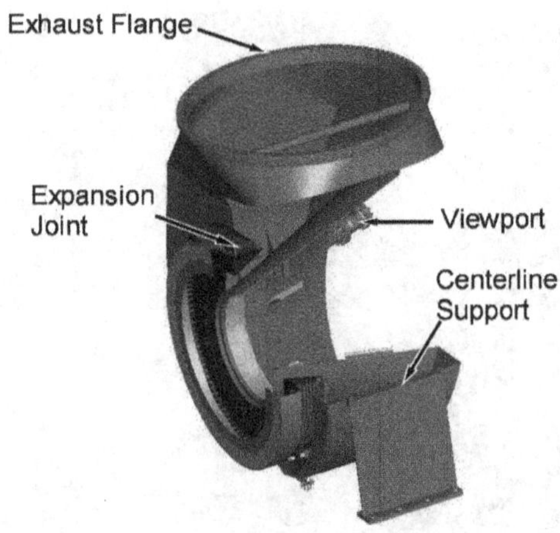

FIGURE 2.48 Single-piece exhaust casing with expansion joint.

Exhaust Casing Structural Design Considerations

The flue gas exiting the expander is typically 300°F (149°C) less than the inlet temperature due to the energy extracted in the expander flow path. Most exhaust casings operate at temperatures between 900 and 1,100°F (480 and 593°C). At these operating temperatures, high-temperature creep and rupture are not a major consideration in the design.

Expander exhaust casings are attached on the outlet to the flue-gas ducting. This ducting can be very large in size (often over 80 in [2 m]) and constructed as a hot or cold wall design. The weight and thermal movements are high and often result in significant loadings being applied to the expander exhaust flange. These loadings can easily cause excessive force and moments to be imposed in the exhaust casing and can lead to high casing stresses and distortions that often affect the rotor-to-casing operating clearances.

Exhaust Casing Materials of Construction

Expander exhaust casings are typically fabricated from 300 series stainless-steel plate and forged rings. Typical construction materials include ASTM A-240/A-182 Types 304, 304H, 321, and 347. Whereas the 300 series stainless steel provides good oxidation and sulfidation resistance, they can be susceptible to shutdown from aqueous corrosion such as polytheonic acid. The fabrications typically receive a stabilizing stress relief after fabrication and prior to machining.

Casing Fasteners

CCU expander casing fasteners have been a problematic area in many expanders over time. The most common problem related to fastener issues is flue-gas leakage from the casing flanges. Flue-gas leakage poses significant safety hazards to plant personnel because the flue gas is hot, toxic, and dust laden. Flange leakages can result from both fastener relaxation and failures.

The fasteners are subject to high loadings primarily from the flange preload, casing pressure loads, and in-service thermal stresses. The thermal stresses result from casing flange differential temperatures and differences in material thermal properties. Fastener operating temperatures can be 50 to 200°F (28 to 110°C) cooler than the casing wall or flange temperature. This temperature differential cases additional loadings in the fasteners. The problem is further accentuated by the fastener material. Fastener materials that have comparable coefficients of expansion to the casing having lower strengths and experience more creep and load relaxation. High-strength fasteners with good creep resistance tend to have lower coefficients of expansion than the casings. For example, most casings are manufactured from 300 series stainless steel that have a coefficient of expansion of 10.3E-6 in/in/°F. A stainless-steel fastener would have the same coefficient of expansion, and the differential growth would only be the result of the operating temperature differences. In contrast, an Inconel fastener would have a coefficient of expansion of 8.8E-6 in/in/°F. The differential growth is much larger owing to the operating temperature and coefficient of thermal expansion differences. This can result in the fastener loadings increasing 20% to 50% under normal operation.

A variety of flange designs have been used as part of the expander casing design. A majority of the flanges are metal-to-metal joints and use a joint compound. Some manufacturers include a high-temperature packing in the face of the flange to reduce the potential for leakages.

A number of design upgrades have been incorporated into many expander casings to eliminate flange leakages and fastener failures that include one or more of the following design features:

- Upgrading the fastener material to a high-strength, more creep-resistant material.
- Changing the material heat treatments to improve the ductility.
- Incorporating more fatigue-tolerant designs that include large undercuts and necked-down bodies. (See Figure 2.49.)
- Utilizing round radius root forms (UNR Per ASME B1.1).
- Rolling the threads to increase strength and corrosion resistance.
- The use of Inconel Belleville washers to absorb transient loadings and maintain preload.

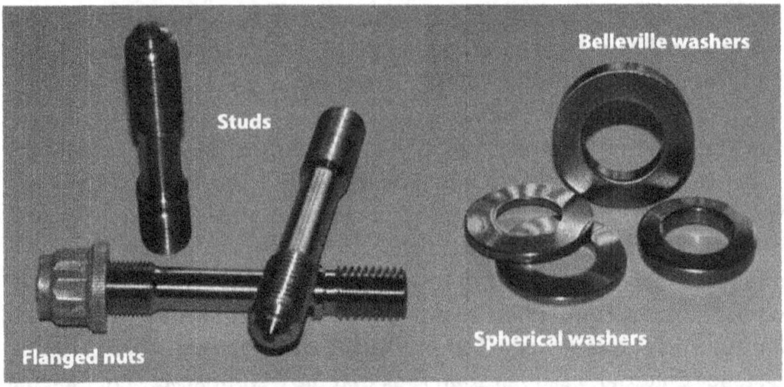

FIGURE 2.49 Expander specialty fasteners.

- The use of spherical washers to reduce fastener bending.
- Shot peening the threads to increase fatigue resistance.
- The judicious use of casing and fastener insulation to reduce thermal gradients.

Critical to the successful application and trouble-free operation of the casing fasteners is the method of tightening. The fasteners must be tightened in a specified pattern that uniformly pulls the casings together with uniform flange loadings and contact pressure. Some fasteners require "hot" retightening after the expander is operating under steady-state conditions.

Expander casing fasteners would be considered critical components that are highly loaded. Typical design criteria for the casing fasteners are:

- The bolt preload stress should be less than 90% of the yield strength of the fastener.
- The force required to produce 80% of the fastener preload stress must be sufficient to maintain the seating load of the gasket.
- The fastener stress must not relax by more than 20% in 50,000 hr of operation.

Fastener Materials of Construction

Fastener material selection needs meet the typical design criteria as well as consider the environment effects. Depending on the location and use, the fasteners can be subjected to both the corrosion and erosive effects of the flue gas.

A large variety of materials have been used for expander casing and piping connections. The variety of materials result because there

Material	Specification	Advantages/Disadvantages
Low alloy steels	ASTM A193	High strength
Cr Mo Steel	Grade B7	Low cost and ready availability
Cr Mo Vn Steel	Grade B16	Poor thermal compatibility with casing materials
		Poor high-temperature corrosion resistance
		Low creep strength accelerates relaxation
300 series SS	ASTM A193	Good corrosion resistance
304	Grade B8	Good thermal compatibility with casing materials
321	Grade B8M	Moderate costs and good availability
316	Grade B8T	Low strength at high operating temperatures
		Low creep-strength accelerates relaxation
Iron-base superalloys	High strength	
A-286	ASTM A453	Good creep strength
Grade 660		High cost and fair availability
Nickel-base alloys		High strength
Inconel 718	AMS-5662	Good creep strength
Waspaloy	AMS-5709	Good corrosion resistance
		Good thermal compatibility with casing materials
		Highest cost and poor availability

TABLE 2.1 Materials for Expander Casing and Piping Connections

is no "ideal" fastener material. The materials listed in Table 2.1 have commonly been used in CCU expander applications.

Users need to take extreme care when using all of the above materials as their material properties are very sensitive to heat treatment.

2.3.9 Rotor Assembly

An overwhelming majority of CCU expander rotor assemblies are built as single, overhung disc designs such as that shown in Figure 2.50.

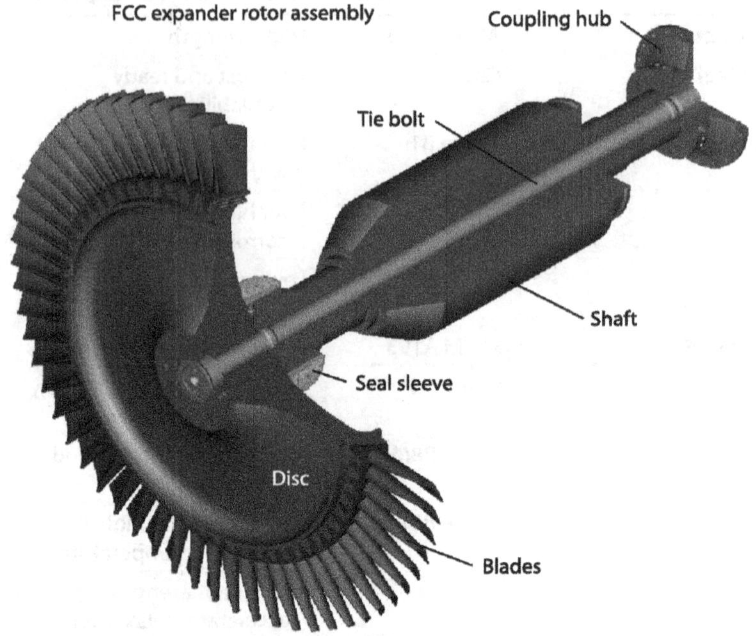

Figure 2.50 Single-stage, single-tie bolt rotor assembly. (courtesy of the Dresser-Rand Co.)

The CCU application is considered a low-pressure, high-temperature, and high-flow application. The expander operates at high power densities. The single-stage design reduces the complexity of the flow path and makes the expander less susceptible to catalyst erosion and deposition. Another significant advantage of the overhung rotor design is the lack of a need for a high-pressure seal. The only shaft-end seals are on the low-pressure exhaust of the expander.

The major components include the rotor blades, disc, tie bolt(s), shaft seal sleeve, shaft, and coupling hub.

Rotor Blades

CCU expanders use tapered and twisted reaction blade designs such as those shown in Figure 2.51. Unique to the CCU application is that the blades need to be designed for both high efficiency and long-term erosive life. The CCU expander application requires the flow-path components to handle high levels of erosive catalyst that is entrained in the flue gas. A typical CCU expander is subjected to more than 100 lbs/hr (45 kg/hr) of highly erosive aluminum-silica catalyst. Over a five-year operating period, that equates to 2,200 tons (2,235 metric

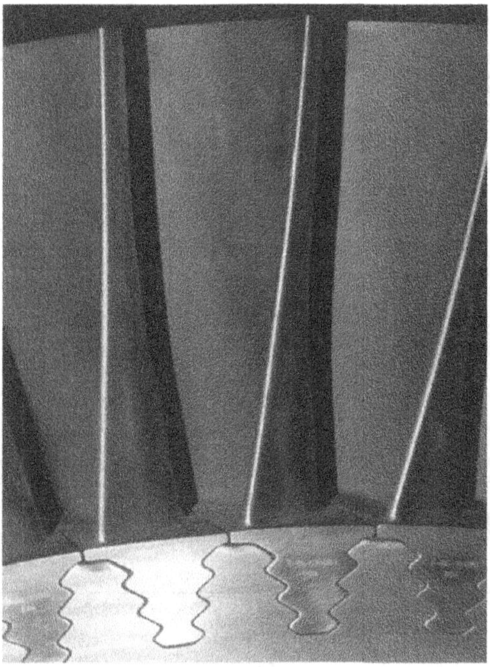

FIGURE 2.51 Typical CCU expander rotor blades.

ton) of a sandlike abrasive material passing through the expander blading. Very few pieces of equipment are capable of handling these levels of abrasive material for the long operating periods.

Early applications of the CCU expander showed that the expander blades would experience excessive wear in as little as a few months of operation. Improved particle separation equipment, new expander blade designs, and improved metallurgies and coatings have allowed modern expanders to operate continually for five or more years while maintaining high performance levels. The rotor blades need to be very robust to account for the erosive wear.

To achieve the high efficiencies accommodate process condition changes and long operating-periods, the CCU expander blade design has had to change over time. The initial applications were designed for lower pressure ratio and energy levels. In the 1960s, pressure ratios of 1.7 along with inlet temperatures of 1,200°F (649°C) resulted in enthalpy drops of 60 BTU/lbm (63 k-joules). As the process designs have changed for increased process throughput and to run a wide range of feed stocks, the operating pressures and temperatures have increased significantly. Today, expander pressures of 3.0 to 3.4 are common along with inlet temperatures in the 1,350°F to 1,400°F (732°C to 760°C) range.

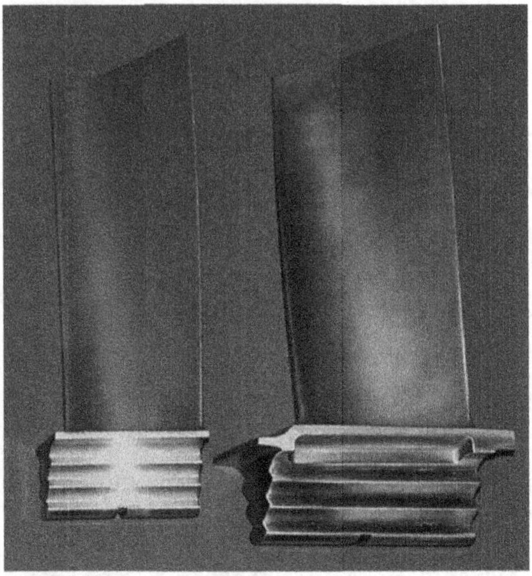

FIGURE 2.52 Early versus modern-day CCU expander blade designs.

These higher inlet conditions result in energy drops as high as 135 BTU/lbm (142 k-joules) or more than twice the energy levels of the early expanders. Power densities have increased from 400 hp/blade (300 kW/blade) to over 1,000 hp/blade (745 kW/blade) under these newer operating conditions. These higher energy applications have required newer blade designs, as can be seen in Figure 2.52. The original blade design is on the left, while the more modern blade design is on the right.

Rotor Blade Structural Design Considerations

The dirty gas environment of the CCU expander affects how the expander blade is designed structurally. Because the blades may experience erosive wear, corrosion damage, and even heavy tip rubbing, the blades must be designed with high structural margins. The normal operating stresses in the expander blades are the result of centrifugal, aerodynamic, and vibratory loadings.

To accommodate these loadings, the CCU expander rotor blades are conservatively designed with low normal operating stresses. Nominal/average cross-sectional stresses are generally designed to be less than 30,000 psi (2,110 kg/cm^2). Generous attachment fillet radii allow the blade attachment concentrated stress to be less than 120,000 psi (8,435 kg/cm^2). Blade gas bending stresses are typically less than 7,500 psi (527 kg/cm^2).

The rotor blade operating temperatures are very dependent on expander inlet conditions, the flow-path design, and rotor cooling. Metal temperatures in the 1,000°F to 1,200°F (538°C to 649°C) range are common. The elevated gas temperatures dictate the need for special creep- and corrosion-resistant materials. A convenient method for examining the adequacy of a given material at a given temperature under known stress levels is by the use of a modified Goodman diagram. The steady-state stresses are evaluated against the tensile capabilities of the materials. At metal temperatures less than 1,000°F (538°C), the steady-state stresses are usually evaluated against the ultimate tensile strength. Above 1,000°F, the lesser of the ultimate strength or the 100,000-hr stress-rupture strength is used for the evaluation. Steady-state stress safety factors of 1.25 or higher are often used. The vibratory or alternating stresses are evaluated against the material endurance strength at the operating temperature. The combined stresses are evaluated against a line drawn between the endurance limit and the ultimate stress-rupture strength. Alternating stress safety factors of 3.0 are often used for the vibratory stress margin.

Vibration Design

All expander blades have a vibration analyses performed to minimize the potential for a blade failure from the excitation of a nature frequency. Because all CCU expander blades are freestanding, only the fundamental frequencies need to be considered.

Typical vibration design criteria for the rotor blade encompass the following:

- The first five natural frequencies (based on the mean blade length) of the blade should be evaluated on a Campbell or SAFE diagram.
- The first fundamental frequency should be above the three-times running-speed excitation. The fundamental frequency should have a safety margin of at least ±5% from the three-times running speed, inlet strut, and stator vane excitations throughout the running speed range of the expander.
- The second- through fifth-mode frequencies should have a safety margin of at least ±10% from the inlet strut and stator vane excitations throughout the expander running speed range.

A more detailed fundamental discussion of vibration design is included in Chapter 6.

Rotor Blade Materials of Construction

The CCU expander rotor blades have been manufactured in a variety of materials depending on the operating temperature of the blades. Both forged and cast materials have been commonly used.

Iron- and nickel-base materials are the primary materials selected for the rotor blades. The most commonly used rotor blade materials include the following: iron base A-286 manufactured per AMS-5735 and nickel-base material such as Inconel X-750 per AMS-5668, Waspaloy per AMS-5704, and IN-738 per AMS-5410.

Expander rotor blades are subject to a number of operating conditions that are life-limiting include:

- *High temperature*. With the rotor blades operating at metal temperatures in the 1,000 to 1,200°F (537 to 649°C) temperature range, the blades can experience high-temperature material issues such as creep, rupture, and aging. When considering extended or multiple operating campaigns, the blades should be evaluated for high-temperature material degradation. Generally, rotor blades are removed from service because of environmental degradation such as erosion or corrosion as opposed to in-service aging.
- *Erosion*. The expander rotor blades both experience significant amounts of erosive catalyst on a continuous basis and even large amounts of catalyst during process upsets. The erosion potential of the blades is reduced by judicious airfoil and flowpath design, modification of the airfoils to accommodate erosion, and by use of blade coatings. The blade coatings are critical to long-term blade life. Tungsten and chromium carbide coatings using 80% tungsten or chromium carbide and 20% nickel chromium binder are commonly used. The coating composition and the method of application are critical to both the erosion resistance and the adhesion of the coating to the blading.
- *Corrosion*. The rotor blades are also subject to a variety of in-service and shutdown corrosion.
    - *In-service*. The unique environment of the CCU expander can promote nickel and chromium oxidation and sulfidation. The corrosive attack can be quite significant and has been a contributor to more than 40% of all CCU expander blade failures. More detailed discussions on CCU expander corrosion can be found in Chapter 3.
    - *Stress corrosion cracking*. Many of the nickel-base alloys used for expander blades have been found to be susceptible

to stress corrosion cracking (SCC). The blade materials become crack-sensitive in the CCU environment when the stresses of the blade exceed a given level. Further discussion of the SCC phenomena can be found in the rotor disc material section.

- *Shutdown.* While it is not common, some CCU expanders have experienced shutdown corrosion. When the flue-gas temperature falls below 400°F (204°C), a number of sulfur-bearing acids can form. These acids are very aggressive and will cause a corrosive attack of both the rotor blade and blade coatings. Keeping the flue-gas system warm and dry is critical in preventing shutdown corrosion. Some installations also use caustic washes during shutdown to neutralize the formation of the acids.

- *Component rubbing.* Another unique aspect of the CCU expander is that of catalyst deposition. Catalyst fines deposit throughout the flow path. The thickness of the deposit can exceed the rotor-blade clearances and result in blade-tip rubbing. The tip rubbing can be hard enough to cause blade-tip overheating. The tip operating temperatures can actually exceed the melting point of the base material. Blade-tip rubbing and cracking is a common CCU expander operational problem.
- *Attachment fretting.* Some expanders have exhibited blade attachment fretting. Fretting is generally associated with component movement in regions where movement is not expected. Fretting damage in the blade attachment region has been found to make the area more susceptible to in-service corrosion.
- *Fatigue.* Expander blades can be subject to both high-cycle fatigue and low-cycle fatigue. The high rotational speeds make the blades susceptible to high-cycle fatigue if the natural frequencies of the blades are excited. Low-cycle fatigue can be an issue if the blades are subjected to rapid temperature changes. Rapid temperature changes from process upsets or the ingestion of a liquid are causes of high-transient thermal stress. The thermal mass of the blade varies significantly within the airfoil as well as from the airfoil to the attachment region. With a rapid temperature change, the airfoil strain is very high and cracking can occur. The result can lead to cracking of the airfoil.

Rotor Blade Aerodynamic Design

As previously noted, the CCU expander blade evolved over time. The original airfoils were designed for the lower energy levels of

60 to 90 BTU/lbm (29 to 42 k-joules/kg). Very conventional 50% reaction (one-half of the energy drop occurs in the stator and one-half occurs in the rotor blade) designs provided for high-efficiency. The airfoils were developed using conventional velocity triangle technology. As the erosion potential of the CCU expander became more evident, the blade designs were modified to accommodate the erosive wear. Thicker blade profiles have been used to reduce the operating stress levels and to accommodate the erosive wear. This is particularly true in the area of the leading and trailing edges.

With the increasing inlet pressures and temperatures over time, the energy drops have steadily increased to levels in excess of 120 BTU/lbm (55 k-joules/kg). For many years, new aerodynamic technology was slow to be accepted by expander operators since operating campaigns of two to three years were achievable with improved blade materials, coatings, and "the tweaking" of the existing airfoil geometries. As the process technologies improved and allowed longer operating campaigns for the process equipment, the need for the expander time between turnarounds also needed to increase. To increase the erosion life of the expander, new stator and rotor-blade geometries were eventually developed in the 1980s and 1990s. Some improvements in both aerodynamic efficiency and erosive life were achieved with conventional aerodynamic design techniques that optimized the airfoil geometries. In the later part of the 1990s, computational fluid dynamics (CFD) technology began to mature and become an integral design tool for modern turbomachinery. Computational fluid dynamics is used extensively today in the design of expander flow-path components. The CFD analyses are three-dimensional, compressible-flow, viscous, multiframe analyses that allow full analysis and visualization of complex flow phenomena. Figure 2.53 shows typical CFD models for a single-stage CCU expander.

In more recent times, CFD analysis capabilities developed significantly in the area of erosion prediction when the analysis could include the injection of particles and tracking of the particle trajectories. The technology advanced further when erosion models were incorporated into the CFD codes.

CFD analysis has also been found to be an excellent tool in reducing the secondary flow-performance losses and its effects on erosion. Figure 2.54 shows the surface flows on an early CCU expander blade design. Figure 2.55 shows an investigation into the effects of expander cooling flow on the blade-platform boundary-layer flow. Analyses of this type have been able to assist in the improvement of the expander efficiency and the reduction in secondary flow erosion. Today, CFD analysis has proven to be a valuable tool in improving

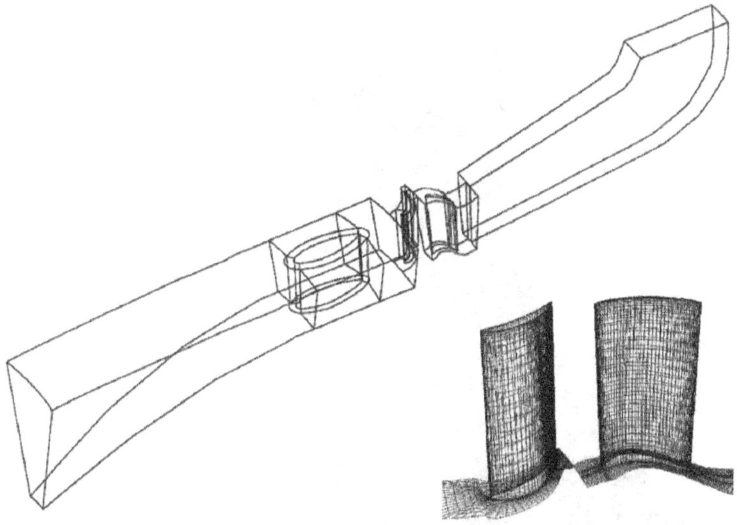

Figure 2.53 Computation fluid dynamics models of a CCU expander flow path.

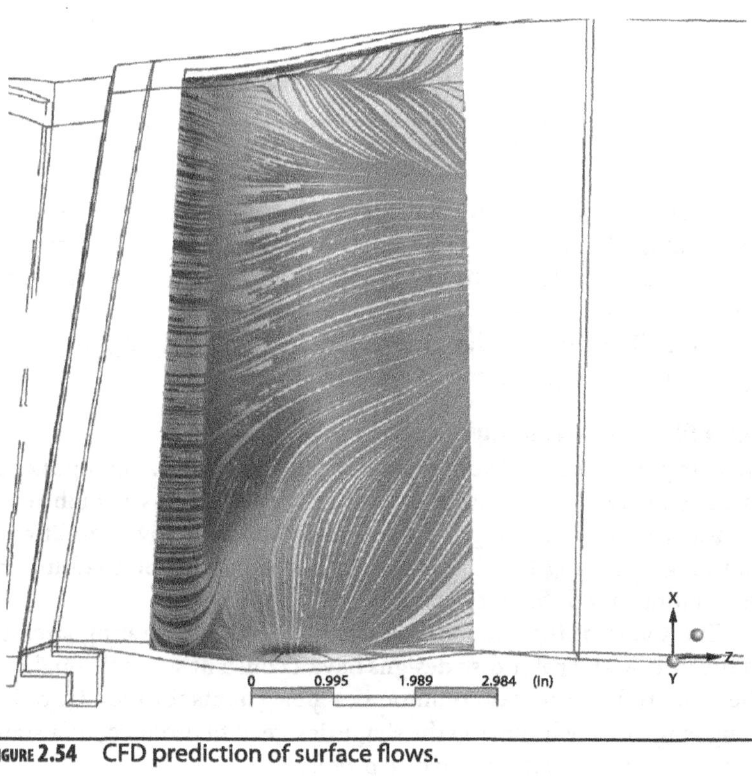

Figure 2.54 CFD prediction of surface flows.

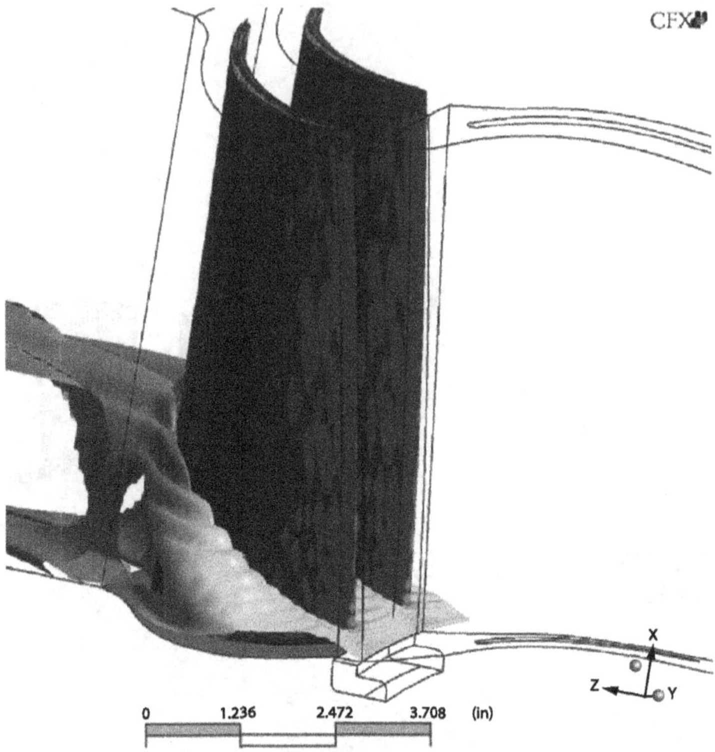

FIGURE 2.55 CFD prediction of expander cooling flows.

the performance and erosion characteristics of CCU expanders. Excellent qualitative agreement exists between the analytical predictions and the field operating experience. From an erosion perspective, expander service lives have been increased between 50% and 100% using the CFD tools. Modern expanders are now achieving consistent operating campaigns of four to six years.

Rotor Discs: General Design

CCU expander rotor discs are one of the most critical components in the expander. Rotor discs are subject to high stresses, high temperature, in-service aging, erosive conditions, corrosive conditions, and long operating time periods. The consequences of a failure are devastating and difficult to contain.

The shape of the rotor disc is dictated by the operating stresses of the disc. Many rotor disc designs have a constant stress throughout the disc profile. Another common design accounts for the disc operating temperature by having the shape designed to produce a constant stress margin throughout the disc profile.

The shape of the rotor disc is also dependent on the type of tie bolt(s) being used in the rotor design. In a single-tie-bolt design such as the rotor shown in Figure 2.56, the missing center material increases the bore stresses. To maintain the low stresses throughout the disc, the disc profile generally bulges out in the center. Multiple tie-bolt designs such as that shown in Figure 2.57 tend to be thinner overall, which often reduces the overhung weight of the disc. Reduced

FIGURE 2.56 CCU expander bladed rotor disc: Center tie-bolt design.

FIGURE 2.57 CCU expander rotor assembly: Multi tie-bolt design.

overhung weight is very beneficial from a rotor dynamics perspective. Multiple tie-bolt designs need to be designed cautiously because the tie bolt through holes or threaded regions when stud bolts are used, can cause high local stresses/stress concentrations. These regions are often the life-limiting region of the disc owing to low-cycle fatigue considerations.

Rotor discs often incorporate seal lips/lands on the face of the rotor disc to improve the effectiveness of the steam cooling. These extensions of the disc are integral to the forging. Using seal lips/lands must be done judiciously as there is a strong tendency for catalyst to deposit and build up on the underside of the seal lip. Seal lip deposits can cause rotor unbalances and be difficult to remove while the expander is in service.

Many rotor discs include integral balancing rings or drilled and tapped balance weight holes to assist in the rotor balance.

Structural Design Considerations

The expander rotor discs are generally very conservatively designed from a structural standpoint. Typical structural design criteria for an expander rotor disc are as follows:

- The average or nominal stresses should be less than 60% of the material yield strength and less than 75% of the 100,000-hr stress-rupture strength.
- Any local point stress in the disc should be less than the yield strength or else a low-cycle fatigue analysis should be run.
- The low-cycle fatigue life should be greater than 10 times the expected life. The minimum number of cycles should be 1,000 cycles (based on a start-up and shutdown once per week for 20 years).

Unlike the expander rotor blades that are likely to be replaced due to erosion or corrosion damage in one or two operating campaigns, the rotor disc is returned to service multiple times with new blades. Rotor discs are generally run until their service life is depleted. The service life of the rotor disc is highly dependent on the operating stress levels, temperatures, and time of service. Most expander manufacturers use a design life of 100,000 hr (11.4 years). Since design criteria of this type are the basis of the maximum-length rotor blade, operating at the maximum continuous operating temperature and speed, the actual design life of the disc may be significantly higher than the 100,000-hr limitation. In addition, the actual operating conditions of the expander will produce an effective life. The effective life can be less than the design life if the rotor has been subject to above-normal operating speeds and temperatures and can be greater than the design life if the rotor has operated at lower speeds, temperatures, and power.

Disc Materials of Construction

CCU expander rotor discs are precision machined from high-alloy, high-strength steel forgings. Closed-die forgings are preferred over pancake forgings due to increased mechanical properties that are the result of the additional work and grain flow imparted into the forging. CCU expander rotor discs have been manufactured from iron-based superalloys such as A-286 (AMS-5735) and nickel-based superalloys such as Incoloy 901 (AMS-5661) or Waspaloy (AMS-5707). The various expander disc materials are chosen for their elevated temperature mechanical properties including:

- Yield strength
- Ultimate strength
- Ductility
- Endurance strength
- Creep
- Stress rupture
- Low-cycle fatigue
- Thermal conductivity
- Thermal expansion characteristics
- Hot corrosion resistance
- Hot erosion resistance

Expander rotor disc is subject to a number of operating conditions that can limit its functional life, including:

- High temperature
- Stress
- Creep
- Erosion
- Corrosion
- In-service aging
- Component rubbing
- Attachment fretting
- High-cycle fatigue
- Low-cycle fatigue

Stress Corrosion Cracking

A significant number of expander blade and disc failures have been related to stress corrosion cracking (SCC). SCC is the initiation and growth of cracks in a corrosive environment. It can lead to rapid failure of normally ductile metals that are subject to tensile stresses and a corrosive environment. The susceptibility to SCC can be accelerated at elevated temperatures. The corrosive environment can be one that is only mildly corrosive to the base material. It is the combination of tensile stress, a corrosive environment, and elevated temperature that causes this cracking phenomenon to exist in the expander. The problem is quite complex and difficult to predict. For example, many of the iron-based superalloys do not experience SCC, whereas virtually all of the nickel-based alloys do. There appears to be a threshold stress level in which the SCC does not occur. Operation of the components with stresses below the threshold value do not experience the SCC phenomena. The threshold stress level in the CCU expander applications has been identified to be approximately 40,000 psi (2,812 kg/cm^2) nominal stress in the rotor component. The SCC potential is a strong argument for the designing and operating of low-stressed rotor components.

Disc Life

With rotor discs operating multiple operating campaigns, it is common for a rotor disc to operate in excess of 100,000 hr. With these long operating periods, the material mechanical properties can change. This is referred to as in-service aging. These changes have a negative impact on the structural margins of the rotor disc. Common life-limiting material issues include high-temperature creep and embrittlement. Periodic life-assessment should be conducted to assess the progress of the material degradations and to ensure safe operation. A life-assessment analysis includes:

- Determination of the actual disc operating life
- Detailed operational review
- Determination of equivalent operating hours (EOH)
- Evaluation of disc erosive wear, corrosion, thermal fatigue
- Nondestructive examination of the disc

Rotor Cooling

CCU expanders are designed and constructed to operate at the elevated flue-gas temperatures ranging from 1,100°F (593°C) to 1,415°F (768°C). These flue-gas temperatures result in expander component metal temperatures between 900°F and 1,382°F (482°C and 750°C). The high metal temperatures of the expander require the use of specialty materials to withstand the resulting normal and transient operating stresses. Additional stress margins in many components can be achieved by using a cooling media in the gas expander to cool the critical components such as rotor discs, tie bolts, and blades. Stationary component cooling is generally not incorporated owing to the lower operating stress levels in the stationary components and the difficulty in cooling the large-size casings. In many cases, the cooling of certain components can be detrimental since large thermal gradients can develop and actually cause a significant increase in component stress. This is particularly true for the expander stationary components such as outer casings, nose cones, and seal housings.

The effect of lowering metal temperatures in an expander component is quite significant with a doubling of the creep or stress-rupture life for every 25°F (14°C) reduction in metal temperature.

Expander Internal Cooling Arrangements

Normal Cooling Normal cooling is the most common form of cooling used in CCU expanders. This cooling refers to cooling media being introduced into the cavity created between the expander disc and the nose cone. The cooling media is most commonly steam, although cooling air has been used in some expander installations. The cooling media lowers the local operating temperature, which then lowers the rotor disc operating temperature. The cooling media escapes by exiting the cavity in the annual area between the rotor disc and the upstream inner stator or shroud ring. The expander normal cooling flow will typically decrease the operating temperature of the rotor disc by more than 50°F (28°C), which will increase the disc life by a factor of four.

In early versions of lower operating temperature expanders, the normal cooling steam entered the cavity between the inlet casing and the nose cone outer liner. The cooling steam provided cooling to the casing and acted as a purge to the cavity formed by the nose cone outer liner and inlet casing. The stream flowed through the nose cone inlet struts and into the bullet region of the nose cone. A baffle plate on the downstream side of the nose cone directed the steam to the hub region of the rotor disc. The various expander manufacturers had different normal cooling steam-flow rates that ranged from 100 to 2,000 lb/hr (45 to 907 kg/hr).

As CCU expander inlet temperatures increased from 1,100 to 1,350°F (593 to 732°C), the need for additional and more effective cooling increased. Normal cooling steam-flow rates have been increased to rates that range between 3,000 and 10,000 lb/hr (1,360 and 4,500 kg/hr). The cooling steam-flow rate is dependent on the expander size, operating temperature, and speed, as well as the steam supply pressure and temperature. To keep the cooling steam as cool as possible, newer expander construction introduced the cooling steam directly into the nose cone to prevent over-cooling of the inlet casing and nose cone outer liner. The normal cooling steam supply tube should be directed toward the rotor disc for more effective disc cooling and to prevent direct impingement of the cooling steam on the nose cone bullet or baffle plate and causing excessive thermal stresses. Many expander nose cones are insulated on the inside of the bullet to prevent nonuniform or excessive cooling.

Impingement Cooling As expander inlet temperatures rose above 1,300°F (704°C), a second cooling system known as impingement cooling was incorporated into many expander designs. Impingement cooling is used to lower the expander rotor disc rim temperature by impinging high-velocity steam onto the disc rim area. Cooling steam is impinged directly onto the rotor rim via a series of six to nine impingement tubes and nozzles. The impingement steam cooling is very effective in lowering the operating temperatures of the high stressed blade and disc attachment regions. Temperature reductions of more than 120°F (67°C) in the disc rim are common and greatly increase both the structural margins and lives of the critical rotor blades and discs.

The relatively cool operating temperature of the impingement cooling steam piping with respect to the nose cone requires a significant amount of flexibility in the system to prevent excessive supply piping loadings. As can be seen in Figure 2.58, the impingement

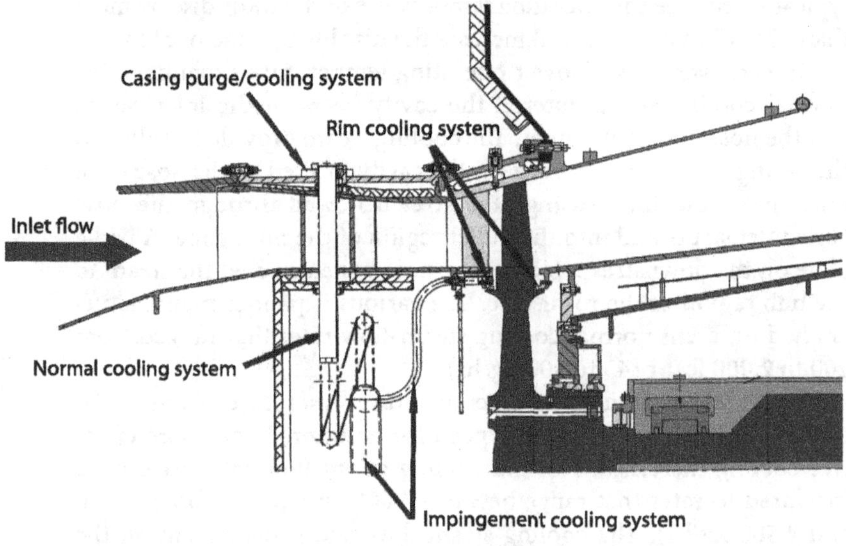

Figure 2.58 Various expander cooling steam designs.

cooling steam system does create added complexity to the expander nose cone design.

Rim Seal Cooling More modern expanders operate with rotor disc rim-seal cooling to accommodate operation at inlet temperatures above 1,385°F (751°C). The rim seals have been shown to be the most effective cooling arrangement for lowering the disc and blade-rim operating temperatures. The rim seals are designed to minimize the amount of hot gas recirculation around the disc faces and keep more of the cooling steam in the bore regions. The rim-seal cooling is very effective and can lower the rotor disc temperature another 100°F (38°C).

While described individually, any of the three cooling system designs can and have been used singularly or in combination in many CCU expanders.

Different expander manufacturers have used different amounts, degrees, and cooling schemes. Both steam and air have been used for cooling in the CCU expander. Steam is the most commonly cooling media owing to its higher heat-transfer characteristics and its inability to cause or contribute to a flue-gas afterburn. In addition to the internal cooling systems described above, several external cooling systems have also been used.

Wind Milling Cooling Expanders that are started up using auxiliary driver(s) and have tight shutoff valves in the inlet line have the potential to heat up from blade windaging. The blade windage will heat up the internal air just as that which occurs in expander shop hot recirculation tests. This situation is unusual as few expanders in conventional trains have tight shutoff valves. Swing-through, clipped, or seatless butterfly valves are commonly used in those applications to prevent regenerator over-pressurization. The wind milling cooling is used under these start-up conditions to prevent excessive heat buildup in the expander. The wind milling cooling steam is generally introduced into the expander via the normal cooling steam system. Over time, the need for wind milling cooling has decreased and is not used in most new installations today.

Afterburn Cooling Afterburn cooling steam is a carryover from the early expander installations. CCU units that operate without complete combustion have CO in the flue gas. The introduction of air can cause ignition, both internal to the process and external to the process, including directly in the expander. The afterburn potential is dependent on flue-gas temperature, with the higher the temperature, the greater the afterburn potential. The afterburn potential decreases greatly under 1,000°F (538°C).

The afterburn potential leads to the need for afterburn cooling steam. Afterburning can produce flue-gas temperature in excess of 1,600°F (871°C) and can last from minutes to several hours. The afterburning can be difficult to detect or identify when or where an afterburn is occurring in the system due to a limited amount of flue-gas line instrumentation. The combustion of CO outside of the regenerator has become less frequent as fewer CCU units operate in an incomplete combustion mode and with the improvement of the process controls and instrumentation.

Most new CCU units operate with complete combustion and have excess oxygen levels between 0.5% and 2.5%. Although the potential for an afterburn is virtually eliminated in the original sense of an external CO afterburn, it should be noted that process afterburning can still occur. Today's afterburning is more likely to be an uncontrolled burning of choke or oil in the regenerator. This type of afterburning tends to produce even higher flue-gas temperatures that can exceed 1,800°F (982°C). Fortunately, the lengths of these afterburns are of short duration (minutes). To protect the expander from these high flue-gas temperatures, most expanders today operate with a shutdown trip at a temperature of 1,625°F (885°C).

The protection of the CCU downstream equipment (including the expander) has typically been via regenerator plenum or overhead line-quenching with steam, water, or a combination of both. Expander protection was supplemented with the use of an afterburn cooling connection that introduced larger amounts of cooling steam into the expander. Whereas some over-temperature protection was afforded to the rotor, the benefit was only short term. The amount of afterburn steam is so small in comparison to the flue-gas flow, the gas flow overwhelms the short-term benefit. This is particularly true for the rotor blades that are bathed in the flue gas. Higher levels of cooling steam have actually been found to unfavorably increase the rotor disc operating stresses due to increased thermal gradients.

Afterburn steam has also been a major problem to the expander casings. The introduction of large quantities of steam into the casings can cause large temperature gradients. High-thermal stresses are generated in components that simultaneously experience major material property degradation from the afterburn temperatures. Many casing cracks and excessive distortion have resulted from the use of expander internal afterburn steam. The problem is accentuated in the older expander designs that introduce the cooling steam in the inlet casing cavity between the inlet casing and the nose cone outer liner.

Another noted field problem associated with afterburn steam has been the ingestion of water into the expander. Since the afterburn supply lines are normally closed until they are needed during an afterburn, the collection of condensate in the line often carries over into the expander causing extensive damage.

Today, many expander manufacturers are recommending the elimination of both the wind milling and afterburn provisions on the expander in favor of an afterburn-only capability upstream of the expander. All installations have some quench capability upstream of the Third-Stage Separator or in the regenerator. Improved or increased upstream afterburn capabilities at these locations is more effective in reducing the magnitude of the afterburn to Third-Stage Separator, inlet duct work, inlet and bypass valves, as well as the expander.

Start-up Steam In more recent years, expander start-up steam can be found in many installations. The start-up steam is usually associated with expander generator sets or in conventional trains where the expander is put online after the process has been brought up to full capacity. The start-up steam is an additional cooling steam flow for controlling the expander heat-up rate. The start-up cooling steam is also used in some installations to preheat the expander prior to introducing full-temperature flue gas into the expander. This mode of operation protects the expander from transient process start-up conditions that might have above-normal operating temperatures or catalyst carryover rates. It also allows the operators to concentrate on the process needs without being concerned about bringing another piece of equipment online.

The amount of steam needed during the expander start-up can be significant since the flue gas is high and at its normal operating temperature. The start-up steam is injected into the expander inlet line to mix with the flue gas and control the expander inlet temperature. A temperature controller is used to vary the amount of start-up steam in use and produce a prescribed expander heat-up rate. The start-up cooling steam-flow rates can be significant with typical flow rates between 10,000 and 50,000 lbs/hr (4,500 and 22,680 kg/hr).

Expander Cooling System Operation

Expander operators need to be careful about the commissioning and decommissioning of the expander cooling steam system. The cooling flows are significant and will produce significant levels of condensation throughout the expander and ducting during cool-down periods. The expander is often the low point of the system in which condensation will collect. The following precautions need to be taken to prevent cooling steam condensation:

- Prior to commissioning, the cooling steam systems must be thoroughly blown down.

- The expander and expander duct drains should be open whenever the expander is shut down and/or cool (less than 250°F [121°C]).
- Prior to commissioning the expander cooling system, the expander seal system needs to be commissioned.
- The expander cooling steam should be commissioned when the expander flue-gas temperature equals the steam temperature. This will prevent the formation of condensation and the thermal shocking of the expander or ducting.
- The expander cooling steam system should be decommissioned when the expander flue gas drops below 800°F (427°C) and before it drops down to the steam supply temperature. The expander does not require cooling in this lower temperature range.
- The expander cooling system should be decommissioned if the expander is operating below 1,000 rpm.
- The expander cooling system should be decommissioned if the expander trips offline.

Shafts: General Design

Modern CCU expanders use large stiff shafts. These rotor assemblies generally operate below their first mode shaft bending critical speeds.

The shaft includes the shaft-to-disc interface, provisions for the rotor tie bolts, a flue-gas seal region (often a shrunk-on sleeve), stiff journal regions, a center counterweight region, active and inactive thrust-bearing collars, overspeed mechanical or electrical provisions, and the coupling hub. The shaft is designed to counterbalance the overhung weight of the rotor disc and blades. The characteristic of the CCU expander shaft is the large-diameter body between the journal bearings. Larger than normal journal-bearing sizes are used to reduce the amount of shaft bending from overhung loadings and to improve the rotor dynamics operating characteristics. This generally leads to very low stress levels in the shafting. The best rotor dynamics characteristics have been achieved when significant downward loadings occur at both journal bearings and with short overhung weights.

The active thrust-bearing surface is generally an integral part of the shaft and part of the large center section. The inactive thrust surface may be either an integral part of the shaft or a shrunk-on collar. Similarly, the coupling hub may be an integral part of the shaft or shrunk on.

CCU expander shafts are designed in accordance with the ASME code and are generally manufactured from forged low-alloy steel material such as AISI-4140 or 4340.

Shaft-to-Disc Interface

Critical to the successful operation of the rotor assembly is the mounting of the rotor disc to the shaft. The disc must remain properly centered on the rotor under all operating conditions or else very large rotor unbalances will develop. In addition, the torque generated by the rotor blades is transmitted through the disc and into the shaft via the disc-to-shaft interface.

The interface must also accommodate significant different thermal growth that occurs between the rotor disc and the shaft. The rotor disc operates at temperatures significantly above those of the shaft and has a higher coefficient of expansion. Conventional disc-mounting methods such as the disc being shrunk on or piloted to the shaft do not work because the interfaces either become loose or can be overstressed. The differential thermal expansion is best accommodated by using spline-type interfaces.

The two most common methods for mounting the expander disc is the use of a Gleason curvic coupling as shown in Figure 2.59. The second mounting method is through the use of radial pins mounted between the interfaces. It is common to see eight radial pins used at the interface. The purpose of the radial pins or curvic splines is to transmit the stage torque and to maintain the disc position to the shaft as the hotter disc grows radially with respect to the shaft. Owing to the large size and heavy weight of the rotor, large amounts of rotor unbalance will develop if the disc moves only 0.001 to 0.002 in (0.025 to 0.05 mm) even small amounts off center. The interface torque transmission capability must be able to accommodate the maximum

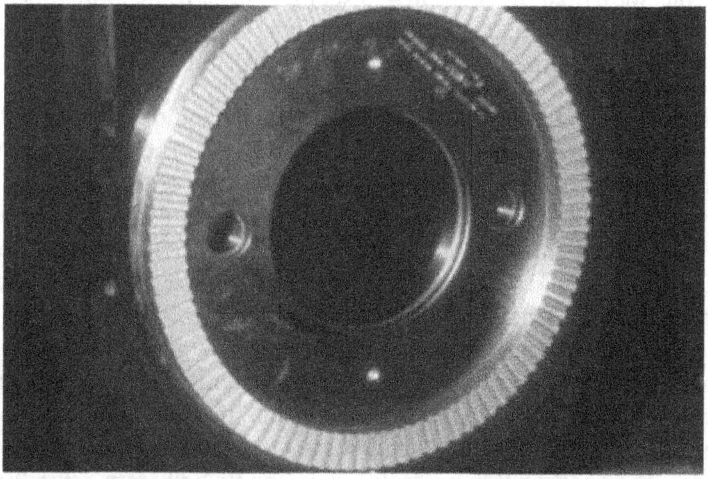

FIGURE 2.59 Curvic coupling spline ground into the face of a shaft.

expander power output at the maximum inlet pressure and temperatures as well as at any transient torques that might be generated during motor/generator synchronizing or loss of load.

Tie Bolt(S)

CCU expanders can be manufactured with either a single-through tie bolt or with multiple tie bolts. Both designs have provided satisfactory service, and it is unusual to have problems with tie bolts. Each design has its own advantages and disadvantages.

Single Tie-Bolt Design In the single tie-bolt design, a single long tie bolt goes through the length of the rotor from the disc to the coupling hub region. An integral tie bolt head is usually located in a recess of the rotor disc. The opposite end uses a spanner-type nut which is also recessed within the shaft end. The single tie bolt is gun-drilled throughout its length to allow a heater rod to be used for stretching and for the measurement of the tie bolt stretch during tensioning.

Very long tie bolts running the length of the rotor need to have the natural frequency of the tie bolt checked to ensure that the tie bolt natural frequency is not coincident with the running speed of the expander. Locator lands or bushings are often used both to center the tie bolt and to limit in-service bending and deflections.

Single tie bolts generally operate at maximum temperatures of less than 700°F (371°C) and can be manufactured out of high-strength specialty stainless steels. The bulk of the tie bolt is located in the rotor shaft and operates at temperatures of less than 150°F (66°C). Tie-bolt relaxation and creep is minimal in the single tie-bolt design. Early expander tie bolts were manufactured in 420 stainless steel, while more modern expanders use high-strength alloys such as 17% chromium and 4% nickel duplex stainless steels conforming to ASTM A564.

Multiple Tie-Bolt Designs Many expanders use multiple tie bolts for the disc assembly as can be seen in Figure 2.60. The number of tie bolts can vary from 6 to 12 depending on the size of the expander. A variety of bolt designs have been used including studs, through studs, stud bolts, and conventional bolts. Generally, tie bolts threaded into either the shaft end or disc are less desirable owing to the higher stress concentrations associated with the threads.

Since the multiple tie bolts are located at a larger diameter in the rotor disc, the multiple tie bolts operate at significantly higher temperatures than those of the single centerline tie bolt. When tie bolts are not centered in the disc and shaft such as in the multi tie-bolt design, the additional bending loads on the tie bolt from centrifugal loading must be limited and accommodated in the design.

FIGURE 2.60 Multi tie-bolt rotor assembly.

Tie-bolt tensioning is very critical to assure that the proper bolt loadings are achieved. This is more critical in multi tie-bolt applications to ensure proper seating of the shaft-to-disc faces or curvic couplings. Nonuniform tightening can actually cause the disc to be mounted cocked on the shaft. The tensioning of multi tie bolts is generally accomplished via torque wrenches or hydraulics. Regardless of how the fasteners are tightened, the measurement of the stretch is the only way to ensure proper bolt loading.

Tie Bolt Structural Design Considerations

Regardless of the tie bolt design, the design requirements are the same. Typical design requirements for the tie bolts are as follows:

- The tie bolt(s) should be sized to provide sufficient clamping force to transmit the maximum torque that the expander can develop.
- The tie bolt should be designed to operate at two-thirds of the material yield strength.
- The tie-bolt preload stress should be less than 80% of the yield strength of the fastener.
- The tie bolt stress must not relax by more than 20% in 50,000 hr of operation.

Very important to the design of the tie bolt is the accommodation of actual operating temperatures and material differences, such as coefficients of expansion and modulus of elasticity. Tie bolts often tighten or loosen in service as a result of these differences.

Tie Bolt Materials

The tie bolts are generally manufactured from high-strength nickel-base alloys such as Inconel 718 conforming to AMS-5662 and Waspaloy conforming to AMS-5709.

Couplings

Flexible couplings are used to transmit the expander power through to the driven equipment. In addition, the coupling is used to accommodate shaft-end misalignments and axial growths.

Early expander installations typically used gear-type couplings. Despite providing acceptable service, the gear couplings often experienced gear-tooth wear and degradation that often led to major expander rotor radial and axial vibrations. Virtually all expanders today use dry couplings. A variety of dry-coupling designs have been used, including diaphragm and various flexible-element couplings. Figure 2.61 shows a typical high-performance coupling that has been designed to reduce the overhung weight and moments of the coupling as well as accommodate two different hub designs. Integral and hydraulically fit coupling hubs are used on the expander shaft ends.

CCU expander couplings tend to use long coupling spacers with typical lengths between 48 and 90 in (122 and 229 cm). The long spacers are required to allow for expander and driven equipment assembly and disassembly.

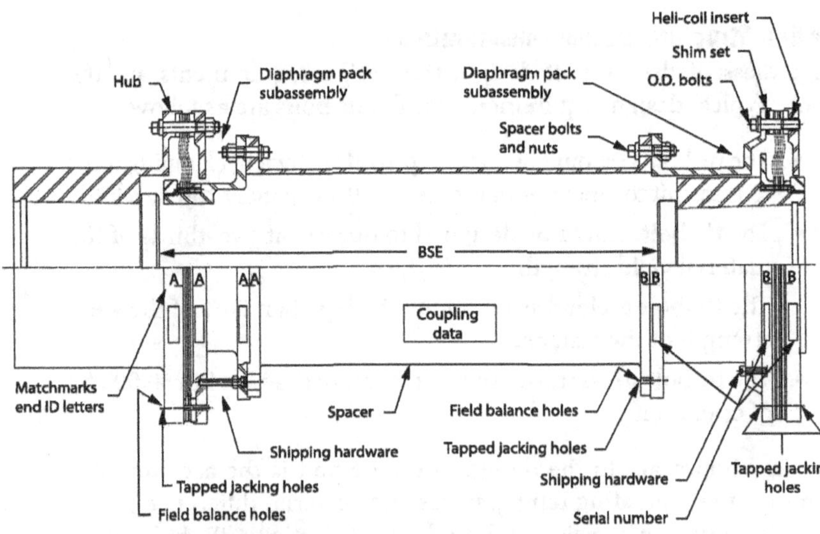

FIGURE 2.61 Typical high-performance power recovery flexible element coupling.

Coupling Structural Design Requirements

It is very important that the coupling manufacturer be provided the design requirements of the coupling. Design requirements should include:

- Operating speed range
- Continuous power and torque limitations
- Maximum transient limitations
- Axial displacements
- Continuous misalignment requirements
- Maximum transient misalignment requirements

The type of power recovery train (conventional train with an air blower or generator set) greatly changes the service requirements of the coupling. In a conventional train, the coupling is sized to accommodate the maximum power generated by the expander and small transient torques. Service factors of 1½ are common. In the generator applications, the generator short-circuit torque can be very high (greater than 7 times the normal torque) and the couplings need to be designed to accommodate these high-transient torques. Whereas the peak torque will be dissipated through the couplings and gear-box shafts, the expander's transient torque can still be more than three times the normal loading.

In all cases, the coupling assembly for the CCU expander should be designed to minimize both the overhung weight and moments on the shaft ends.

Expander couplings are typically designed for infinite life when operated within the design ratings and environment guidelines. All components of the coupling should be centerd by piloted or rabbeted fits. These fits should be designed to tighten under centrifugal loading.

All shim packs used in the coupling assembly should be made from corrosion-resistant stainless steel. When the shim packs are used to adjust the axial float, the shim packs should be located on pilot fits of sufficient length to allow the pilot fit to engage onto the mating coupling component with the full complement of shims installed.

Vibrational Design Requirements

Because of the expander's sensitivity to coupling unbalance, CCU expander couplings are typically balanced to residual unbalances of half that specified in most API equipment standards. Component and assembly balances are typically required to be 2W/N (in-oz), where W is the component weight in pounds and N is the maximum continuous operating speed in rpm. The coupling final disassembly and reassembly requirement is typically 10W/N.

Special sets of coupling bolts are used to ensure that the coupling fasteners are of equal weight within 0.0035 ounces (0.1 grams).

The coupling actual lateral critical speeds should be designed to be a minimum of 20% away from the operating speed range. The actual critical speed accounts for the drop in frequency due to shaft-end flexibility and bearing-fluid film flexibility. The lowest lateral critical speed of the coupling spacer should be a minimum of 20% above the maximum continuous operating speed over the range of coupling axial stretch. Gear-type couplings should have a minimum critical speed margin of 200% over the operating speed range to account for gear-tooth oil film flexibility. Disc or flexible element natural frequencies over the allowable range of coupling stretch should be a minimum of 10% away from the operating speed range of the expander.

Coupling Materials of Construction

Gear-type couplings are typically manufactured from low-alloy steel forgings conforming to ASTM A-470. Most flex elements of dry-type couplings are designed with the use of specialty stainless steel such as ASTM A564/A705 (15-5) and ASTM A564 (17-4 or 13-8). The spacer pieces and hubs are low-alloy steel conforming to ASTM A-470.

Coupling Guards

The operating speeds of the expander require that the guards be designed to prevent excess heat buildup in the guard. Figure 2.62 shows

Figure 2.62 Expander-to-gear coupling installation.

a typical CCU expander coupling guard. Windage plates and baffles are commonly incorporated to reduce the coupling windage and heat buildup. Coupling operating temperatures should be maintained below 180°F (82°C). The coupling guard operating temperature should be maintained below 120°F (49°C) for personnel safety and to prevent instrumentation overheating.

All coupling guards should have baffled vent and oil drain provisions regardless of a wet or dry coupling design used. Horizontally split, flanged, cylindrical shell guards are the most commonly used designs for all expander applications. With the long couplings used in the expander applications, the guard must be designed to accommodate the thermal growth of the guard between the expander and the driven equipment. Slip fits or even expansion joint designs have been used to accommodate the differential growth.

Coupling Guard Materials of Construction

All coupling guards should be manufactured from spark-resistant materials. Aluminum guards are most commonly used.

2.3.10 Rotor Bearing and Seal Assembly

Virtually all modern CCU expanders are built with an overhung rotor design. The rotor is supported in the journal bearings in a cantilevered bearing pedestal. The bearing pedestal also supports the flue-gas seals and part of the expander exhaust casing. The rotor and its support are commonly referred to as the rotor, bearing, and seal assembly. A typical construction is illustrated in Figure 2.63.

Bearing Support

Rotor alignment to the driven equipment is accomplished by moving the expander rotor, bearing, and seal assembly as a complete unit. Vertical alignment is accomplished with the use of stainless-steel laminated shim packs that are installed between the baseplate and the subbase. Axial and transverse alignment is made via jacking screws. The bearing pedestal is then doweled to the baseplate.

The bearing housing can either be an integral part of the bearing support or bolted in. The attachment of the bearing housing to the support requires careful attention since it must be designed to accommodate high-dynamic loads and the shock loading of a major rotor failure (blades or disc). The bearing housing must be precision machined to ensure proper alignment of bearings and seals. The bearing housing also contains the lubrication oil supply, discharge provisions, and supervisory instrumentation.

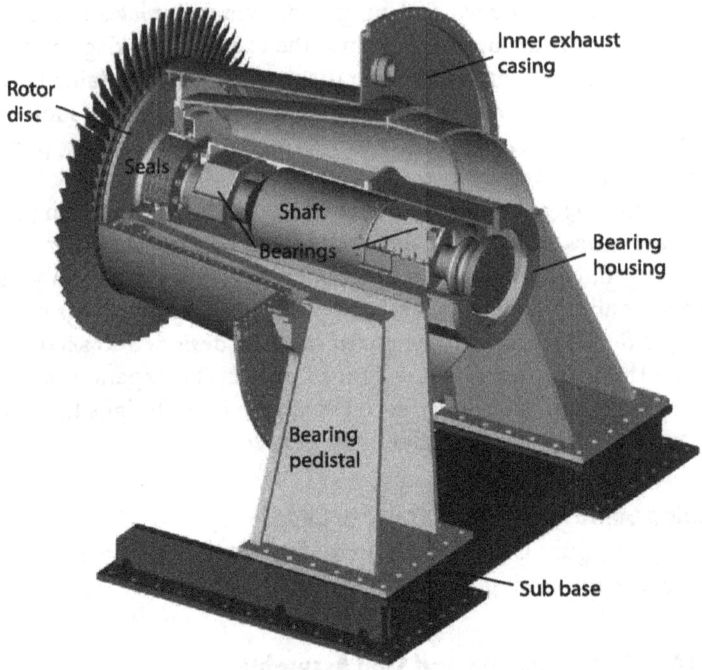

FIGURE 2.63 Rotor, bearing, and seal assembly. (courtesy of the Dresser-Rand Co.)

Bearing Support Structural Design Considerations

The bearing support is a critical component in the support of the rotor assembly and a controlling factor in the rotor's dynamic operating characteristics. The bearing support stiffness plays a major role in determining the rotor critical speeds, system dampening, and response to unbalance or other rotor stimuli. The stiffness of the actual bearing support differs greatly between the disc end and the coupling end of the support. With the main vertical supports under the coupling end bearing, the coupling end stiffness would be expected to have a high stiffness. Coupling end-support stiffnesses have been measured with stiffness values ranging from 3 million to 5 million lb/in (3.5 million to 5.8 million kg/cm). The cantilevered disc-end-support region has a much lower stiffness with values ranging from 300 thousand to 1 million lb/in (350 thousand to 1.1 million kg/cm).

The bearing-support structural margins are generally very high because of the need to maximize the stiffness and minimize deflections resulting from the rotor loadings. Bearing supports are generally problem free and operate for multiple campaigns without rework.

Bearing Support Materials of Construction

The bearing supports are fabricated from carbon steel plate. Bearing supports are painted with high-temperature paints to prevent oxidation.

Bearings

CCU expanders use standard-design pressure-lubricated fluid film journal and thrust bearings (see Figure 2.64). The bearing design and selection is important as these expanders operate at high speeds and have complex rotor dynamics characteristics. Chapter 5 provides detailed discussions of the rotor dynamics characteristic of CCU expanders.

Journal Bearings

CCU expanders have used a variety of journal-bearing designs over time. Both sleeve and tilting pad journal bearings are commonly used. Many early expanders were designed with tapered land or lemon bore sleeve-type bearings. These bearings provided high-load capabilities along with a high degree of fluid damping. However, many of these early designs experienced operational vibration problems with sub-synchronous oil whirl. Successful operation of the sleeve bearings depends on proper alignment of the bearing bores. Misalignment can lead to bearing edge loadings. More modern expander designs use self-aligning tilting-pad journal bearings. Four- and five-shoe tilting-pad bearings with load between shoe designs are most commonly used. The tilting-pad bearings have greatly reduced the incidence

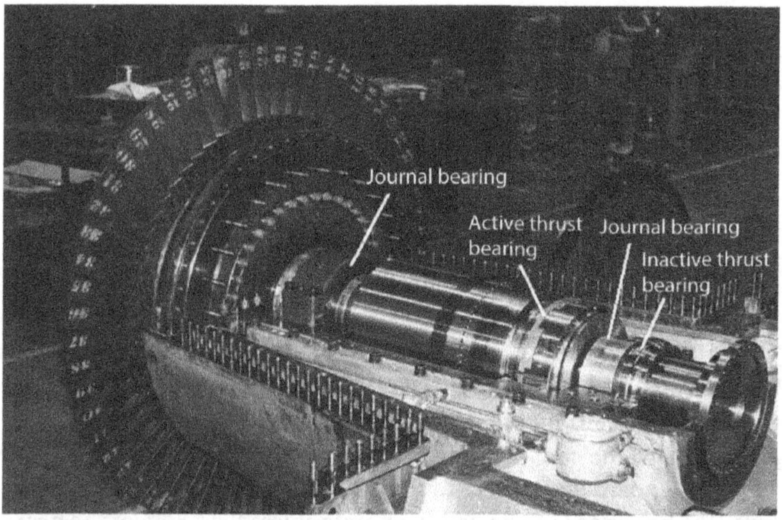

Figure 2.64 CCU expander bearings.

of the subsynchronous vibration. Although the tilting pad bearings often have less dampening than the sleeve bearings, the other benefits of the tiling-pad bearings outweigh this one disadvantage.

All expander bearings are force fed lubricated with typical turbine mineral oils. An oil viscosity of 150 SSU at 100°F (ISO VG 32) with an inlet oil temperature of 120°F (50°C) is typical for this application. Bearing oil flows are regulated with supply-side orifices. The lubrication oil should be cooled and filtered to a 10 micron or finer level.

Journal-bearing loadings are unique to the CCU expander. Because of the overhung rotor design, the disc-end journal bearing accommodates a majority of the rotor weight. It is common for the disc-end bearing to accommodate 80% to 90% of the total rotor weight. The coupling-end journal bearing accommodates the remaining rotor and coupling weights. The coupling-end journal is usually sized to transmit the expander's generated power, and therefore is often fairly large in diameter. The journal-bearing unit loads are conservative and typically fall between 100 and 150 psi (7.0 and 10.5 kg/cm^2) on the disc-end journal and only 30 and 75 psi on the coupling end. Surface speeds are typically 150 to 175 ft/s (46 to 53 m/s).

The journal-bearing clearances are critical in assuring reliable expander operation. The expander rotor dynamics are highly influenced by the bearing clearances and fits. Whereas most turbomachinery operates best with diametral bearing clearances of 0.0015 in (0.037 mm) per inch of bearing diameter, CCU expanders are often designed with much tighter clearances. Clearances of 0.001 in (0.025 mm) per inch of bearing diameter plus 0.001 in (0.025 mm) are typical. For example, an 8-in (20.3-cm) bearing journal would operate with 0.009 in (0.229 mm) diametral clearance. These reduced bearing clearances have been found to be helpful in the reduction of rotor radial vibration sensitivity to stimuli such as rotor unbalance and blade-tip rubbing from catalyst deposits.

The disc-end bearing of the expander is also unique in that the journal operates with a high shaft temperature typically in the 130 to 160°F (54 to 71°C) temperature range. The journal thermal growth needs to be accommodated in the bearing clearance sizing.

Thrust Bearings

The overhung design of the CCU expander results in a unidirectional thrust from the high-pressure inlet toward the discharge. This generally is toward the coupling or drive end.

The standard design uses multipad self-leveling active thrust bearings. The bearings are selected to operate at less than 80% of the bearing manufacturer's calculated load capacity. Bearing loadings range between 300 and 400 psi (21 and 28 kg/cm^2). Mean surface speeds between 200 and 300 ft/s (61 and 91 m/s) are common.

For some highly loaded or rerated expanders, thrust bearing upgrades have been required. Copper chrome shoes, offset shoe pivots, and directed lubricated bearings are commonly used.

Inactive thrust bearings are commonly a reduced-size tapered land bearing since the inactive thrust bearing is little more than a bumper bearing and sees small loadings only during unit start-up or shutdown.

Seals

Flue-Gas Seals CCU expanders generally use two main flue-gas seals to prevent leakage of flue gas to the atmosphere. Flue-gas leakage poses significant safety hazards to plant personnel. The gas is very hot and poses a burn hazard, is laden with catalyst fines that pose a dust hazard, and can be potentially lethal owing to its lack of oxygen and the presence of sulfur compounds.

Labyrinth Seals Labyrinth seals are used for the flue-gas seals. Some manufacturers position the seal teeth on the seal ring, whereas other manufacturers position the teeth on the rotor. Many designs include a replaceable seal sleeve on the rotor as can be seen in Figure 2.65.

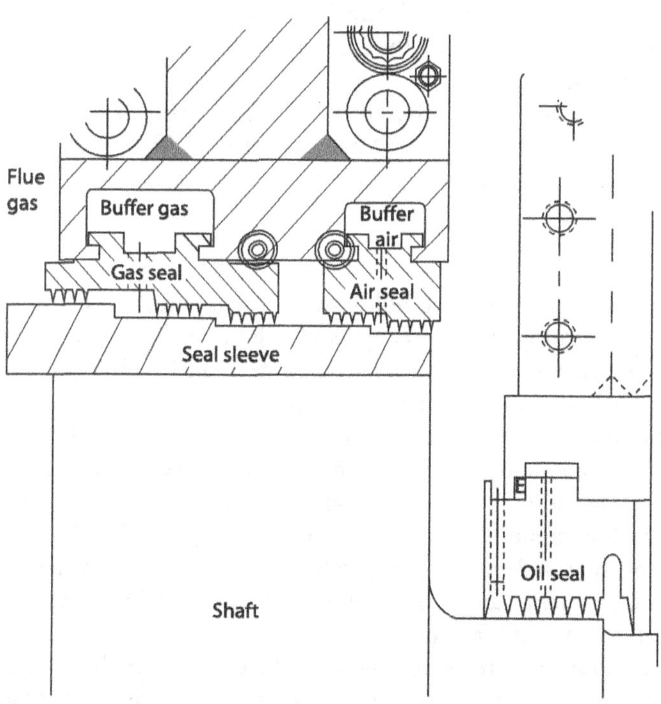

Figure 2.65 Expander flue-gas seals.

FIGURE 2.66 Seal rings and bearing assembled in housings.

The seal rings are manufactured as half rings and rolled into the retaining grooves of the seal housing. Low-alloy-steel seal rings are often used to accommodate the high operating temperatures and corrosive environment.

The flue-gas seal is buffered with either steam or air. A second backup seal is often used and buffered with air. The region between the two seals is often educted when steam is used as the buffer media. (See Figure 2.66.)

Seal Housing

The expander seal housing holds the gas seal rings in position, centered to the rotor. The housing is attached to the bearing support either by being integrally welded or bolted on. In addition, the seal housing is bolted to the exhaust casing.

Seal housings are actually among of the most difficult regions to design of the expander. The seal housings operate with larger radial thermal gradients. The outside diameter of the seal housing is operating near the expander exhaust temperature of 1,000°F (538°C). The inside of the seal housing is typically 400°F (204°C). This large temperature gradient causes high-thermal stresses in both the housing and horizontal joint fasteners.

Seal housings are generally manufactured from high-strength, low-alloy steels. They may be produced as a casting or welded fabrication. High-strength, high-alloy fasteners are used to hold the halves together. Seal housings often have a number of piping connections that make the bottom half fairly congested. Typical seal housing connections include:

- Expander exhaust pressure
- Gas seal buffer gas
- Gas seal buffer gas supply pressure
- Eductor cavity pressure
- Air seal buffer
- Air seal buffer supply

Information on setup, control, and operation of the flue-gas seals can be found in section 2.5.

Internal Flow-Path Seals

Many expanders use flow-path seals that both reduce the amount of flue gas and catalyst that enters the nose cone cavity and improve the effectiveness of rotor cooling (air or steam). Owing to the pumping action of the rotor disc, a slightly negative pressure develops in the nose cone cavity. This reduction in pressure causes flue gas and the entrained catalyst to be pulled down into the flue-gas cavity. The flue gas is then pumped up the face of the rotor disc due to the friction between the disc face and the gas. The pumped gas will both reenter the flow path and recirculate in the cavity.

Recirculating flue gas will have a tendency to heat up in the cavity, and catalyst will deposit throughout the cavity and on the rotor disc. Deposit thickness in excess of 0.375 in (9.5 mm) has been found in the nose cone cavity and on the rotor disc. Whereas the deposits are fairly uniform in thickness, they can spall off at times and cause major rotor unbalance and vibration.

To minimize the amount of flue gas and catalyst pulled down into the nose cone cavity, reduce the cavity temperature, and increase the effectiveness of the rotor cooling, nose cone seals are often included in the expander design. The use of a rotor disc seal reduces the flow area between the inner flow-path components to rotor by 75% to 85%.

Figures 2.67 and 2.68 show two commonly used nose cone cavity seal designs. In Figure 2.68, a radial seal is used with a rotor disc seal lip that is integral to the rotor disc. The actual seal ring has been integrally machined as part of the flow-path components, with separate wiper plate(s), J strips, or a bolted-on multitooth labyrinth design. The design of the seal is critical as contact with the rotor can and has caused major rotor damage. The design of the seal lip on the disc is also critical as catalyst buildup under the seal lip has occurred and

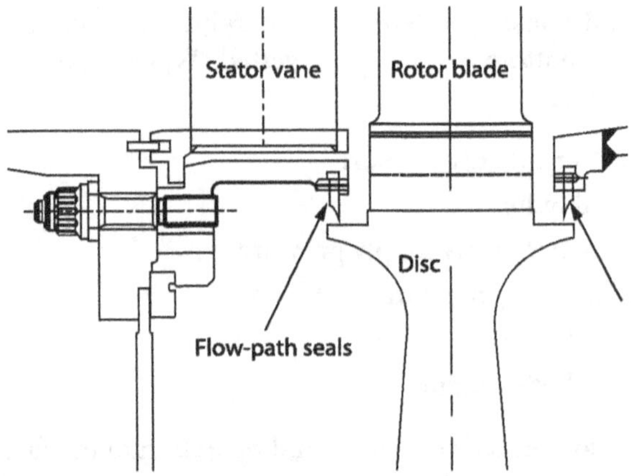

FIGURE 2.67 Flow-path seal rings—disc seal lip design.

caused significant rotor unbalance. With the use of a cooling media, the seal lip designs can significantly reduce the operating temperature of the rotor disc and blade attachment regions.

In more recent years, new technology designs have been integrated into a number of expanders. These new designs use axial seal designs similar to those depicted in Figure 2.68. The designs often include blade platform extensions sometimes referred to as "angel wings." Axial seal rings can also be integral or separate wiper plates, J strips, or metallic brush seals. Disc and blade attachment temperatures can be

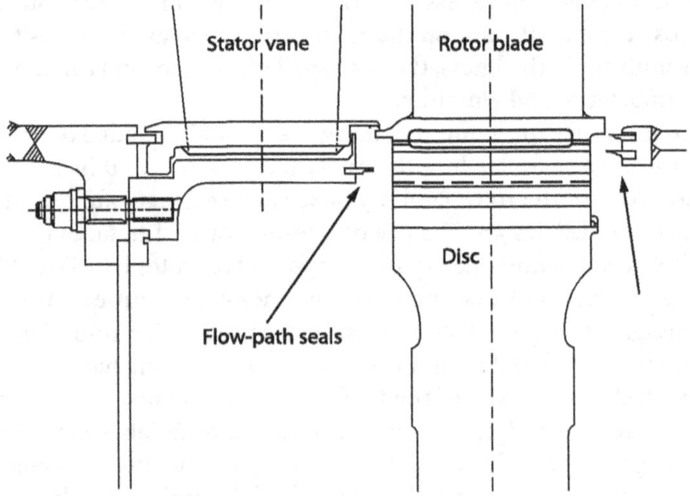

FIGURE 2.68 Flow-path seal rings—axial design.

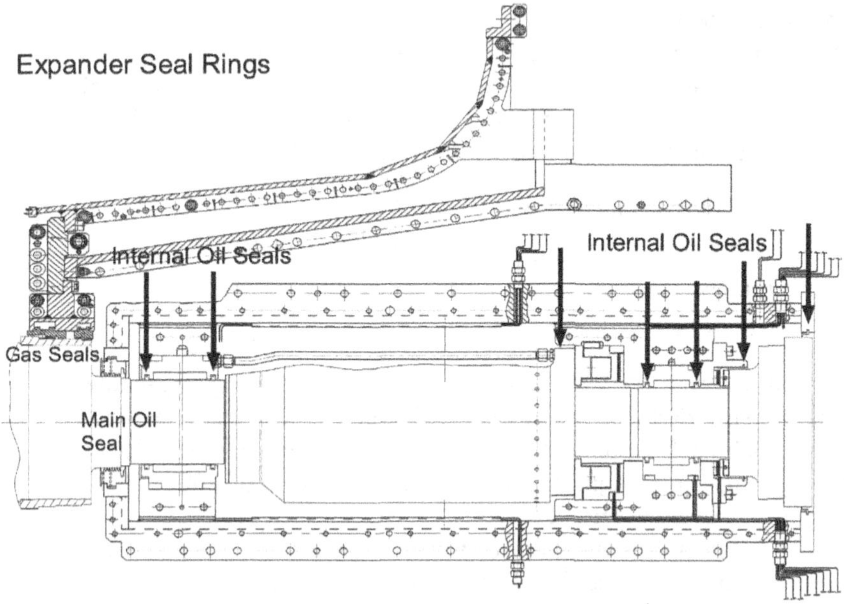

FIGURE 2.69 Bearing housing internal oil seals.

further reduced with the axial seal design. Another major advantage of the axial seal design is that the entire blade attachment region can be shielded from the flue gas. This can be very helpful in the reduction or prevention of blade and disc attachment corrosion.

Main Oil Seal

Figure 2.69 shows the various oil seals used in the bearing housing. One of the most critical seals in the expander is the bearing-housing end wall/main oil seal. This seal is intended to prevent the leakage of oil out of the bearing housing and onto the hot casings. Oil leakage poses a major fire hazard around the expander. The main oil seal is a buffered labyrinth design with the labyrinth teeth on the stationary seal ring. Clean and dry air (instrument) is used for buffering to prevent the introduction of contaminants into the bearings and lube oil system.

The oil seal is usually made up of a rub-tolerant material such as a bearing bronze.

In the operation of the expander, it is important to remember that the expander lube oil system should not be turned on without first commissioning the bearing buffer air.

Internal Oil Seals

A number of oil seals are used in the bearing housing to control the flow of oil to and from the expander bearings. Figure 2.69 shows typical oil seals used in the bearing housing of an CCU expander. In this particular example, note the following oil seals:

- Journal-bearing end-wall seals (4)
- Active thrust-bearing control ring
- Inactive thrust-bearing control ring
- Coupling hub wiper plate

The internal oil seals can be of the oil wiper or spring-loaded floating ring designs. Wiper-type seals operate with seal clearances of approximately 0.002 to 0.003 in per inch of shaft diameter, whereas the floating-ring designs operate with seal clearances of 0.001 to 0.002 in per inch of shaft diameter.

The seals are manufactured from bearing bronze materials that are forgiving of a rub and will not damage the shaft.

Design Validation

As with most critical machinery used in the refinery business, Hot Gas Expanders go through extensive quality-assurance preassembly and assembly checks and testing to ensure successful installation and start-up. All pressure-containing and high-stress components are full-penetration welds or high-quality forgings or castings of radiographic quality. The high-pressure casings receive a temperature compensated hydro-tested in accordance with the ASME pressure vessel code. Similarly, the expander exhaust casing receives a temperature-compensated air test.

Hot Recycle Mechanical Test Run One unique test performed on the CCU expander is a hot mechanical test run. The mechanical test run is intended to meet the 4-hr mechanical run test delineated in API Standard-617. The expander mechanical run test requires the expander to be driven by either a shop driver or one of the auxiliary drivers used in the power recovery train. During the test, the expander is heated up in a recirculation loop that is heated by the blade windage power consumption. Figure 2.70 shows a typical expander hot mechanical run test setup. The expander should be tested on the job baseplate supported in a similar manner to that in the field. Either a shop or the job lubrication system is used for the test. If possible, the job coupling should be used for the hot mechanical test.

The expander test includes running the expander for four hours at the maximum continuous operating speed of the expander and the

FIGURE 2.70 New CCU expander receiving a hot recirculation test.

higher of either the design exhaust temperature or 1,050°F (565°C). The expander heat-up and cool-down rates are controlled by both an inlet and exhaust valve that can be used to bleed off hot air or introduce cooler ambient air.

Typical operating parameters monitored in the hot test include:

- Radial vibration
- Any housing accelerators
- Axial position
- Speed pick ups
- Key phasor
- Overspeed system parameters
- Journal-bearing operating temperature
- Active thrust-bearing operating temperature
- Inactive thrust-bearing operating temperature
- Bearing and casing support temperatures
- Lube oil pressure
- Lube oil supply temperature
- Lube oil drain temperature
- Lube oil flow

- Bearing buffer airflow
- Flue-gas seal airflow
- Test loop temperatures (inlet, crossover pipe, and exhaust)

Most expander mechanical runs include testing both the main and spare rotors in an identical manner and receive posttest bearing and seal inspections. The expander overspeed protection is fully tested, calibrated, and verified during the mechanical run test. When required, a rotor unbalance response test may be run.

The hot mechanical run test does not include any performance testing of the expander nor does the expander generate any power. Expander performance testing is usually conducted in the field by performing a power recovery train power balance.

Maintenance

Most CCU expanders operate for many years with minimal online maintenance. Major overhauls and turnarounds are generally set by the CCU maintenance schedule. Today, turnaround times of three to six years between each unit major overhaul are not unusual. The actual time between unit overhauls is very site-dependent. Rarely does the expander set the plant turnaround time.

Expander Components

Rotors CCU expander rotor discs and blades can be expected to have component lives of 100,000 hr (11 years) or more. Detailed rotor inspections need to be conducted after each operating campaign. A remaining-life analysis of the rotor is often conducted at each major turnaround to determine its suitability for further operation. Generally, the rotor blades are replaced at each major turnaround due to blade deterioration from erosive wear, corrosion, or mechanical deterioration. The rotor shaft and tie bolt are not life limited.

Flow-Path Components Although parts of the rotor are life limited, the stationary components such as the shroud rings and exhaust diffuser are not life limited. These components are often operated for multiple campaigns if excessive creep deformation or erosive wear is not noted.

The stator vanes generally wear at a rate less than that of the rotor blades. It is common to get two operating campaigns out of a set of stator vanes.

Casings The expander outer casings are not life limited but do require periodic rework. The most notable form of distress is casing bore and flange distortion. Remachining of the casings is time consuming and needs to be accounted for in any major overhaul schedule. Most CCU expander operators find that the outer casings need to have a major rework (solution annealing, replacing flanges, remachining, etc.) or replacement after 12 to 15 years of operation.

Wearing Parts Generally all wear parts (bearings and seals) are replaced at each major turnaround of the expander.

CCU expanders are manufactured with numerous tight and critical clearances as well as many thermal expansion design features. The maintenance or refurbishment of these units requires significant machinery repair experience, precision machining and assembly techniques, and/or equipment that is not typically found in the field or in refinery facilities. These capabilities are best found in a turbomachinery repair shop and make it common practice to remove the entire expander as an assembly during a major overhaul. The casings generally are not rigid enough to allow the unit to be removed in a single lift. Since the expander overhaul requires significantly more time than that of the other blower train machinery, some CCU expander users have found it to be economically viable to spare the complete expander. Maintaining a spare expander assembly is driven by process shutdown time. The spare unit typically can be swapped out in 14 days or less, whereas a unit rebuild in an outside shop generally requires four to eight weeks.

Spare Parts Most users of CCU expanders maintain a typical array of spare parts that include the following:

1. Class I Spare Parts. Wearing parts commonly replaced at each unit turnaround.
    - Journal-bearing insert or tilting pads
    - Thrust-bearing pads
    - Thrust-bearing shim packs or locator rings
    - Bearing housing oil seals
    - Flue-gas seal rings
    - Flow-path seal rings
    - Supervisory instrument probes
        - Vibration probes and proximeters
        - Axial position probes
        - Speed pick ups
    - Flange gaskets and sealants

2. Class II Spare Parts. Assemblies that might reduce turnaround times and the replacement of flow-path components that might be damaged.
   - Journal-bearings shells and retainers
   - Thrust-bearing retainers
   - Complete set of rotor blades
   - Complete set of rotor-blade locking hardware
   - Complete stationary vane assembly
   - Shroud ring
   - Spare coupling
   - Spare critical casing fasteners
3. Class III Spare Parts. Parts that provide the maximum degree of reliability and backup protection owing to major machinery problems.
   - Complete rotor assembly
   - Inlet casing assembly

Some CCU expander users have chosen to spare the complete rotor, bearing, and seal assembly. This would include the bearing support, seal housing, and inner exhaust casing as well as the normal spares.

2.4 Expander Performance

There are approximately 400 CCUs around the world, but only one-third of those units currently operate with a Hot Gas Expander. New expanders are being installed every year, though some CCU installations operate at conditions that do not make an expander economically viable. Other barriers to installing expanders are incompatibility with existing equipment or a lack of real estate necessary for an expander train and its supporting equipment. Virtually all new CCUs are built with a Hot Gas Expander.

The FCC expander is a custom-built turbine for each specific installation. Catalytic cracking units operate with different technologies and under a wide range of conditions owing to different designs, feed, and intended end products. Consequently, the operating conditions of the expander are nonstandard. Table 2.2 shows the wide range of conditions that current FCC expanders have been designed to accommodate. It can be clearly seen that FCC expanders can be categorized as a low-pressure, high-temperature, high-flow turbine. The high flue-gas flow and temperature makes the amount of recoverable power quite significant. Although all expander manufacturers

	Minimum Values	Maximum Values	Minimum Values	Maximum Values
Inlet temperature	1,060°F	1,415°F	571°C	768°C
Inlet pressure	10.5 psig	50.0 psig	0.74 kg/cm^2	3.52 kg/cm^2
Discharge temperature	940°F	1,140°F	504°C	616°C
Discharge pressure	0.5 psig	5.0 psig	0.035 kg/cm^2	0.35 kg/cm^2
Pressure ratio	1.5	3.4	1.5	3.4
Flow	115,000 lb/hr	1,700,000 lb/hr	52,000 kg/hr	771,000 kg/hr
Isentropic enthalpy drop	60 Btu/lbm	135 Btu/lbm	63 kJ	142 kJ
Power	3,350 hp	72,000 hp	2,500 kW	53,690 kW
Rotational speed	3,000 rpm	7,800 rpm	3,000 rpm	7,800 rpm

Table 2.2 FCC Expander Range of Operating Conditions

provide a number of standard frame sizes that cover the range of FCCU operating conditions, each individual expander flow-path design is unique.

Early expanders tended to operate at lower pressure ratios, temperatures, and overall energy levels. A single-stage turbine design was well suited and provided good efficiency for these early expanders. Over time, the CCU regenerator pressures and temperatures increased and multiple-stage expanders were developed to maximize the recovered power. Although a majority of the multistage expanders were two-stage expanders, a number of units were built with as many as four stages. Unfortunately, many these multistage expanders experienced poor reliability due to high deposition of catalyst in the interstage areas. This was particularly true under upset conditions when high levels of catalyst were being carried over into the expander. Even though several multistage expanders remain in service and have been successful, many multistage expanders have been converted to a single-stage design.

Flue-Gas Chemistry

While there are a variety of FCCU designs, a majority of the operations can be described as being either an incomplete combustion unit

	N_2	CO_2	CO	O_2	H_2O	SO_2	MW
Incomplete combustion	70.4	11.6	6.4	0.15	11.5	0.10	28.8
Complete combustion	72.1	14.6	0.02	1.72	11.8	0.08	29.4

Table 2.3 Average Flue-Gas Chemistry for Incomplete and Complete Combustion

(ICU) or a complete combustion unit (CCU). As shown in Table 2.3, in the ICUs, not all of the carbon monoxide is burned in the regenerator vessel and the flue gas has virtually no excess oxygen. In the CCUs, essentially all of the carbon monoxide is burned in the regenerator. A small amount (0.5% to 3.5%) of excess oxygen will exist in the flue gas.

Although the combustion mode causes small changes in the flue-gas properties such as k value and molecular weight (MW), it does significantly affect the flue-gas temperature and pressure as well as other significant issues such as the corrosion potential in the expander. ICU units tend to have flue-gas temperatures in the 1,100°F to 1,285°F (593°C to 696°C) range, whereas CCUs tend to operate in the 1,300°F to 1,415°F (704°C to 768°C) range and under the higher pressures. The lack of excess oxygen greatly increased the expander corrosion potential in the expanders. For example, a combustion turbine operates with excess oxygen of approximately 22% (more than ten times the amount).

Performance Map

Expander manufacturers provide the owner/operators an expander performance map that allows them to predict the expander performance under various operating conditions. As can be seen in Figure 2.71, the expander performance map is actually a series of curves at various expander inlet pressures and temperatures. The expander power can be determined knowing the expander inlet pressure, temperature, and flow. The typical expander performance map is generated assuming constant flue-gas properties, discharge pressure, and rotational speed.

FCC expander performance maps generally show over-temperature operation as these turbines can experience significant over-temperature operation known as afterburning. Afterburning is an uncontrolled burning of carbon monoxide outside of the regenerator vessel. It can occur in the expander and/or expander ducting and is a very dangerous condition as it greatly degrades the materials the system is constructed of. While afterburn temperatures can exceed 1,800°F (982°C), generally the expander is tripped offline if the inlet

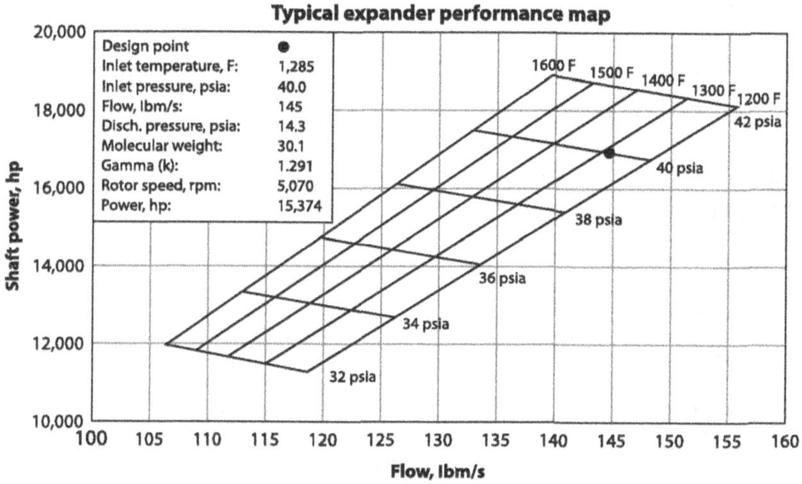

Figure 2.71 Typical FCC expander performance map.

temperature exceeds 1,625°F (885°C). It is important that the expander-driven equipment and auxiliary systems be designed to accommodate these upset conditions.

Flow-Path Design

Expander flow-path design has evolved with the technological developments over the more than 50 years of their application. Initial flow-path designs were the basis for the early FCC expander applications that operated at lower inlet temperatures and pressures. A single-stage turbine was appropriate for these initial expander applications. Conventional stator vanes and rotor blade shapes were developed for the FCC expander.

Conventional Two-Dimensional Turbine Aero Design

Early FCC expanders were designed with basic aerodynamic design techniques that had been used in turbine designs for many years. The techniques involved maintaining the basic conservation principles of mass, linear momentum, energy, and annular momentum in a steady one-dimensional form to calculate flow, efficiency, reaction, and so forth.

First the nozzle flow was calculated based on the stator area, stage pressure ratio, the degree of reaction, gas pressure and temperature, and stator size. The theoretical flow was calculated and then corrected for using nozzle area, critical flow, gas property, geometric conditions, and leakage corrections that were obtained from laboratory and field testing.

The efficiency of the stage was calculated in a similar manner with a baseline efficiency obtained from laboratory and field-testing data.

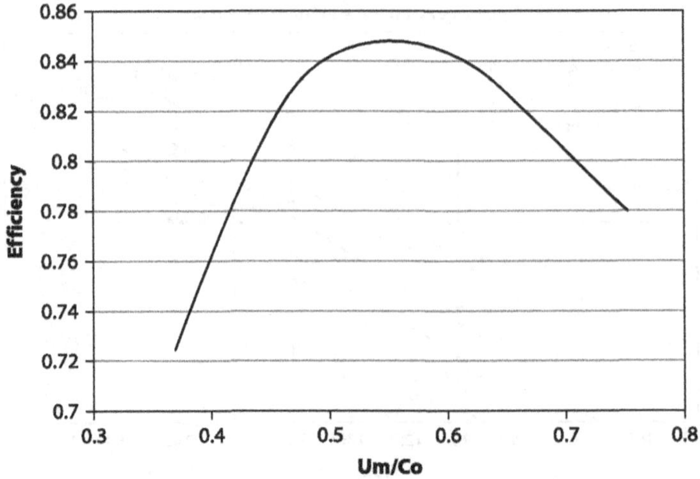

FIGURE 2.72 Typical state efficiency curve.

The baseline efficiency was then corrected for various geometric conditions (aspect ratio, vane size, expansion number, stage leakage, stator trailing edge thickness, etc.). Baseline efficiency is generally expressed in terms of the velocity ratio defined as blade mean-line velocity (U_m) divided by the theoretical nozzle gas discharge velocity (Co).

Each stage design has a different performance characteristic similar to that shown in Figure 2.72. It can clearly be seen that as the stage loading changes, peak efficiency decreases. This is particularly true of higher stage loadings (small Um/Co).

Expander Efficiency

Blade Mean-Line Velocity Calculation

$$Um = \frac{DN}{229}$$

Um—Blade mean-line velocity (ft/s)
D—Mean diameter of the blade (in)
N—Rotational speed (rpm)

Theoretical Nozzle Discharge Velocity Calculation

$$Co = 223.76\sqrt{\Delta His}$$

Co—Nozzle theoretical sonic velocity (ft/s)
ΔH_{is}—Isentropic enthalpy drop (Btu/lbm)

Isentropic Enthalpy Drop

$$\Delta His = T1\left(\frac{k}{k-1}\right)\left(\frac{R}{778}\right)\left[1-\left(P2/P1\right)^{\frac{k-1}{k}}\right]$$

R—Gas constant = 1,545/MW
MW—Molecular weight
k—Ratio of specific heats
T_1—Inlet temperature (degrees R)
P_1—Inlet pressure (psia)
P_2—Discharge pressure (psia)

Isentropic Gas Horsepower

$$HP_{gas} = \frac{m\Delta His\ 778}{550}$$

$HPgas$—Isentropic gas power (hp)
m—Mass flow (lbm/s)

Expander Power The expander power was then calculated based on the mass flow, isentropic enthalpy drop, and efficiency:

$$HP_{shaft} = HP_{gas\ n\ -\ brg\ loss}$$

$HPshaft$—Expander shaft power (hp)
$HPgas$—Isentropic gas power (hp)
N—Stage efficiency
$Brg\ loss$—Bearing losses (hp)

Example of Expander Performance Calculation

Expander Data

Ambient pressure = 14.7 psia
Inlet temperature = 1,325°F $T1$ = 1,785°R
Inlet pressure = 32 psig $P1$ = 46.7 psia
Discharge pressure = 1.5 psig $P2$ = 16.2
Pressure ratio = 46.7/16.2 = 2.88
Flue gas MW = 29.3 R = 1,545/29.3 = 52.7
K value = 1.31
Flow = 750,000 lbm/h
Mean diameter = 57.0 in
Rotor speed = 3,600 rpm

$$Um = (58)(3{,}600)/229 = 912 \text{ ft/s}$$

$$\Delta His = \left(\frac{1.31}{1.31-1}\right)(46.7)(1{,}785)\left[1-\left(16.2/46.7\right)^{\frac{1.31-1}{1.31}}\right] = 113.3 \text{ Btu/lbm}$$

$$Co = 223.76\sqrt{113.3} = 2{,}382 \text{ ft/s}$$

$$Um/Co = 912/2{,}382 = 0.383 \qquad n = 0.735 \text{ basis efficiency curve}$$

$$HPgas = (750{,}000/3{,}600)(113.3)(778/550) = 33{,}390 \text{ hp}$$

$$HPshaft = (33{,}390)(0.735) - 125 = 24{,}417 \text{ hp}$$

Velocity Diagrams

Flow-path velocity diagrams have been used extensively in the blade design process and continue to be used today to evaluate blade aerodynamic loadings. Figure 2.73 shows the typical mean-line velocity diagram layout used in Hot Gas Expander design. For a typical expander operating at an inlet temperature of 1,300°F (704°C) and a pressure ratio of 2.6 would be as follows:

Stator absolute = 1,760 ft/s	536 m/s
Stator relative = 770	235
Blade speed = 1,100	335
Rotor relative = 1,330	405
Rotor absolute = 584	178

The velocity diagram changes from the hub to the tip of the blading due to changes in the flow and the change in blade tangential

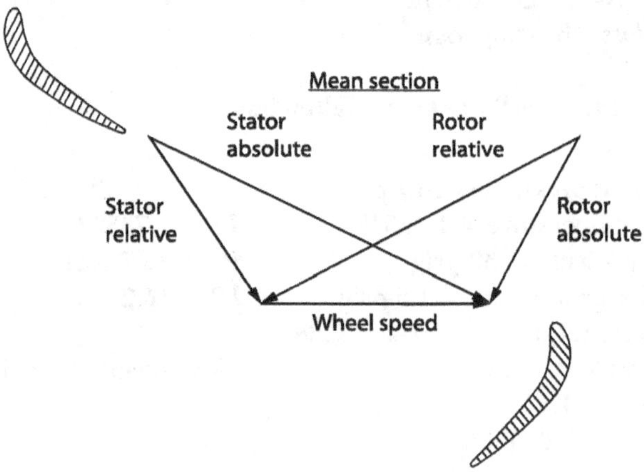

FIGURE 2.73 Typical mean-line velocity diagram.

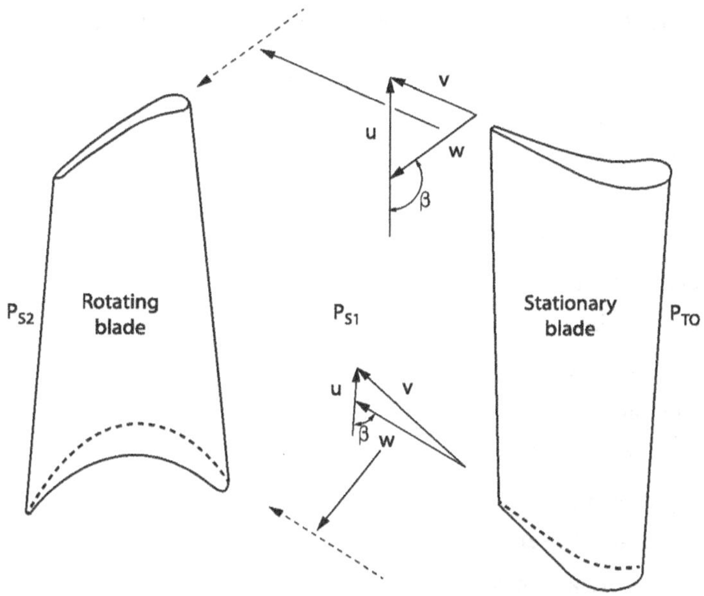

FIGURE 2.74 Velocity diagram—hub versus tip.

velocity (Figure 2.74). Changes in the airfoil are required to accommodate the radial flow-path velocity variations.

Quasi-Three-Dimensional Turbine Aero Design

In time, it was recognized that predicting the performance of turbine stages using two-dimensional analyses was not as accurate as it should be. The problem was exasperated when the turbine was operating in off-design operating conditions of speed and pressure ratio. It was recognized that the complex three-dimensional flows within a turbine flow have powerful influences on stage capacity and performance and needed to be accounted for in the performance prediction process.

New calculation methods were developed that make use of data derived from the analysis of a very large number of turbine tests and test rigs. The calculation methods first determined the characteristics of pressure loss and gas efflux angle for each blade row in the turbine. This can be done over a wide range of turbine inlet conditions. The second step is to take a fixed set of conditions at the inlet of the turbine and follow the course of a fixed mass flow from one blade row to the next. New gas conditions are calculated at each step to create a complete definition of the flow at each point within the turbine. With detailed flow-path definitions, the stage and overall efficiencies and mass flow can be determined. Accounted for in the row-by-row calculations are the following losses:

Gas compressibility

Profile loss

Surface friction

Entry and incidence

Overexpansion and shock loss

Leaving loss

Secondary passage loss

Clearance

Tip loss

Trailing edge loss

Exhaust diffuser

Disc wind age losses

The outputs of these analyses are quite extensive and include complete stage flow definitions. The development of these more sophisticated analytical tools and better computing capabilities have evolved turbine performance design and performance prediction programs to a much higher degree of accuracy. The estimated tolerance on the absolute values of gas mass flow and peak efficiency are estimated to be ± 1 1/2% for efficiency and ± 2% for mass flow. These developments and their refinements occurred between the 1960s and the 1990s and are still in use today by most expander manufacturers.

Computational Fluid Dynamics

In recent years, computational fluid dynamics (CFD) analysis has matured and become a major analytical and design tool for turbo machinery. CFD analysis is a fluid mechanics calculation procedure that uses numerical methods and algorithms to solve and analyze problems that involve fluid flows. The CFD analysis basis is governed by the three fundamental fluid state principles: conservation of mass, Newton's second law, and the conservation of energy. CFD works by dividing the domain of interest into a series of small discrete volumes using a mesh or grid. The physical properties of the fluid, such as temperature, pressure, or velocity, are calculated at a single point with each of these volumes as a solution to the governing equations. CFD in many ways is similar to FEA for structural analysis of components.

The analysis capabilities have evolved to a level where the interactions of gases within a defined space can be accurately simulated.

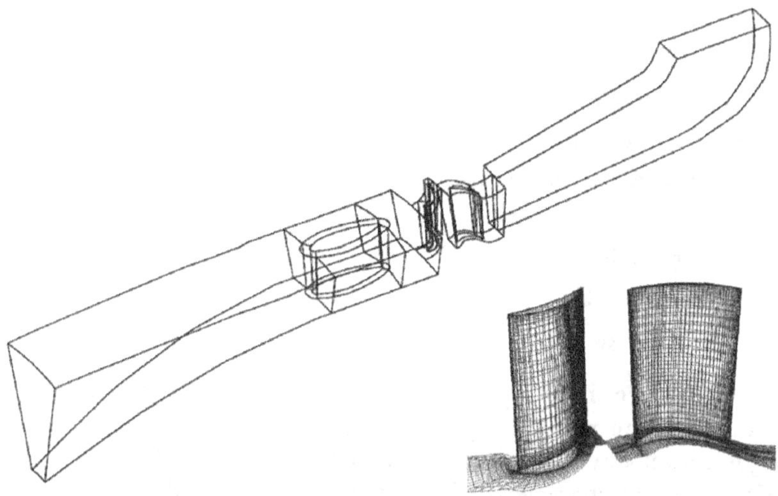

FIGURE 2.75 CFD analytical model—full flow path and detailed stator and rotor blade.

Complex flows including transonic and turbulent flows can also be modeled.

To perform a CFD analysis, a complete model of the flow path needs to be created. The model or mesh fully describes the full expander flow path and includes the expander nose cone, stator, shroud, rotor blade, and exhaust diffuser. Figure 2.75 shows a typical CFD analysis model for an FCC expander. The CFD code allows for the solution of the three-dimensional, turbulent, high-speed compressible flows in multiple reference frames. Particle tracking capabilities allow for the flow field solution to be affected by the catalyst particle paths. In addition to improved accuracy and capabilities, the CFD analytical tools have been upgraded and advanced to include the effects of particle-laden flows such as the catalyst in the flue gas of the FCCU. Understanding the catalyst particle dynamics allows new expander flow-path designs to be optimized for improved efficiencies, reduced erosion, and deposition. On a three-dimensional basis, a large amount of data and information can be obtained from the CFD analysis, including:

- Mach number
- Relative Mach number
- Absolute velocities
- Relative velocities
- Velocity vectors

- Static pressure
- Absolute pressure
- Flow angles
- Flow incidence
- Particle trajectories including rebounding
- Particle impaction
- Particle impact velocity
- Erosion rate
- Erosion density

Figure 2.76 shows typical CFD analysis outputs. The figure on the left is a contour plot of the gas Mach number, while the figure on the right is a velocity vector plot. These tools allow detailed visualization of the gas-flow characteristics throughout the flow path.

Figure 2.77 shows how CFD analysis can be used to study the surface flows within the expander flow path. The large flow separation seen at the base of this airfoil has been associated with localized leading-edge erosion. This analysis also shows a large leakage flow over the top of the airfoil. Airfoil redesign resulted in significantly reduced secondary flows and the reduction in leading-edge erosion.

Surface flow analyses such as that shown in Figure 2.77 allow the blade designer to see the localized secondary flows and to improve the blade design to minimize these flows. Secondary flows are often

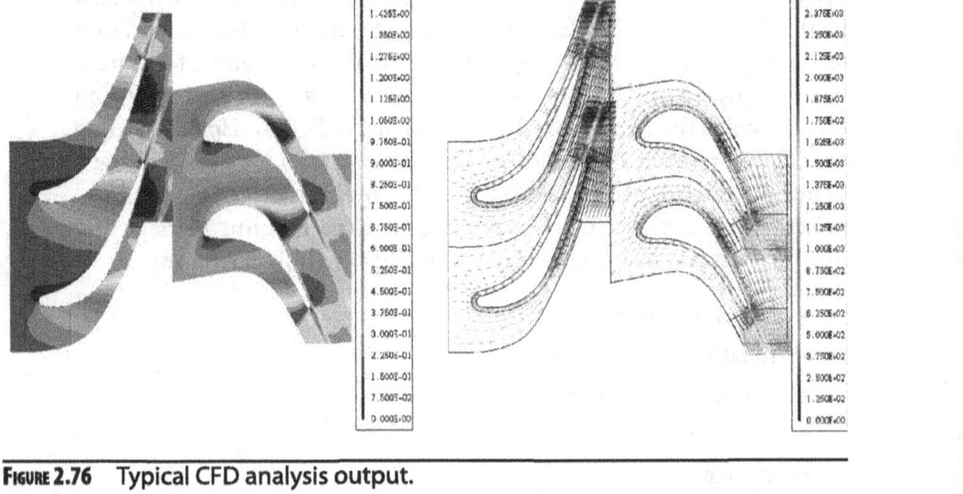

FIGURE 2.76 Typical CFD analysis output.

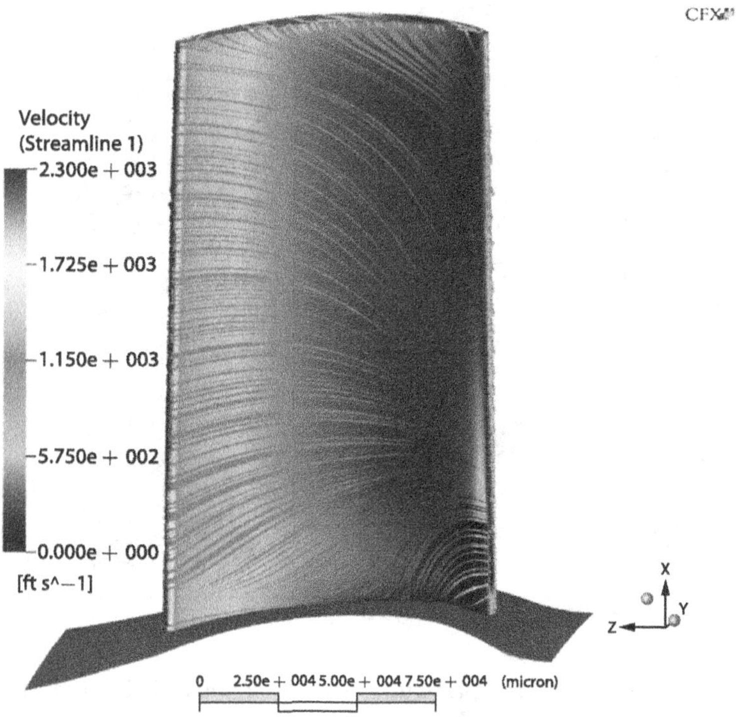

FIGURE 2.77 CFD analysis results—surface streamlines.

associated with localized turbulent flows around the blade profile and increase the airfoil profile losses, cause flow blockages, and cause local airfoil erosion and catalyst deposition.

Particle-trajectory analysis can also be conducted in a CFD analysis. Particles of any size or size distribution can be injected into the flow field and followed throughout the entire flow path. Figure 2.78 shows typical fine particle (2 micron) trajectories at the 90% span. The analysis shows that the fine catalyst particles follow the flow streams very closely except for those that get caught up in the blade-tip leakage flow.

Since the catalyst particles have both mass and momentum, many of the particles will strike the flow-path surfaces. Erosion models have been developed and are coupled with empirical test data to allow erosion damage assessment and the prediction of particle behavior both on impact and post-impact. The particle momentum loss due to a collision is categorized as the restitution ratio (ratio between the particle velocity before and after the collision). The direction of the particle after a collision is determined from the rebound-to-incidence angle ratio. These critical parameters are determined from

FIGURE 2.78 Particle trajectory analysis.

erosion test rigs for the particular particle characteristics and blade materials. Depending on the size and mass of the particles, many particles will have multiple impactions throughout the flow path.

The CFD predictions have gotten quite good over time as can be seen in Figure 2.79. The left figure is the analytical prediction of the erosion, the right figure is a blade removed from service after almost four years of operating experience. Note the mid-chord and leading-edge erosion in both figures.

Even though the analytical tools have proved to be a very useful tool in erosion prediction, they can only provide qualitative predictions. The accuracy of erosion prediction obviously is very sensitive to the catalyst loadings into the expander. In reality, the catalyst loadings change over time with various process operations and deterioration of the separation equipment over time.

CFD analysis has shown that most aerodynamic improvements that are good for increased expander efficiency are also good for flow-path erosion reduction. Efficiency improvements generally involve making the flow path less torturous. Flow paths with smooth transitions and minimized steps, along with lower velocities, slower gas turning, and improved airfoil shapes have all helped to improve the expander's overall efficiency and reduce the potential for erosion.

While the CFD analysis has developed into a major design tool for the expander manufacturers, it should be noted that additional

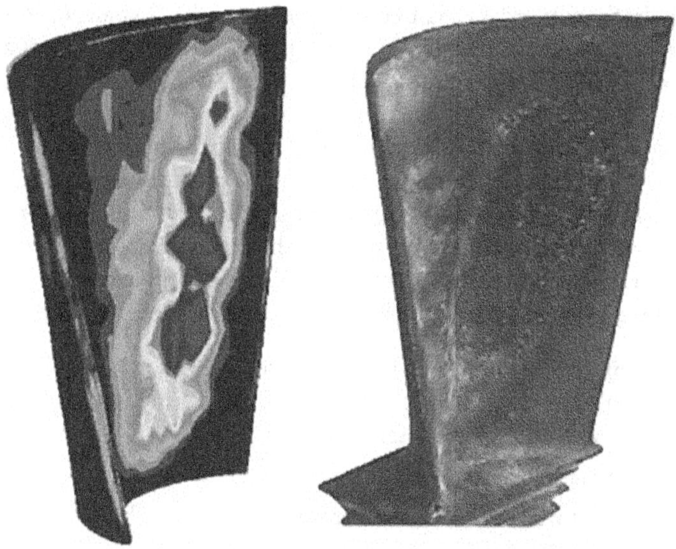

FIGURE 2.79 Analytical erosion prediction versus field experience.

work is required to correlate the analytical work to the actual field-operating experience. This is particularly true in the area of flow predictions. Flow deviations between the analytical predictions and the actual field experience are the result of the flow-path model not accurately reflecting the actual hot geometry of the expander flow path.

Performance Measurement

The measuring and monitoring of the FCC expander performance can be difficult. This is because the measurement of expander flow is difficult and not performed on a continuous basis. Flow meters have not been developed for large-sized ducting. Flow has been measured on a sampling basis via iso-kinetic probes and with the use of radioactive isotopes. The accuracy of these measuring methods is limited, and they represent only a single point in time.

Expander flow is often calculated from regenerator flow determinations:

- *Main air-blower flow.* The regenerator flow is estimated by adding regenerator combustion makeup to the air-blower train. The combustion makeup is generally fairly well known. The makeup is generally 10% of the air-blower flow.
- *Nitrogen balance.* With a known flue-gas composition and the air rate, a regenerator nitrogen balance can be conducted to determine the flue-gas flow rate.

In both expander calculation methods, the expander flow is arrived at by correcting the regenerator flow for any mass-flow additions such flue-gas quenches and losses such as TSS underflow (1% to 3%) and bypass valve leakage or flow.

Expander performance is generally measured using a PRT power balance. The power balance measuring method involves determining the expander output via measuring or calculating the driven equipment power. Obviously, power put into the train must equal the power consumed in the train. In conventional PRTs, the measurement of the consumed power is accomplished by measuring the following:

- *Main air-blower performance.* The blower performance can usually be determined fairly accurately by measuring the normal operating parameters, such as:
    - Ambient conditions
    - Inlet pressure and temperature
    - Discharge pressure and temperature
    - Flow
    - Variable stator position

Testing is best performed without the inlet throttle louvers, discharge antichoke and antisurge valves in service.

- *Auxiliary steam turbine performance.* If applicable, the turbine performance is determined by measuring the turbine inlet and discharge pressures and temperatures and flow. If a condensing turbine is used, the performance is more difficult to measure without a determination of the quality of the discharge steam.
- *Gear box.* If a gear is used in the PRT, the losses can be calculated using factory test data or OEM performance data. The gear power losses are a function of speed and power. Generally, approximately one-third of the gear box losses are related to the bearing losses and are fixed despite the box power loading. The other two-thirds of the gear losses vary proportionally to the power level being transmitted.
- *Motor.* If a motor is used in the train, the motor input or output can be measured electrically. Motor/generator performance data are as follows:
    - Amperage
    - Voltages
    - Power factor
    - Power (watts)

- *Expander inlet and bypass valve.* The expander inlet and bypass valves can often be used to check the system flows. Both valves generally operate similarly to an orifice, and if the inlet pressure and temperature, discharge pressure, and position are known, a flue-gas flow can be calculated.

For an expander/generator train, the consumed power can be measured directly at the generator.

The expander power is calculated based on the measured inlet and discharge pressure and temperatures and estimating the expander efficiency and flow per the equations shown below the performance-monitoring section. The calculated expander power is compared to the PRT consumed power, and if the data are good, the two valves will be very close in value. A power balance is generally considered acceptable if the two calculated expander powers match within 2%.

A number of PRTs have used torque measuring couplings between the expander and the driven equipment to measure the output of the expander directly. Unfortunately, the accuracy and long-term reliability of the torque meters have not been at the levels required.

Performance Monitoring

Monitoring the condition of the expander is critical to successful and long-term reliable operation. In addition to monitoring the expander supervisory instrumentation and typical process conditions, the monitoring of the expander performance can also be very informative. Performance monitoring is accomplished by measuring and tracking the expander efficiency and flow functions.

The measurement and monitoring of the expander efficiency are possible since the parameters required to calculate the expander efficiency (expander inlet and discharge pressure and temperatures) are usually continuously monitored parameters. The trending of the expander flow capacity and efficiency is an excellent tool for determining when the expander flow path requires cleaning or refurbishment. Figure 2.80 shows an example of operating data from an FCC expander. It can be seen that the expander efficiency and flow factor are fairly steady for the first 48 days of operation and, over the next 40 days, there is an increase in the expander efficiency and a reduction in flow factor. The condition of the flow path is confirmed by looking at the rotor radial vibrations and the blade operating photographs shown in Figure 2.81. After fairly steady operation over the first month of service, change can be noted in both the flow function and radial vibration over the next month of operation. These changes are consistent with catalyst buildup on the rotor causing a reduction in flow capability and rotor unbalance. Toward the end of the second month of operation, all of the monitored parameters make

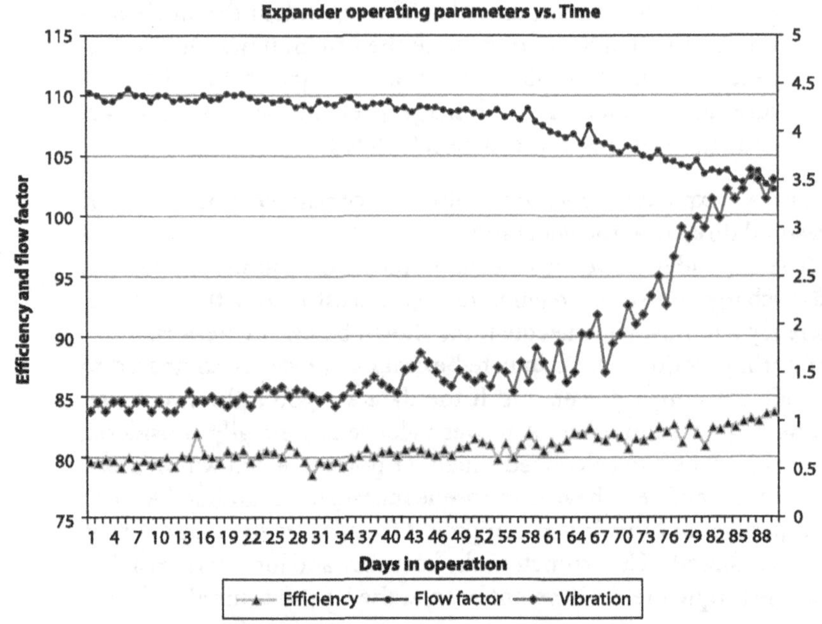

FIGURE 2.80 Expander performance over time.

FIGURE 2.81 Expander viewport photographs after 30 and 90 days of service.

another change in value and operating characteristics. These changes are more consistent with major shroud deposits causing the flow function to decrease further and an increase in expander efficiency owing to the loss of the blade-tip clearance. Blade-tip rubbing is occurring as evident by the blade photographs and the vibration becoming more erratic and the magnitude increasing significantly. Flow-path wear also can be detected via monitoring the expander performance. Erosive wear will show as a reduction in expander efficiency and as an increase in flow factor.

Expander-performance monitoring is often done on a continuous basis using the basic formulas for turbine efficiency. Either polytrophic or adiabatic efficiency can be monitored.

Polytrophic Efficiency

$$\eta poly = \frac{\ln\left(\frac{T2}{T1}\right)}{\ln\left(\frac{P2}{P1}\right)}\left(\frac{k}{k-1}\right)$$

Adiabatic Efficiency

$$\eta ad = \frac{1-\left(\frac{P2}{P1}\right)^{\eta poly\left(\frac{k-1}{k}\right)}}{1-\left(\frac{P2}{P1}\right)^{\frac{(k-1)}{k}}}$$

Expander efficiency calculations are generally calculated and used on a qualitative basis for monitoring the expander efficiency. To calculate a more exact efficiency such as that needed for machinery-acceptance testing, one must account for the effects of expander cooling and sealing flows. These flows will typically reduce the expander discharge temperature 10°F to 20°F (5°C to 11°C), which artificially makes the expander efficiency appear higher than the true value. A simple gas mix calculation can be used to estimate the effect of the cooling and seal flows if a more accurate efficiency calculation is required.

Expander flow function monitoring can be done with an abbreviated version of the flow function. The full version of the expander flow function is calculated with the following formula:

$$FF = \frac{m\sqrt{RT1}}{\sqrt{\left(P1^2 - P2^2\right)}}$$

Because the flue-gas composition and discharge pressure are essentially constant throughout the expander run, the flow function calculation formula reduces to:

$$FF = \frac{m\sqrt{T1}}{P1}$$

The expander flow can be calculated from the flow function accordingly:

$$m = \frac{FF(P1)}{\sqrt{T1}}$$

If there is a need to monitor expander power, the formulas present in the two-dimensional flow-path design can be used.

CHAPTER 3

Dirty Gas and Particle Separation

3.1 Dirty Gas

The Fluid Catalyst Cracking process is one of the most important and commonly used processes used in petroleum refining. The FCC process vaporizes and breaks the long chain molecules of the high boiling point hydrocarbon liquids referred to as heavy gas oil or vacuum gas oil (HVGO). The cracking produces shorter molecules by contacting the feedstock with a fluidized powdered catalyst at high temperature (1000°F [538°C]) and under a moderate pressure (25 psig [1.75 kg/cm^2]). This process occurs in a large vessel referred to as the reactor. The catalyst particles are spherical in nature and range in size between 10 and 150 microns in size. The average particle size is typically 80 microns. Figure 3.1 shows typical equilibrium FCC catalyst particles. Over time, the catalyst particles break down and become more irregular in shape and more varied in size. The CCU catalyst is typically an alumina- or silica-based ceramic particle.

The cracking reactions produce some carbonaceous material (referred to as coke) that deposits on the catalyst and very quickly reduces the catalyst reactivity. To effectively reuse the catalyst, the catalyst needs to be regenerated. The regeneration of the catalyst occurs in a separate vessel called the regenerator where the coke deposits are burned off of the catalyst in a fluidized bed. The catalyst is transferred back to the reactor vessel for reuse in the cracking process. The combustion of the coke produces a combustion by-product referred to as *flue gas*. It is this flue gas that is directed to the Hot Gas Expander to generate mechanical power.

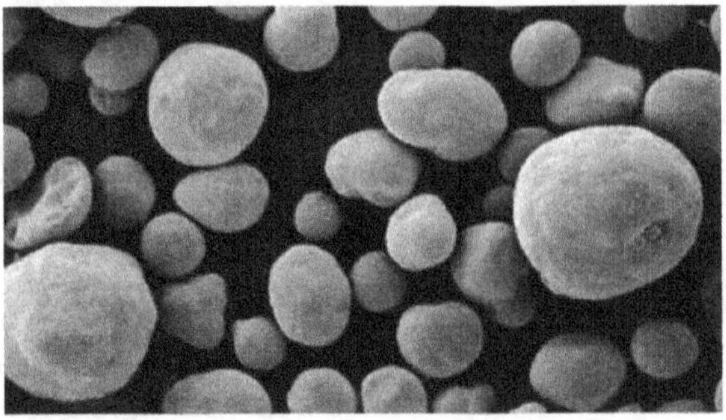

FIGURE 3.1 Fresh CCU catalyst.

As the flue gas exits the regenerator, two stages of cyclone separators are typically used to remove a majority of the catalyst entrained in the flue gas. Although the regenerator cyclones are quite efficient and typically remove more than 97% of the entrained catalyst, regenerator discharge particles loadings are typically 400 ppm (weight %). For a typical 100,000 BPD regenerator, this would equate to 400 lb (181 kg) of catalyst exiting the process and entering the Hot Gas Expander. This amount of catalyst entering the expander is a problem because the catalyst particles are quite abrasive. Early FCC expander operating experience showed that the expander blades would be eroded beyond use in as short as four months of operation at these higher catalyst levels (see Figure 3.2). This excessive wear caused

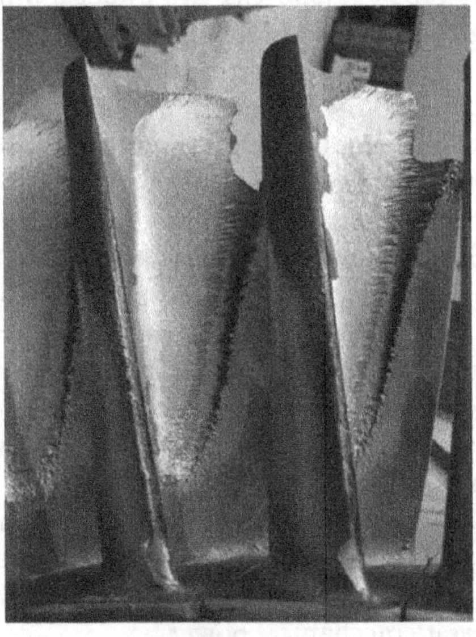

FIGURE 3.2 Heavy worn expander rotor blades from excess catalyst carryover.

both major losses in expander performance, as well as reduces the structural integrity of the blades.

3.2 Particle Separation

During the early application of CCU expanders, it quickly became clear that additional flue-gas cleanup was required to improve the operating time of the FCC expander. A number of development programs by various operating companies and particle separator manufacturers over the years showed two technologies could provide separation efficiencies that would greatly extend the life of the CCU expander and that were also cost-effective. The two most commonly used technologies included the use of conventional cyclones or axial swirl vanes. These different technologies are illustrated in Figure 3.3. All of these technologies work on the same basic principle. The particle-laden flue gas is accelerated to a higher tangential or swirling velocity, and the particulate is thrown to the outside via centrifugal force. Additional particle separation occurs when the flue gas quickly

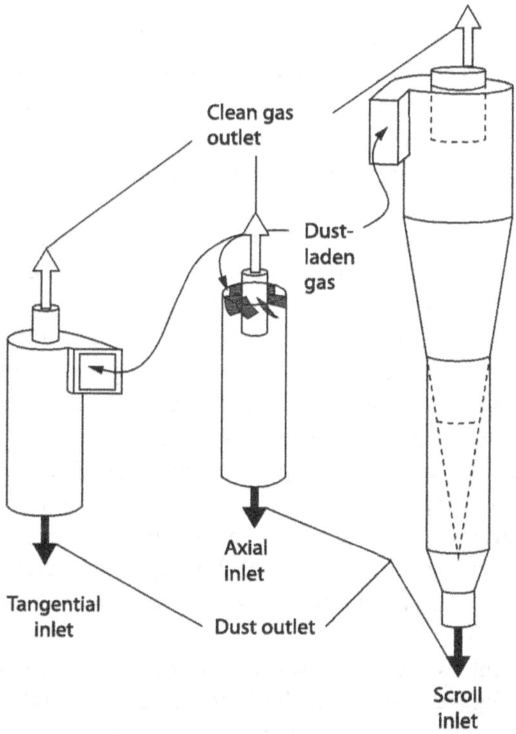

FIGURE 3.3 Catalyst separation technologies.

changes flow direction to exit the separation device. The particles drop out of the bottom of the swirl device, while a majority of the flue gas exits out the top of the cyclone. A small amount (1% to 3%) of flue gas typically is used to transport the separated particles out of the bottom of the separator. Over time, various improvements to these technologies have evolved to improve their separation efficiencies.

An added benefit of this separator technology is that it can be used to reduce the process environment discharge as well.

Early TSS designs were capable of removing more than 95% of all catalyst particles greater than 10 microns in size or larger. Expander erosion has been found to be greatly reduced for particles less than 10 microns in size. With the use of the TSS, the amount of catalyst entering the expander was reduced from the original level of 400 ppm to less than 120 ppm. This corresponds to approximately 100 lb of catalyst per hour for a 100,000 BPD FCC expander. Over time, the efficiency of the TSS has improved to what it is today, the TSS designs are capable of removing more than 97% of all catalyst particles greater than 5 microns in size or larger. This reduces the amount of catalyst entering the expander to less than 50 ppm.

3.2.1 Third Stage Separator Construction

Third Stage Separators are the assembly of multiple cyclonic separation devices. As previously noted, the TSS is used to reduce the amount of catalyst entering the Hot Gas Expander (for increased blade life) and to reduce the amount of catalyst being emitted to the atmosphere.

TSSs were developed after the initial operational trials of FCC expanders showed that without additional particle separation, the expander flow-path components were heavily eroded in operating times of less than 1 year. TSS were initially developed in the 1960s and provided satisfactory service for 2- to 3-year operating periods. This was consistent with typical FCC turnaround schedules at the time. Shell Oil developed a multi-swirl vane design that became the industry standard. The Shell TSS design uses multiple 10 in (25 cm) diameter axial inlet swirl vane assemblies. Depending on the volume flow of flue gas generated by the FCC process, the number of swirl vanes can vary from 80 to more than 120. Flue-gas flow rates are proportional to FCC capabilities and can vary from 300,000 lbm/h to over 1,400,000 lbm/h.

Flue gas enters the top of the vessel and is directed between two tube sheets (Figure 3.4). From the tube sheets, the catalyst-laden flue gas flows downward into the swirl vane assemblies. The swirl vanes impart a tangential force to the gas and catalyst that causes the catalyst

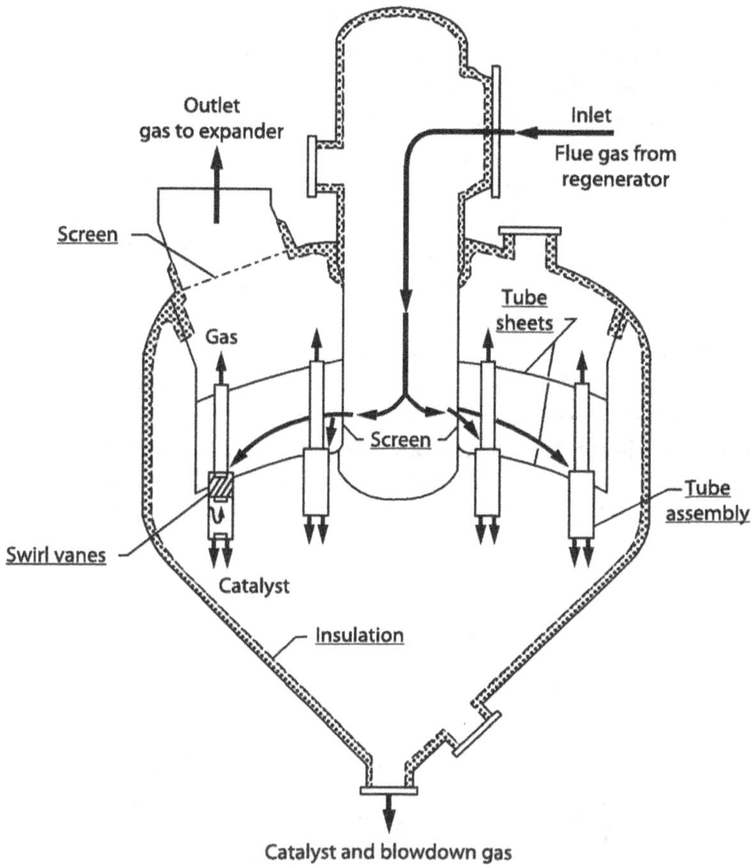

FIGURE 3.4 Shell TSS design with multiple axial swirl vane assemblies.

to collect on the outside of the assembly tubes. The catalyst drops out of the gas and into the bottom of the TSS vessel, while the flue gas is quickly turned vertically upward. Finally, the flue gas exits out of the top of the TSS vessel. The separated catalyst exits out of the bottom of the TSS vessel. The separated catalyst is transported to downstream separation equipment such as an electrostatic separator or Fourth Stage Separator.

The multiple-swirl vane or cyclones used in the TSS are generally mounted in a single cold wall vessel. The vessel is refractory-lined carbon steel with stainless-steel internals. Early TSS installations developed some reliability issues that included swirl vane plugging and some mechanical issues associated with the tube sheet support, attachments to the outer vessel walls and refractory reliability.

Over time, these issues have been resolved and today's TSS assemblies are very reliable and operate for many years with minimal maintenance. It should be standard practice to inspect the internals of the TSS at every major FCC turnaround.

Numerous product improvements and new designs have developed over time to reduce discharge and stack emissions. These lower emissions have obviously been beneficial in reducing expander flow-path erosion and deposition. TSS upgrades include improved mechanical designs, optimized geometries to improve separation efficiencies, and designs that are more tolerant of process upset conditions (afterburning and catalyst dumps).

Throughout the history of power-recovery expanders, several conventional cyclone TSS designs have been developed similar to that shown in Figure 3.5. Separators designs using conventional cycles have been constructed by a number of companies including Buell, Emtrol, Marsellex, and Van Tongrin. The conventional cyclone TSS

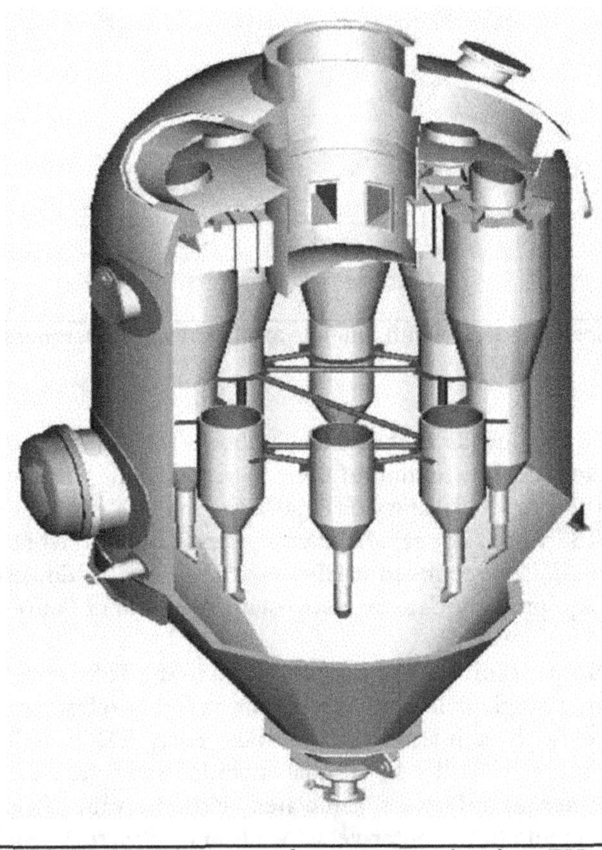

Figure 3.5 Typical cutaway view of a conventional cyclone TSS.

designs use as few as 6, to as many as 25 individual tangential or scroll inlet cyclones.

In addition to the Shell swirl vane and convention cyclone technologies, several other TSS designs have been successfully applied on a limited basis. These include:

- *KBR/Mobil TSS.* The KBR TSS uses a large quantity of smaller-sized (10 in) vertically mounted cyclones in a similar arrangement as the Shell swirl vane design. The design is still actively marketed.
- *Polutrol TSS.* The Polutrol TSS design was a unique design that used a large number of smaller-sized horizontally mounted cyclones in a vessel.
- *UOP TSS.* The UOP (formerly known as Universal Oil Products) TSS design uses a large number of vertically mounted swirl vane assemblies. Flow enters the top of the TSS, is directed vertically downward through swirl vane assemblies, and finally exits the swirl vane assembly in a downward direction. Unique to the UOP design is that the flue gas leaves the TSS out the bottom of the vessel. This separator is also actively marketed today.

Critical to the successful operation of the TSS is the underflow at the bottom of the TSS vessel. The separated particles are carried out of the vessel via an underflow line. The underflow line directs the separated catalyst around the Hot Gas Expander and can return the catalyst back into the flue gas or into an additional separation device such as an electro-static precipitator, flue gas scrubber, or Fourth Stage Separator (Figure 3.6).

Plugging of the underflow will allow separated catalyst to collect in the bottom of the TSS vessel. Eventually, when the vessel capacity is reached, all of the catalyst entering the TSS will be carried over into the Hot Gas Expander. Plugging is best detected by monitoring the operating temperature of the underflow line. Plugging of the underflow will reduce the underflow rate and a corresponding reduction in operating temperature can be noted. The most common causes of underflow plugging are TSS vessel refractory failures and catalyst dumps from the CCU process. The plugging often occurs in the critical flow orifice/nozzle (CFN). The CFN is used at the end of the underflow line to control the amount of underflow. CFNs are sized to provide a flow rate equal to 1% to 3% of the TSS full flow. The required amount of underflow is an area of technical debate. Lower underflow rates will increase the amount of power recovery in the Hot Gas Expander. While a small amount of underflow is required to move the normal amount of separated catalyst, the lower flow rates tend to make the underflow line

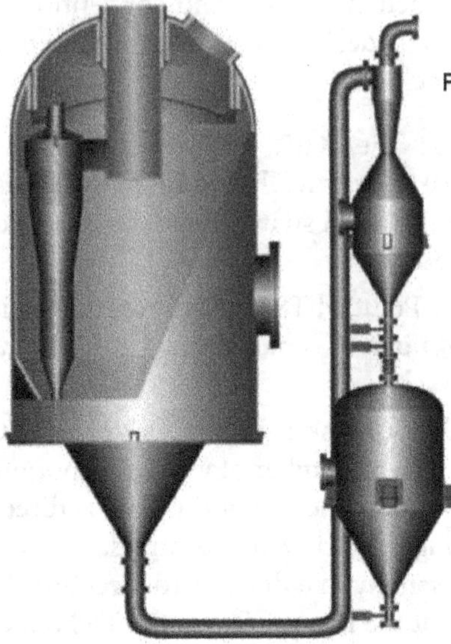

FIGURE 3.6 Fourth Stage Separation equipment.

more prone to plugging. This is particularly true during periods of abnormally high catalyst carryover from the FCC process. Increasing the underflow has been found to help prevent underflow line plugging and has been found to improve the separation efficiencies of the finer (<10 micron) particles. These finer catalyst particles tend to increase the amount of catalyst deposition in the Hot Gas Expander.

The overall condition of the TSS is difficult to monitor while in service. A number of instruments can be installed to assist in continuous monitoring, including:

- TSS inlet temperature
- TSS inlet pressure
- TSS outlet pressure
- TSS tube sheet differential pressure
- Catalyst-level indicators—bottom cone
- Underflow line temperature

Third Stage Separator performance is critical to successful Hot Gas Expander reliability and blade life. As previously noted, expander blades can wear out in as short as a few months of operation or experience virtually no wear in more than 8 years of operation. The TSS is the single-most important item that controls the expander blade life.

3.2.2 Fourth Stage Separator

Initial TSS installations were designed with underflow lines that bypassed the Hot Gas Expander and dumped the separated catalyst back into the flue-gas line downstream of the expander. As concerns developed and environmental emission requirements became more stringent, additional separation equipment was required. Early CCU units collected the catalyst fines in an electro-static precipitator.

In more recent years flue-gas wet-scrubbing systems have been used to capture both the catalyst fines and SO_x emissions. Although these technologies have been found to have acceptable performances, they are expensive to install, operate, and maintain.

With the ever-increasing need to reduce FCC emissions, the use of Fourth Stage Separators (FSS) has become more common. With the use of a FSS, the TSS underflow is carried over into a single fourth-stage cyclone that separates the catalyst from the underflow gas. The separated catalyst is collected in a fines collection hopper as can be seen in Figure 3.6. When the fines collection hopper is full, the collected fines are transferred to a fines receiver vessel. The fines collection hopper allows the catalyst to cool down and provides continuous separation while the fines receiver is being unloaded. The fines collector is often sized to hold a single day of catalyst recovery, whereas the fines receiver is usually sized to collect a week of separated catalyst. Level indicators are used to determine when the collection vessels are full and weight sensors on the fines receiver can be used to monitor the amount of separated catalyst.

3.2.3 Isokinetic Testing

Testing the performance of particle separation equipment is very difficult. It requires extracting the catalyst-laden flue gas at high temperature and pressure from large-diameter ducting. Experience has shown the most accurate testing is accomplished by the use of Isokinetic testing. Isokinetic testing involves the insertion of Pitot tube–type probe into the flue-gas duct and extracting samples of the catalyst-laden flue gas. The extraction must be done with the extracted gas velocity equal to the duct-flow velocity as shown in Figure 3.7. If the gas withdrawal velocity is at a velocity less than the duct velocity, some of the catalyst particles will not enter the probe and will bypass the probe. This results in erroneously low test results. If the withdrawal velocity is greater than the duct velocity, the probe will actually suck more particles into the probe and give erroneous high test results.

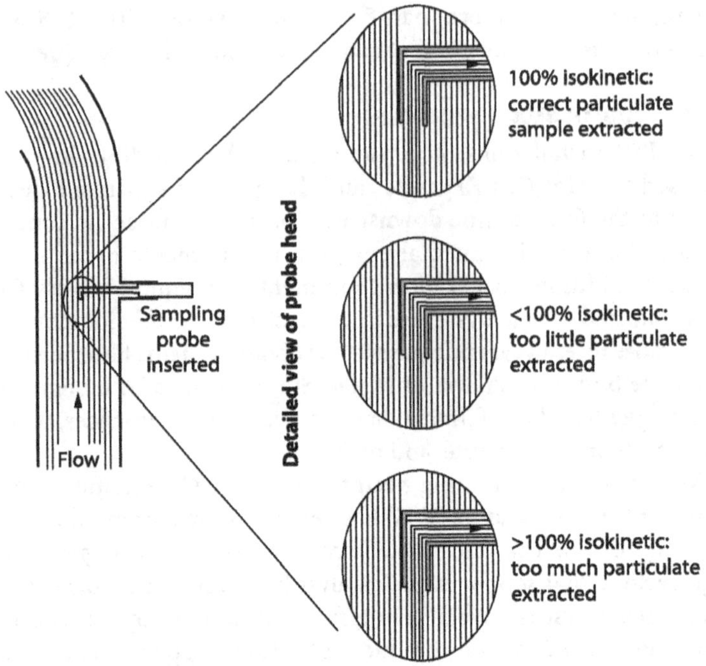

FIGURE 3.7 Isokinetic testing—particle withdrawal.

Isokinetic testing is best performed in long straight runs of ducting away from any geometric changes or obstructions such as elbows and valves. The best test results occur when the flow is fully developed and laminar. Testing is usually performed at various insertion depths into the duct owing to the reduced velocity boundary layer flows that exist along the walls of the ducts. Generally the duct is tested at eight different insertion depths to identify and quantify the flow profile. In addition, isokinetic testing is usually performed in two different directions 90° apart in order to identify any flow disturbances or nonuniformity within the duct.

Isokinetic testing inaccuracies encountered during the testing often include:

- Unsteady flue-gas flow in the ducting
- Varying flue-gas flow over time owing to unsteady process conditions
- Nonuniform flow profiles across the diameter of the duct
- Liquid condensation as the flue gas cools down
- The formation of metallic salts in the samples as the flue gas cools down
- The conglomeration of particles

Special procedures have been developed and adopted by the U.S. Environmental Protection Agency (USEPA). USEPA Method 17 has been found to give the most accurate test results and addresses the problem of condensation and the formation of salts or conglomerates during the gas cooldown. The size and frequency of the particles captured during the isokinetic testing are determined by use of a Coulter counter.

Catalyst particle size distribution changes throughout the FCC system. Figure 3.8 shows a typical catalyst distribution of equilibrium catalyst as it is added to an FCC unit. The average particle size can be seen to be approximately 80 microns in diameter. Figure 3.9 shows a typical catalyst particle distribution exiting the FCC regenerator and entering the Third Stage Separator. The catalyst size data shows a bimodal distribution with both large and very fine particles.

The catalyst fines are generally the result of fracturing and the breaking up of the larger catalyst particles during the cracking process (Figure 3.10). If the Hot Gas Expander were subject to this particle distribution, the expander would likely experience high levels of catalyst erosion and flow-path deposition.

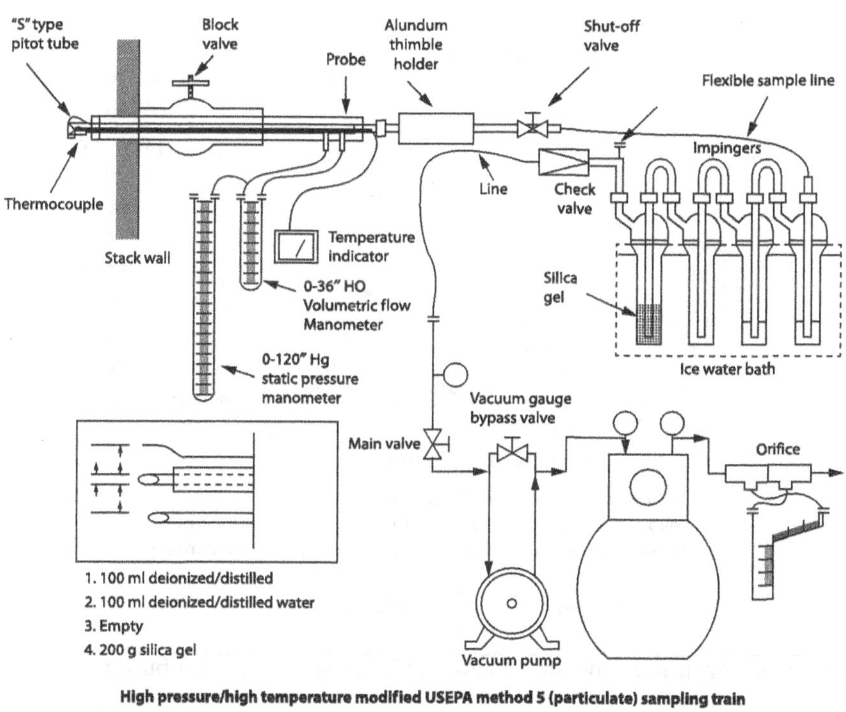

Figure 3.8 Isokinetic test instrumentation and setup.

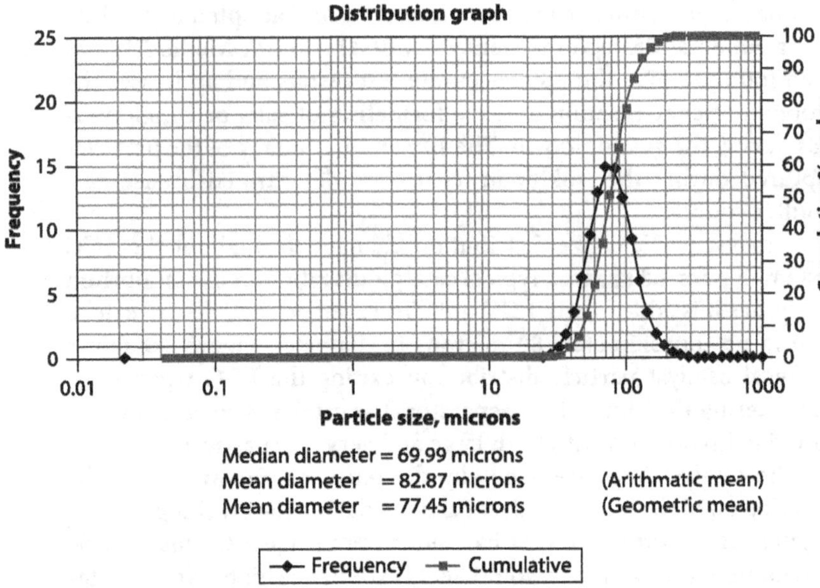

FIGURE 3.9 Typical e-cat catalyst particle distribution.

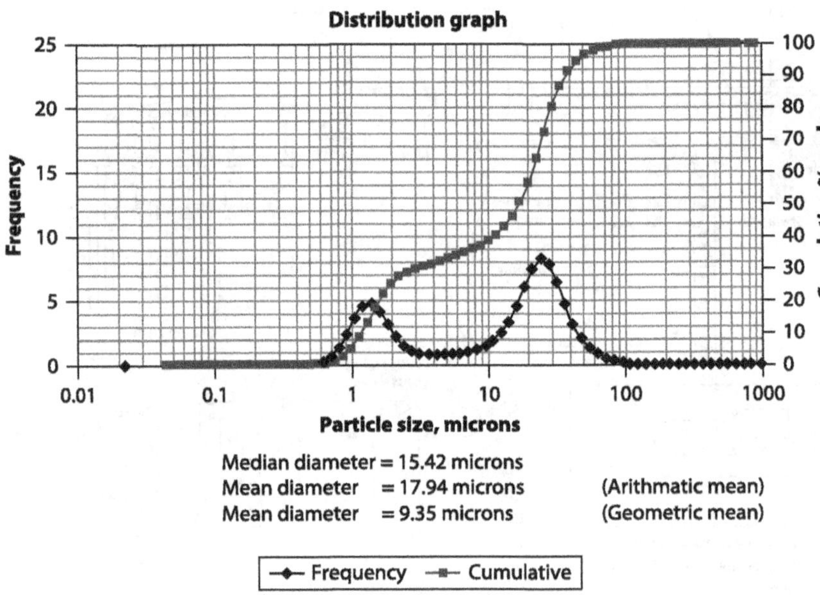

FIGURE 3.10 Typical regenerator discharge/TSS inlet particle distribution.

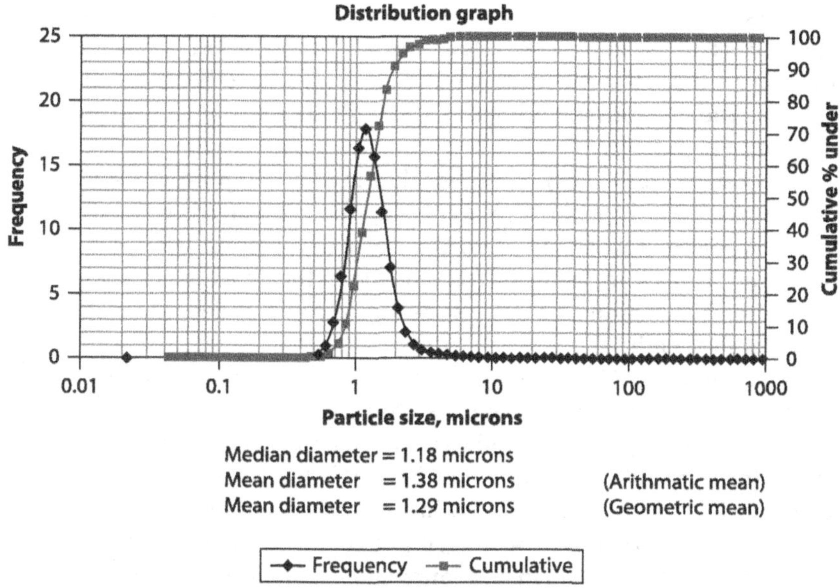

Figure 3.11 TSS discharge/expander inlet distribution.

Figure 3.11 shows a typical catalyst distribution for a properly performing Third Stage Separator. This catalyst distribution would allow for long-term reliable operation of the Hot Gas Expander.

3.3 Expander Erosion

Shortly after the first FCC expander was put into service on a Catalytic Cracking Unit it was evident that the catalyst carried over in the flue gas was very abrasive and would rapidly lead to excessive erosion damage in the expander flow path. Over the last 50 years, CCU expanders have evolved and greatly improved the erosion resistance of the expander. The reduction in erosion potential has been accomplished via technology upgrades in the TSS equipment, expander flow-path design, expander metallurgy, and coating technology.

The unique feature of CCU expanders is the dirty gas environment that is highly erosive. Even though the process particle separation equipment is very efficient (>98%), significant amounts of erosive catalyst is still carried over into the downstream equipment. Typical regenerator catalyst overflow rates are of the order of 400 ppm (parts per million by weight). This loading was found to be excessive for the Hot Gas Expander and excessive blade wear was found to occur in as little as a few months of operation. With the addition of the TSS, expander catalyst loadings have been greatly reduced. Today, catalyst loadings of 100 ppm are normal and can be as low as 50 ppm.

The catalyst is a very abrasive alumina—silica powder. With the use of a TSS, the catalyst size entering the expander is generally under 10 microns in diameter during normal operation. Even with the use of the TSS, the carried-over catalyst to the expander can still be erosive. In addition, the catalyst can be sticky and deposit in the expander flow path.

Since the first CCU expander was put into service over 50 years ago, the erosion characteristics have changed. Many of the erosion characteristic changes are related to the changes in process operation. The major operational changes that have affected the erosion characteristics include residual oil cracking, the cracking of higher sulfur feed stocks, more selective cracking, higher temperature and pressure operation, and the changes in environmental regulations that have led to changes in CCU operation, catalyst type, and the use of emissions control additives.

3.3.1 Erosion Characteristics

Erosion of the stationary flow-path guides has been relatively minor with only the nose cone and shrouds experiencing any significant erosion. The erosion occurs over many years of operation provided there is no major catalyst upset conditions. Virtually no erosion has been experienced in the discharge diffusers or exhaust casing regions.

Duct erosion is generally insignificant as well owing to the low gas velocities. Duct velocities are typically less than 150 feet per second (ft/s) (45 m/s). Local piping erosion has been noted only in the areas of high gas velocities such as in the regions of control valves or flow orifices. Erosion generally is not experienced when the gas-path velocities are maintained less than 300 ft/s (91 m/s).

Nose Cone Erosion

Expander nose cones have experienced nose cone bullet erosion. Nose cone bullet erosion can occur on nose cones that have short axial lengths that rapidly accelerate the gas and catalyst into the stator vanes. These short, blunt nose cone designs also have a tendency to have flow separations and concentrate the catalyst in the boundary layer flows of the cone. These catalyst-laden boundary-layer flows tend to concentrate the catalyst along the flow-path inner wall and at the base of the stator and rotor blades. Nose cone inlet strut, stator vane, and rotor-blade erosion can result from the heavier concentration of catalyst in the nose cone boundary layer flow.

Additional area of nose cone erosion has been noted in support flange seal surfaces, around guide pins, on nose cone seal rings, and around cooling piping. These wear regions are generally related to gas and catalyst leakage in regions where flow is not normally expected. Nose cone distortion or inadequate nose cone-to-casing seals often contribute to this leakage erosion. Newer expanders have improved seal designs that minimize the leakage of flue gas and catalyst out of the flow-path regions.

Inlet Guide Vane/Strut Erosion

Inlet guide vane erosion can be significant enough to reduce the structural integrity of the inlet struts. The inlet strut wear can be characterized as a chamfering or "V"ing of the leading edge of the vane and as localized tracking or grooving in the welded regions where the vanes join the inner or outer liners, as can be seen in Figure 3.12.

The inlet guide vane design is significant to the stator blade erosion. Wide inlet guide vanes can cause significant flow blockage, separations, and shielding of the downstream stators from the catalyst-laden flue gas. The stator vanes adjacent to the shielded vanes often experience significantly higher vane erosion whereas the shielded stators behind the vanes can see little or no erosion. This is illustrated in Figure 3.13, where each guide vane behind an inlet strut shows virtually no erosion.

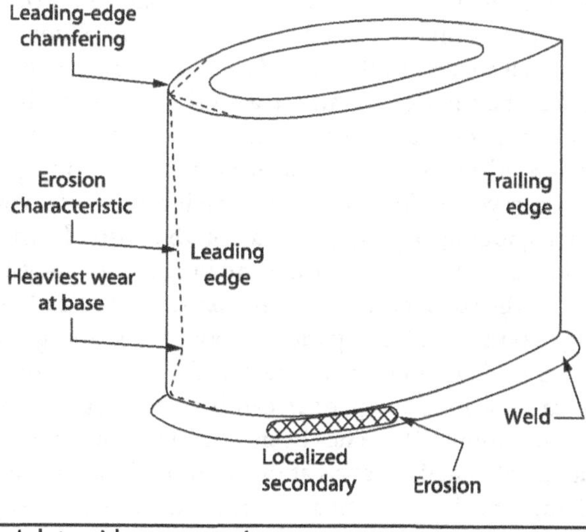

Figure 3.12 Inlet guide vane erosion.

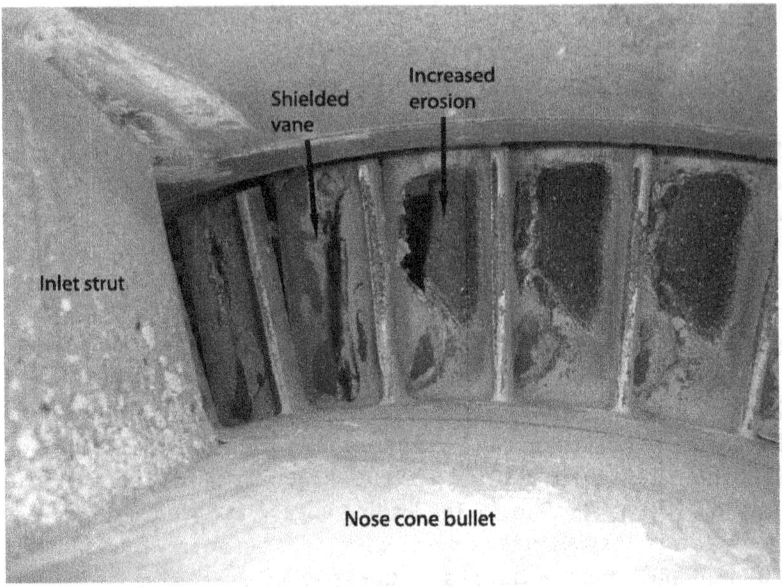

FIGURE 3.13 Stator vane erosion: effect of inlet strut blockage.

Shroud Erosion

Under normal operation, shroud erosion is generally low in CCU expanders. The rate of erosion is generally less than that experienced by the stator vanes and rotor blading. It is common to get multiple operating campaigns from a shroud ring. Erosion-resistant coatings are often applied to the flow-path regions to further extend the life of the shroud. Periodic refurbishment and recoating of the shroud can further extend the life of the shroud ring. Shroud ring wear is gradual although the erosion can eventually penetrate the shroud coating. This generally occurs in the region over the rotor blades where gas velocities are high. Once the shroud coating is penetrated, the base material will erode at a faster rate.

Heavy shroud erosion has been experienced during periods of high catalyst carryover. These periods of high catalyst carryover often lead to axial gouging over large areas of the shroud. The erosion can be accelerated from a grinding and rubbing of the catalyst between the shroud and the rotor-blade tips. This condition can lead to extreme overheating of both the blade tips and shroud. The spalling of the coating on the shroud usually occurs during these rub conditions. The amount of catalyst in the expander can be quite large (tons) during these upset conditions. There has been evidence that some expanders have operated with catalyst levels at the bottom of the casing in excess of 1 ft (0.3 m). Local overheating of a shroud from catalyst grinding or deposition as described above is the most common cause of shroud distortion and distress and is the number-one cause of shroud wear and replacement.

Stator Blade Erosion

Stator blading can experience erosion during long-term normal operation as well as from upset conditions with high catalyst carryover. The erosion patterns from both the normal operation and during the periods of catalyst carryover are similar with the leading edge of the vane suffering significant chamfering of the inlet radius. Pressure side (concave) erosion is also a typical erosion pattern on stator vanes. A typical stator vane erosion pattern is illustrated in Figure 3.14.

The amount of stator-vane coating coverage greatly affects the erosion pattern. Prior to the early 1980s, most stators were coated only along the trailing edge of the vanes. This allowed the catalyst erosion to undermine the coating when the erosion started in the mid-chord region upstream of the coating, as can be seen in Figure 3.15. Severe

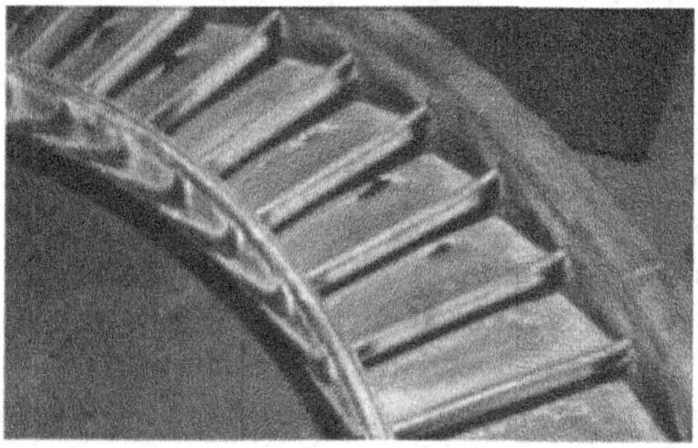

Figure 3.14 Typical stator vane pressure side erosion.

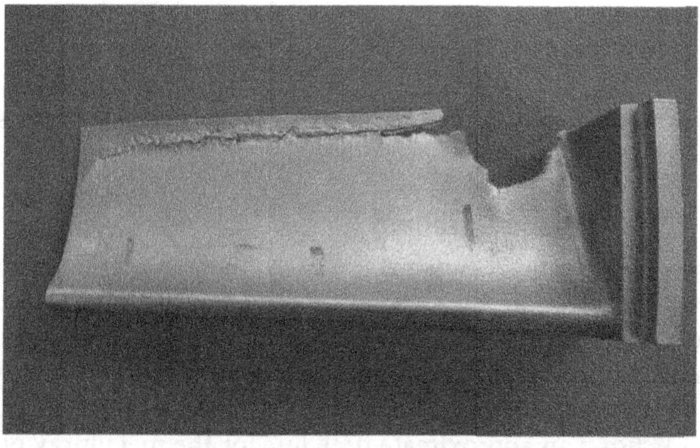

Figure 3.15 Heavy stator vane erosion: pressure (concave) side.

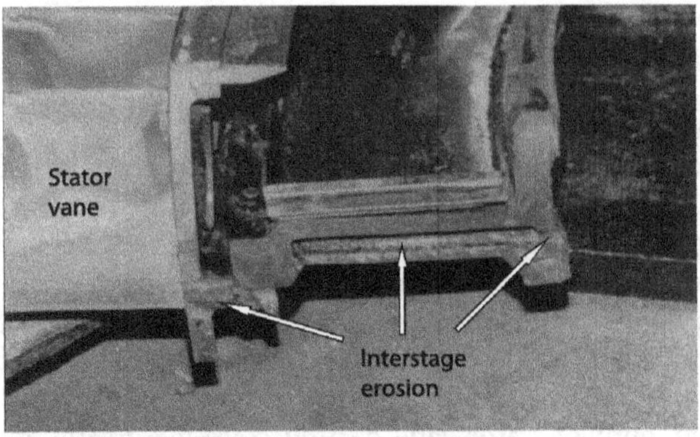

FIGURE 3.16 Interstage erosion in two-stage expander.

erosion of the stators generally shows major mid-chord gouging and a thinning of the trailing edge of the vane. The wearing of holes through the vane cross section is also common on heavily worn vanes.

Erosion of the stator is often concentrated toward the base of the vanes owing to a larger concentration of catalyst at the base of the vane, as was described in the nose cone section above. Stators do not experience significant vane tip erosion, whether the stator is cantilevered from the inside or the outside of the flow path. Although stator erosion does not appear to depend on how the stators are mounted with respect to the inner or outer portion of the flow path, the rotor blades do appear to experience increased erosion in the area of stator leakage flow.

A number of single and multiple-stage expanders use integral stator/shroud assemblies. In general the multiple-stage expanders have experienced extremely heavy erosion in the interstage regions from leakage of gas around the stators. The erosion occurs in every location where the interstage pressure drop is experienced or the gas is rapidly turned. Figure 3.16 shows typical erosion across a stator segment split line from a two-stage expander. Note the wear on both the inner shroud and drop diaphragm. Unique rotor erosion patterns have also been experienced from the interstage leakage flows.

3.3.2 Stator Blade Metallurgy

Stator vanes have been constructed from several different metallurgies over the history of the CCU expanders. The metallurgy has been selected for both high-temperature strength and erosion resistance. Upgraded metallurgies have been incorporated in recent years to eliminate the high-temperature aging effects of Stellite 31 material. Table 3.1 shows an approximate time line for the metallurgy of FCC expander stator blades.

Year of Application	Expander Inlet Temperature		Material	Material Chemistry
1950s through 1960s	1100–1345°F	621–729°C	AISI-304 SS	Iron base—18 Cr-10 Ni
1960s through 1990s	1050–1355°F	566–735°C	Stellite 31	Cobalt-base superalloy
1960s through 1990s	1050–1355°F	566–735°C	A-286	Iron base—15 Cr-25 Ni
1980s through today	1100–1350°F	593–732°C	Waspaloy	Nickel-base superalloy
1990s through today	1100–1355°F	593–735°C	X-40	Cobalt-base superalloy
1990s through today	1325–1375°F	718–746°C	IN-738	Nickel-base superalloy

TABLE 3.1 Stator Blade Metallurgy

The vane metallurgies (Stellite 31, X-40, and IN-738) are the hardest and have provided the best erosion protection and vane life.

3.3.3 Rotor-Blade Erosion

Background

CCU expander rotor blading has experienced a number of different types of erosion rough out the more than 50 years of their use. Early expanders experienced rapid blade wear owing to continuous high levels of catalyst carryover into the expander. Figure 3.17

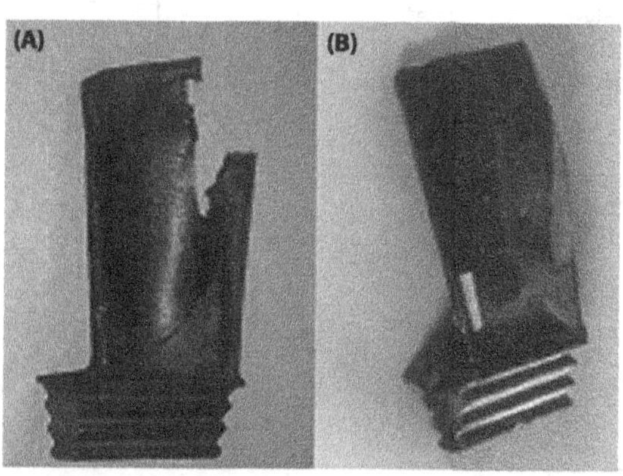

FIGURE 3.17 Early expander blade erosion.

shows the type of blade wear experienced in the early years of the CCU expander. Even with the development of the TSS, the early expander rotor blading lasted less than 2 years before large sections of the airfoils experienced heavy erosive wear. This heavy blade wear caused by larger catalyst particles is called *primary erosion.* Figure 3.18 shows the typical early primary erosion. Throughout the 1960s, with the development of better catalyst-separation technology, the expander blade life more than doubled. Improved expander design and the implementation of erosion-resistant coatings provided for a further increase in expander blade life. Expander operating times increased from the initial run time of less than 1 year to 3 to 4 years.

The 1970s saw little change in TSS or expander design. It is of interest to note that fewer than 10 FCC expanders were built between 1950 and 1978. Over the next 30 years, more than 120 new FCC expanders were manufactured and installed. This larger expander population demanded improved reliability and increased blade life. This need drove the improvement of both the particle separation equipment and the Hot Gas Expander. The equipment designs were improved and the primary erosion characteristics of the past were virtually eliminated in properly operating power-recovery systems.

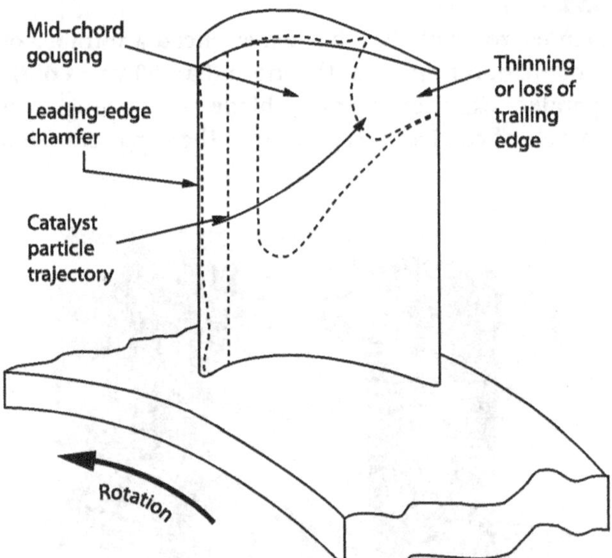

Figure 3.18 Expander primary erosion characteristics.

During this same time period, alternate TSS and expander designs were introduced by a number of manufacturers. Rotor-blade life was now found to exceed the turnaround times of the process equipment (typically 3 to 4 years). The need for rotor-blade life-assessments and refurbishment techniques arose as many blades exhibited minimal or only localized erosion wear.

Over the life of the CCU expander, major changes occurred to the process operating conditions, that dictated further changes to the expander design. Some of the more significant process changes included the following:

- The change to complete combustion operation in many existing and most new CCUs.
- Increased residual oil cracking in existing CCUs. The construction of new residual oil Catalytic Cracking units.
- The changeover and increased production of unleaded gasoline.
- The construction of both larger and small CCUs.
- Increased environment restrictions.
- Greater selectivity changes.
- New catalyst technology.
- Higher pressure operation.
- Higher temperature operation.
- The use of a wider selection of feed stocks.
- The introduction of new separation technology.

To accommodate these process changes, the CCU expander designs also had to change. New designs were required to accommodate operating at higher process flows, temperatures, and pressure ratios. Higher energy costs also drove the need for improved expander efficiency.

The process changes of the 1980s brought new expander operational problems to light. New erosion patterns began to evolve, high temperature corrosion, and most significantly, the problem of catalyst deposition and resulting blade wear became common place.

The most common type of catalyst erosion of rotor blades evolved from the large particle primary erosion into fine particle secondary flow erosion. Figure 3.19 shows the complex secondary flows that exist in all turbine blading. While it is not possible to eliminate secondary flow erosion, it has been reduced and eliminated in some expanders by incorporating more modern blade designs and with the

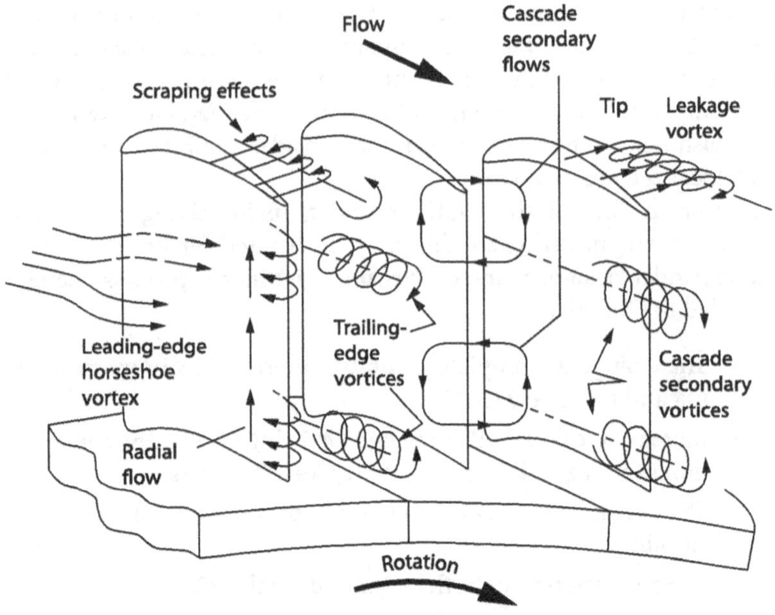

FIGURE 3.19 Secondary flow vortices in a rotor blade.

use of fully coated airfoils. Some examples of severe secondary flow erosion can be seen in Figure 3.20.

Catalyst deposition is the most significant concern of expander users in terms of expander rotor-blade wear. Chapter 4 goes into a more in-depth discussion of catalyst deposition.

The 1990s became a period concentrating on expander reliability. A strong emphasis was placed on maximizing run times of the expander and the elimination of operational problems using existing

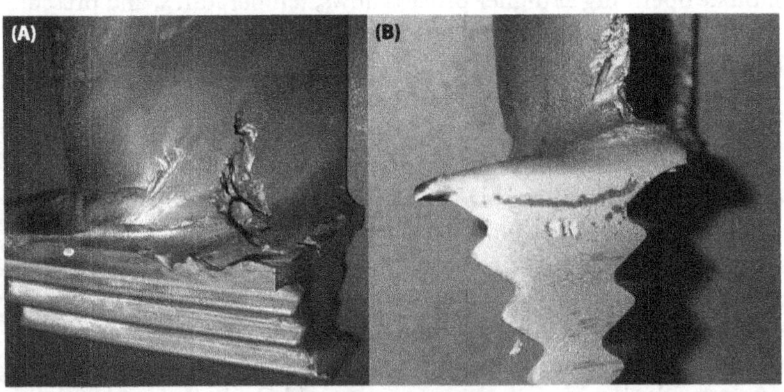

FIGURE 3.20 Examples of severe secondary erosion.

technology. In the 2000s, expander development began to use the many technology tools that have evolved in the turbomachinery analytical world, including erosion-prediction capabilities, improved flow-path design tools, improved performance-prediction program, computational fluid dynamics, and more.

3.3.4 Rotor-Blade Metallurgy

The base material of rotor blades has changed over the years owing to increased operating temperatures of the FCC process and the need for more erosion-resistant blade materials. The higher operating temperatures require materials of higher temperature strength combined with greater corrosion resistance.

Table 3.2 is an approximate time line of the metallurgy of rotor blades.

The development of expander rotor-blade material has paralleled that of industrial and aero derivative gas turbines. As the operating requirements of the service have become more severe, the materials have changed from wrought or forged iron-based superalloys, to wrought or forged nickel-base superalloys. The latest evolution has been to use cast equiaxed crystal nickel-based superalloys.

Increased erosion resistance has been found with higher temperature metallurgies. A comparison of the high temperature erosion characteristics of the various expander blade materials and a commonly used coating can be found in Figure 3.21. It can be seen that the maximum erosion rate of the base materials is at an impingement angle of approximately 30°. In addition, it can be seen that the coating has a very flat erosion characteristic and the erosion resistance can be

Years of Application	Expander Inlet Temperature		Material	Material Chemistry
1960s through 1990	1050–1200°F	566–649°C	A-286	Iron base—15 Cr-25 Ni
1974 to 2000	1200–1285°F	649–696°C	Inconel X-750	Nickel-base superalloy
1978 through present	1285–1380°F	696–749°C	Waspaloy	Nickel-base superalloy
1978 to 1990	1285–1350°F	621–732°C	Nimonic 90	Nickel-base superalloy
1990s	1325–1415°F	718–768°C	IN-738	Nickel-base superalloy

TABLE 3.2 Rotor Blade Metallurgies

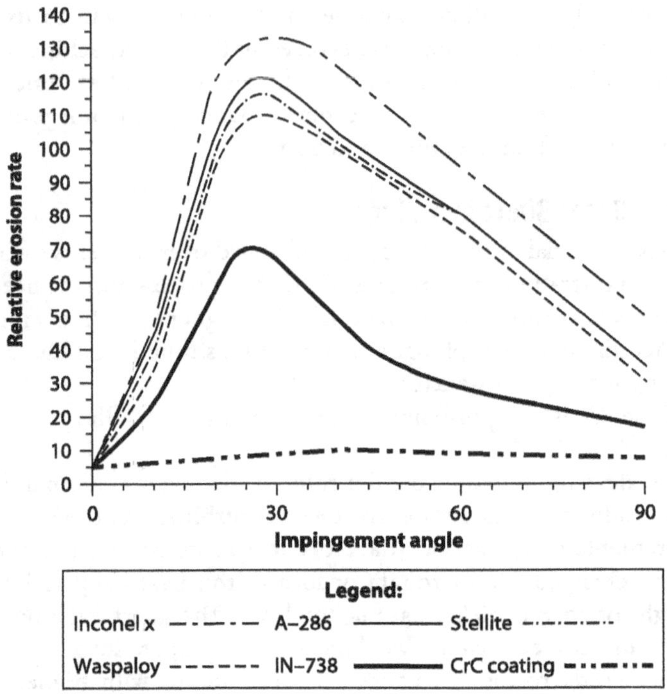

Figure 3.21 Relative erosion resistance of various expander blade materials.

as much as 10 times that of the base materials. The erosion resistance of the rotor-blade base materials has been found to be directly related to the high-temperature strength and hardness.

In addition to the additional erosion resistance, the high-temperature metallurgies give higher structural margins for normal and upset condition operation.

With the reduction in expander rotor-blade erosion over time the reparability of the blades also is of greater interest. A number of weld-repaired blade sets have been successfully repaired and returned to service in the last 5 years.

3.3.5 Flow-Path Coatings

As previously noted, flow-path coatings have been found to significantly increase the life of the expander flow-path components including stator vanes, rotor blades, and shrouds. The flow-path coatings evolved over time and have continued to improve with expander development.

Early expander erosion-resistant coatings were primarily a tungsten carbide coating. The stator vanes and rotor blades had a 1-in (25-mm) band applied along the trailing edge of the vanes and blades. The coating extended along the airfoil from the platform to the outer-most diameter. This type of coating was effective in reducing the primary erosion along the trailing edges of the vanes and blades. Experience showed that the erosion was often occurring in front (upstream) of the trailing edge–coated strip in a typical primary erosion characteristic called mid-chord gouging. This erosion was reduced by extending the coating forward on the vanes and blades until eventually the coating covered the entire pressure (concave) side of the airfoils. With the elimination or reduction of the mid-chord erosion, the leading edge of the vane and blades became the limiting area for component reuse. In more recent years, the wrapping of the leading edge of the stators and fully encapsulating the rotor-blade airfoils with the erosion-resistant coating has further increased component lives.

As the expander inlet temperatures continued to rise (from 1200°F [650°C] to 1350°F [735°C]), the oxidation and spalling of the stator tungsten carbide coating was experienced and became more of a concern. This prompted a change of coating from tungsten carbide to a chromium carbide. The chromium carbide coating has been shown to be very effective in the reduction of flow-path component wear and has been effective for expander inlet temperatures in excess of 1415°F (768°C). Further enhancements to the coating erosion protection were found with the use of fully robotic systems and the use of very high-density coatings achieved with specialty coating guns such as Praxair's D-Gun and Super D-Gun Chromium Carbide coatings. The robotic coating system allows the application of the coating uniformly, covering all areas of the blade including the top of the blade platform.

The chromium carbide coating was selected for its more ductile binder and the coating's ability to withstand high-temperature oxidation. The chromium carbide, fully coated airfoil has been in service for more than 20 years and continues to be the standard coating for essentially all expander installations. Field experience has shown that fully coated airfoils and platform regions have proven to be very effective in reducing both primary and secondary flow erosion.

Several expander installations have attempted to coat the expander blade tips to minimize tip wear caused by catalyst deposition. These applications were not successful in the reduction or elimination of the blade tip wear.

3.3.6 Rotor Disc Erosion

A number of expanders have experienced heavy disc erosion on the outer portion or on the face of the attachment region. This erosion can be the result of catalyst particles being channeled down

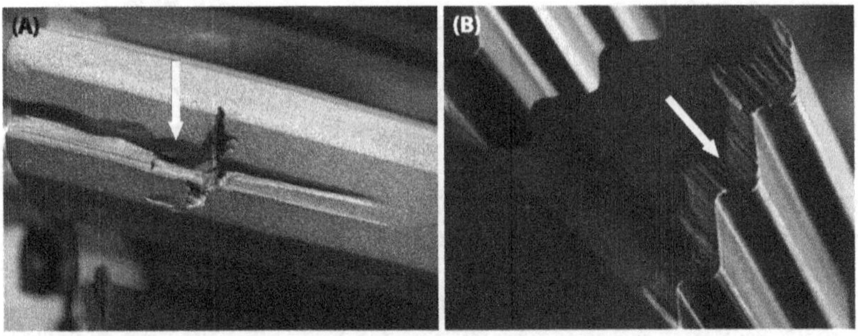

FIGURE 3.22 Rotor disc attachment erosion: OD and face.

the clearance area between the blades, and the rotor disc or in some expanders owing to wet cooling steam. Figure 3.22 shows two cases of heavy rotor disc erosion. Figure 3.22 (A) shows cutting on the outer-most region of the disc attachment. This type of erosion is very sharp and forms a long, narrow erosion channel. The attachment erosion has penetrated deeply into some expander discs by more than 0.25 in (6.4 mm). In at least one installation, the erosion actually penetrated into the disc sufficiently to break through the disc into the outer neck of the blade attachment. Severe erosion of the disc attachment regions can have a negative effect on the structural integrity of the rotor disc.

The second type of erosion shown in Figure 3.22 (B) is face erosion of the rotor discs. Although this type of erosion can occur on either side of the disc, it tends to be greater on the inlet side. The catalyst tends to be caught by the edges of the blade attachment geometry and then channels the catalyst along the attachment faces and flanks. Sharp penetration in the region of the attachment fillet is the greatest concern.

A significant number of expander installations have had to prematurely replace rotor discs solely because of these types of erosion. The disc-eroded regions at times can be reworked by blending the eroded regions and coating the region to prevent further damage. Many expander operators have coated new discs to prevent the noted erosions. Erosion-resistant coatings have been effective in preventing these types of disc erosion. Coatings can be applied to the outer section of the blade root attachment area and/or the disc face to combat the noted erosion damage. A chromium carbide coating typically has been used.

Multiple-stage FCC expanders have experienced a great deal of rotor disc erosion. The leakage of flue gas and catalyst around the downstream stators has caused locally heavy erosion on the faces and

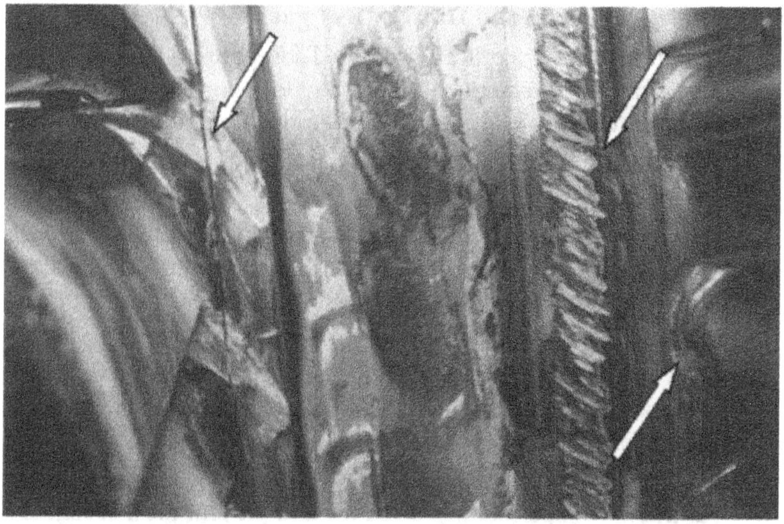

FIGURE 3.23 Multistage expander interstage erosion damage.

seal regions of the downstream rotating row. Drum-style rotors have experienced significant seal and spacer disc erosion as the catalyst and gas leaks under the stator blades. Damage of this type can be seen in Figure 3.23. Coatings have been effective in reducing this type of rotor damage.

In addition, the application of erosion shields and the redesign of the interstage areas to reduce local velocities around the rotor disc have been effective in reducing or eliminating the face and seal region wear.

Erosion Prevention and Reduction

A dramatic reduction in expander flow-path erosion has been accomplished over time by addressing the many contributors that increase the erosion potential. A multi-aspect approach is needed to ensure long-term expander blade life. The process, Third Stage Separator and Expander designs and operation all contribute to the expander erosion potential. The erosion problem cannot be solved by addressing only one aspect of the erosion contributors.

3.3.7 Expander

Over time, many changes have been made to the expander to improve the erosion life of the flow path. Experience has shown that the following design features should be incorporated into all expanders to maximize the unit run times.

Nose Cone, Flow Path and Diffuser Designs

- Use more aerodynamically designed components that minimize pressure drop, reduce boundary lay flows, provide for smooth and uniform gas and catalyst acceleration, and deceleration.
- The nose cone shape should be developed to be continuously accelerating.
- Newer inlet guide vane designs should be used that are more aerodynamically shaped. The new shapes reduce the inlet strut blockage and prevent flow separation of the gas on the downstream side of the vanes.
- Low-angle shrouds (less than 10°) are preferred to prevent flow separations in the flow path and to reduce the catalyst deposition potential.
- Integral stator/shroud designs should be used whenever possible to improve expander performance and prevent local secondary flows and erosion.
- All flow-path guides should be designed with smooth transitions and gradual accelerations or diffusions to prevent flow separations.
- Flow-path angle changes should be maintained to less than 15° to prevent flow separations.
- The exhaust diffuser geometry should be optimized to smoothly and uniformly decelerate the flue gas.

Stators

- Use improved stator vane designs that provide for more uniform gas acceleration and minimize flow separation of the gas and catalyst.
- Use blade path designs with stator reaction in the 40% to 50% range.
- Incorporate airfoil designs that take advantage of today's aerodynamic design tools.
- Use low-aspect ratio (vane length/vane chord) vanes that reduce the rate of gas turning.
- Upgrade and use the most erosion-resistant materials.
- Use stator vane coatings:
  - Fully coat the full-pressure side vanes.
  - Coated the vane-leading edge.
  - Use chromium carbide coatings.
  - Use automated high-density coating equipment such as the Super Detonation Gun.

- Minimize or eliminate stator blade tip clearance and leakage. Integral stator/shrouds should be used.
- Incorporate the following design features:
    - Use low-stress designs that are not structurally susceptible to blade erosion.
    - Use thicker trailing edges to make the vane more tolerant to erosion.
    - Control the inter-platform clearances to minimize channeling of catalyst.
    - Control the platform shape to minimize steps between blade platforms.

Rotor Blading

- Used improved rotor-blade designs that provide for more uniform gas acceleration and minimize flow separation of the gas and catalyst.
- Use blade path designs with rotor reaction in the 50% to 60% range.
- Incorporate airfoil designs that take advantage of today's aerodynamic design tools.
- Use low-aspect ratio (blade length/blade chord) vanes that reduce the rate of gas turning.
- Maintain rotor-blade speed.
- Upgrade and use the most erosion-resistant materials.
- Use rotor-blade coatings:
    - Fully coat the entire airfoil (both sides).
    - Use Chromium carbide coatings.
    - Use automated high-density coating equipment such as the Super Detonation Gun.
    - Increase blade-tip clearances to accommodate catalyst deposition.
- Incorporate the following design features:
    - Use low-stress designs that are not structurally susceptible to blade erosion.
    - Use thicker trailing edges to make the vane more tolerant to erosion.
    - Control the inter-platform clearances to minimize channeling of catalyst.
    - Control the platform shape to minimize steps between blade platforms.

- Control blade to disc fit and location to prevent face channeling.
- Use generous fillet radii.
- Use locking designs that are not prone to erosion.

3.3.8 CCU Operation

The FCC unit operation continues to be the primary factor controlling erosion and deposition in the expander. Start-up and shutdown sequences often cause periods of high catalyst carryover and care must be taken to minimize the expander catalyst loadings. Even small problems during normal operation can lead to expander flow-path wear owing to the high number of operating hours. System upsets also can cause major expander damage in very short time periods (minutes).

Start-Up

Start-up conditions of an FCCU are often very conducive to high catalyst losses. The catalyst loading sequence should be reviewed to minimize the catalyst losses by controlling the flow and pressures at which the catalyst is loaded. Lower catalyst loading rates and velocities should be maintained throughout the catalyst loading periods until the primary cyclone dip leg bottoms are adequately covered and the cyclones are operating at their design conditions.

Starting the process with the expander in a bypass mode has been most successful from an expander-erosion standpoint. This type of operation requires significantly higher auxiliary power to provide adequate power and air flow for FCC unit start-up. Even bypassing the expander can have risk. A number of expander installations have experienced large amounts of catalyst carryover from catalyst collecting in the dead-headed expander inlet-piping system. The geometry of the piping may leave dead legs that collect large amounts of catalyst. On start-up of the expander, all of this catalyst is blown into the expander similar to what occurred during a process catalyst dump. This problem is particularly true for expander/generator sets and conventional trains with full bypass capability. A detailed review of the piping systems should be conducted to assure that the catalyst does not collect during transient or steady-state operation.

Upset Conditions

The most damaging operating condition is when the process has a pressure or catalyst-level upset. Under these conditions, major (tons) of catalyst carryovers can occur. This can occur if there is rapid depressurization of the regenerator from control or control valve failures. In these scenarios, large amounts of catalyst are carried out of the regenerator. The large amount of catalyst carried over from the regenerator often overloads and even plugs the Third Stage Separator. The overloaded Third Stage Separator allows the catalyst to carry over into the expander with catalyst.

Complete control and process system design reviews should be conducted to identify potential causes for process upsets. Dynamic simulations can be performed to identify potentially harmful conditions to both the rotating machinery as well as to process vessels and controls. These hazard and operational studies allow the different failure modes to be identified and the consequences of the event evaluated.

Catalyst Selection

Catalyst and catalyst additive manufacturer, type, and usage should be closely tracked and correlated to the expander flow-path erosion and deposition rates. Problematic catalysts and/or additives can be identified and their use minimized.

Catalyst Particle Separators Primary Cyclones

The primary regenerator cyclones are a critical part of the power recovery system. The cyclones must be properly sized and designed to assure proper particle separation and controlling the catalyst loadings into the Third Stage Separator. In many installations, the primary cyclones have been found to have been inadequately sized for the current operation owing to changes in the process conditions that have occurred over time.

Maintenance of cyclones is also very important to assure long-term reliable operation. Refractory breakdown, anchor failures, support failures, dip leg plugging, dip leg failures, trickle valve failures, and other problems have all been experienced and can greatly affect the cyclone separation performance.

3.3.9 Third Stage Separators

Maintenance of the Third Stage Separator is critical to assuring proper operation. Blocked or worn swirl vane assemblies and tubes, liner wear or degradation, refractory integrity, dip leg wear, tube sheet distortion, worn or blocked critical flow orifices, and more all can significantly affect the Third Stage Separator performance.

Most of the Third Stage Separator designs have been upgraded over time and have incorporated new technology that can greatly increase the separator performance and expander erosion life.

The Third Stage Separator underflow line can be the most critical part of the separator operation and needs extra design and maintenance attention. A blocked underflow line will fully disable the separator and cause continual high catalyst carryover to the expander. The use of large radii underflow lines is critical to prevent plugging of the underflow line. The monitoring of the underflow line to ensure proper operation is also highly recommended. The most effective monitoring of the separator underflow line operation has been the use of skin thermocouples. Plugging of the underflow line causes the metal temperature of the line to decrease. Monitoring of the line allows the detection and the opportunity to take appropriate action prior to a major blockage. The installation of online rodding provisions and nitrogen purging connections at all elbows and bends allows the online clearing of plugged lines.

The critical flow orifice in the underflow line must be properly sized to allow the proper amount of gas and catalyst flow. Sufficient underflow is critical to ensure proper separator performance.

Recent trends have been to decrease the underflow rate from the normal rate of 3% down to 1.5% to 2% of the flue-gas flow. Experience has shown that this may not be desirable as it appears to reduce the fine particle separation efficiency and reduce the ability of the separator to handle a catalyst upset. Increasing the underflow rate to 4% or greater would be recommended by this author. It should be recognized that this will decrease the expander power proportionally to the change over the current underflow rate.

3.3.10 Expander Blade Life Estimate

Many expander users have asked how good the particle separation needs to be to get an acceptable expander run. Expander erosion potential has been estimated using various erosion models and techniques. The best of these estimate an expander's erosion potential by accounting for the expander's gas and catalyst mass flows, catalyst particle distribution, stator and rotor-blade gas velocities, and rotor-blade incidence angle. Most significant in various contributing factors are the catalyst mass flow and gas velocities. Erosion rate varies directly proportional to the catalyst mass flow and the gas velocities rise to a power between 2.5 and 3.0. Using several analytical predictions and field experience, Table 3.3 shows the approximate expander blade life based on different catalyst loadings.

Acceptance Criteria	Loading	Approximate Blade Life
Very good	<75 mg/Nm^3	5 years
Good	100	4
Acceptable	150	3
Poor	>200	2
Very poor	>400	<1

Table 3.3 Expander Blade Life Versus Catalyst Loadings

3.4 CCU Expander Catalyst Deposition

3.4.1 Unique Environment

One of the most unique aspects of operating a CCU Hot Gas Expander is dealing with expander flow-path catalyst deposition. Catalyst deposition can occur throughout the CCU, but is more noticeable and detectable in the gas expander. This is because the deposition can be viewed in real time in the expander while the unit is in operation via the expander viewports. Catalyst deposition can also affect the expander rotor vibrations that are continually monitored.

The problem of catalyst deposition is that it is difficult to understand, predict, and eliminate. Numerous theories for the causes of catalyst deposition have been investigated and found to be a contributor under certain operating conditions. Some of the suspected causes of catalyst deposition include particle inertial impaction, vapor or particle diffusion, catalyst hydration, electrostatic attraction, magnetism, and thermal deposition.

Catalyst deposition occurs in most CCU expanders and is the number one-operational problem throughout the industry. Catalyst deposition causes expander performance losses, blade tip wear, blade tip mechanical damage, and high rotor vibrations. These operational problems often result in CCU production losses and/or power recovery train or process shutdowns. Figure 3.24 shows an expander experiencing severe flow-path deposition. The rotor blade airfoils have catalyst deposits in excess of 0.25 in (6 mm) in thickness, whereas the shroud deposit is more than 0.5 in (12 mm) thick. The blade tips have heavily

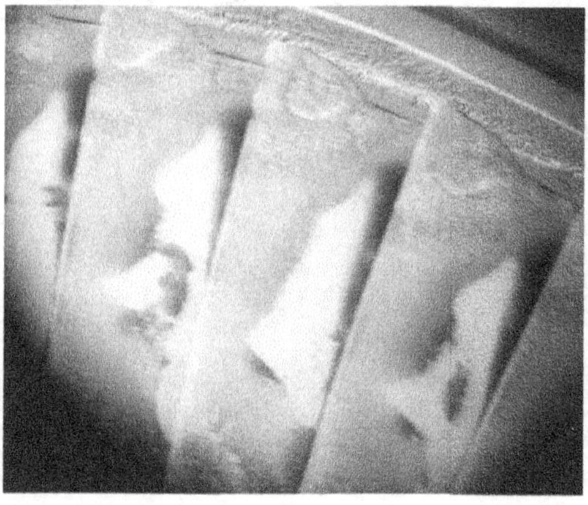

Figure 3.24 Severe expander flow-path catalyst deposits.

rubbed the shroud deposit and the rubbing has caused major mid-chord blade tip wear. The flow capacity of this expander was reduced by 15% and the power output by 20%.

Although these discussions of catalyst deposits concentrate on expander deposits, similar catalyst deposits are also found throughout the CCU flue-gas circuit. The deposits tend to occur wherever there is a pressure reduction or letdown in the flue-gas system. Deposition at these locations is somewhat unexpected at these locations inasmuch as the flue-gas velocities tend also to be highest at these locations. Common deposit locations include the following:

- All of the particle separation equipment including the regenerator cyclones, Third Stage Separator, and Fourth Stage Separator.
- On the gas path side of the expander shroud.
- On the expander rotor blades.
- On the expander stator vanes. Figure 3.25 shows a large stator deposit on the inlet side or leading edge of stator vanes.
- On the seats or in the bearing areas of process control valves. Figure 3.26 shows a catalyst deposit removed from the seat area of an expander inlet control valve.
- In the flow passages of TSS swirl vanes.
- In critical flow orifices. Figure 3.27 shows a catalyst deposit found in the critical flow nozzle. The nozzle area was reduced by more than 50%.

FIGURE 3.25 Expander stator vane deposits—inlet side.

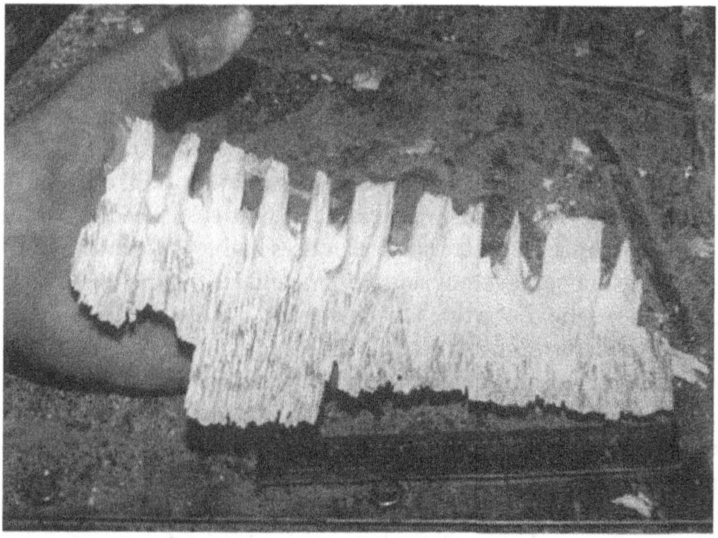

FIGURE 3.26 Expander inlet valve seat deposits.

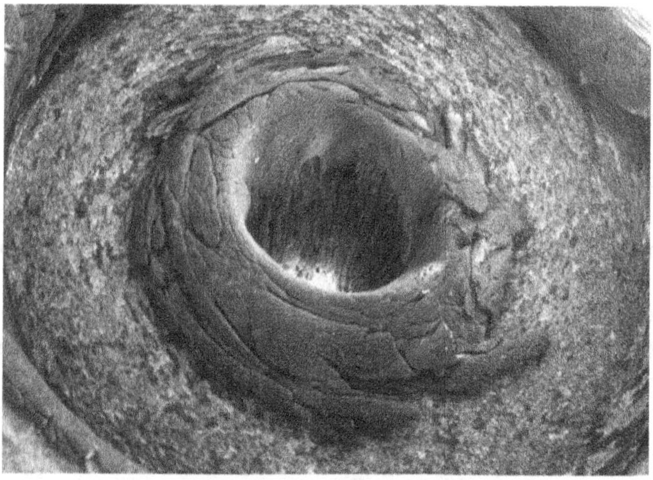

FIGURE 3.27 TSS underflow—critical flow orifice deposits.

Most catalyst deposits typically build up over months of operation, although the most severe deposits have built up sufficiently to cause rotor-blade tip rubbing in less than 1 week of operation. This would correspond to a deposition rate of more than 0.050 in (1.25 mm) per day! More typically, the deposits build up in a 3- to 6-month time period.

Catalyst deposition is not unique to any particular process licensor, separation equipment design or manufacturer, or expander design or manufacturer. Although there appears to be some correlation between particular types of catalyst or catalyst additives, it is more likely that the supposed problematic catalyst is not the problem rather that the catalyst was chosen for certain process conditions that are conducive to catalyst deposition. For example, certain catalysts are used for residual oil cracking. Residual oil cracking produces higher coke rates and usually contains higher levels of salts and metals contamination. The salts and metals are the root cause for the deposit, not the catalyst in use.

Catalyst deposition in the CCU flue-gas system greatly affects the overall system performance and efficiency in many ways. For example, deposition in regenerator cyclones can lead to excessive flue-gas duct wear and Third Stage Separator overloading. Overloading or deposition of the Third Stage Separator can lead to excessive expander wear and increased catalyst deposition.

Deposition in the control valves can lead to both operational efficiency losses as well as potential safety issues. For example, deposition in an expander bypass valve can prevent tight closure of the valve and lead to a significant amount of valve leakage. This leakage causes flue gas to bypass the expander and results in a loss of expander output. Bypass leakage losses have been greater than 15%.

The greatest concern is when the deposits present safety hazards. A control valve deposit can cause sticky valve operation or, more importantly, prevent the operation of a control valve. Poor process control can obviously cause production losses and, more importantly, cause dangerous operating conditions. In the case of an expander generator set, an inoperable control valve can cause a power recovery train over speed condition and/or process flow blockage or bottleneck.

Deposition in the TSS underflow can greatly reduce the amount of underflow and even lead to a complete blockage of the underflow. Reduced TSS underflow greatly reduces the performance of the TSS and obviously affects the erosion potential of the gas expander. Figure 3.28 shows a catalyst deposit building up in the underflow critical flow nozzle.

3.4.2 Catalyst Deposits

Catalyst deposit samples from power recovery systems are often very difficult to recover because a process or expander shutdown usually causes the catalyst deposits to spall off during the shutdown operations. The deposits are carried downstream in the ductwork and break up into fine debris. The slow shutdown times of a normal CCU process clears out a majority of the deposits owing to the change in flow conditions and, more importantly, due to the thermal incompatibility that

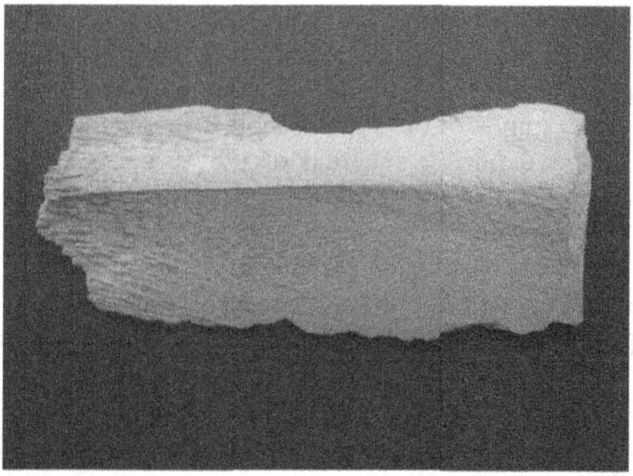

FIGURE 3.28 Type A catalyst deposit.

exists between the ceramic deposit and the base material to which it is adhered. As the process temperature is reduced, the catalyst deposits tend to fracture and spall off. It is this thermal incompatibility that has been used to develop online thermal cleaning procedures.

Expander flow-path deposits have been retrievable when the process equipment is rapidly shut down, such as during an emergency shutdown or if the deposit is of sufficient thickness and mass to cause it to fall into the bottom of an expander casing or flue-gas ducting. Deposits that do not spall off can cause expander rotor seizing and restart problems. This occurs when the expander rotor quickly comes to a full stop prior to the expander flow-path cooling off. With the rotor stationary, the expander casings and shroud reduce in size as the expander cools down. If the deposits are still on the shroud, they become trapped between the shroud and rotor blade tips. This causes the expander rotor to seize up. The seizing can be quite tight and the power recovery train auxiliary driver (steam turbine or motor) may not have sufficient torque to turn and start the train. A disassembly of the expander is often required to allow the frozen deposits to be chipped out from the blade tips.

3.4.3 Types of Catalyst Deposits

Over time, numerous catalyst deposit samples have been collected and classified into four different types:

Deposit Type A

The first type of catalyst deposit is generally a light chalky deposit. This type of deposit occurs in areas of lower pressure and/or an area of flow separation in the expander flow path. These deposits are often seen in the boundary layer flows of both the stator vanes and rotor blades. It is the most common type of deposit and is found throughout every section of the regenerator and downstream equipment. Figures 3.27 and 3.29 are examples of the Type A deposit.

The Type A deposit is generally of a lower concern owing to its fragile nature. After a unit shutdown, any remaining Type A deposit can be easily removed from the component to which it is adhered by brushing, hydro blasting, or mechanically cleaning. These deposits are easily broken up by rubbing it between one's fingers. It leaves a chalk-type residual. The Type A deposits can normally be removed in service via abrasive cleaning techniques and procedures.

Type A deposits can grow significantly in thickness and at times the thickness has exceeded 0.75 in (19 mm). Thick deposits can reduce both the flow capacity and efficiency of the separation equipment, gas expander, and downstream waste heat or CO boilers.

Occasionally, the Type A deposit has black layers throughout the deposit or a black color on the surface. This is usually the result of a carryover of coke covered catalyst during process upset conditions.

Deposit Type B

The second type of catalyst deposit is generally caused by the hydration of the Type A deposit. The normally harmless powdery Type A deposits become very sticky when wet. The most common sources of moisture are from wet steam during steaming or purging operations, the use of steam or water quenches for continuous flue-gas temperature control,

FIGURE 3.29 Type B catalyst deposit.

condensation that occurs during unit start-ups or shutdowns, and even in low flow ducting during normal operation. Liquid condensation can occur in flue-gas lines at temperatures as high as 360°F (182°C).

Flue-gas quenching is the most common source of water throughout the flue-gas system. The highest catalyst deposition rates have been directly correlated to the use of flue-gas water quenching. Low-quality quench water with significant levels of salts greatly increases the amount of catalyst deposition throughout the flue-gas system. The use of demineralized quench water with high-pressure atomizing spray nozzles has been found to reduce the amount of catalyst deposition.

Type B catalyst deposits often have the appearance of a dried out mud or slurry. The Type B catalyst deposits are more tenaciously adhered to the expander flow-path surfaces and therefore are more difficult to remove.

Deposit Type C

The Type C deposit is also commonly found throughout the entire power recovery system, and is one of the most troublesome deposits. This deposit is very hard and appears to collect in the areas of highest gas velocity, such as the blow-down slots or tubes in Third Stage Separators, critical flow orifices, the Hot Gas Expander flow path, and across the regenerator top pressure control valves. Figures 3.27, 3.28, and 3.30 show a typical Type C deposit. The Type C deposit has a much greater density and is often characterized by an irregular surface condition. The Type C deposits are amorphous with no crystalline

FIGURE 3.30 Type C catalyst deposit.

FIGURE 3.31 Type D catalyst deposit.

shape and tend to build up in thickness in a columnar shape that grows in an *upstream* (against) direction.

Expander thermal cleaning is generally required to remove the Type C deposit.

Deposit Type D

The last type of deposit found in the power recovery system is one that basically occurs only in the Hot Gas Expander. This type of deposit is actually a transformation of a Type C deposit into a very hard, tightly adhered, sinter catalyst mass. The Type D deposits form when metallic salts form and make the deposits very sticky or when an existing catalyst deposits are subjected to high temperatures and pressures. A common cause for the formation of the Type D deposits on the expander shroud is blade rubbing. The blade rubbing quickly generates enough heat and pressure to sinter or collapse the catalyst particles into a noncoherent mass without actually melting the deposit. It generally requires deposit temperatures to exceed 2000°F (1090°C) for this transformation to occur. Evidence of the high temperatures can be seen both on the catalyst deposits exhibiting melted blade material (Figure 3.31) and from the melted material on the blade itself.

3.4.4 Macroscopic and Common Characteristics of Expander Flow-Path Deposits

One of the most interesting characteristics of expander catalyst deposits is that the deposits typically build up in discrete layers.

Figures 3.32 through 3.34 are examples of a layered catalyst deposit. In these cross sections, it can be seen that the layers are not uniform and change their shape and characteristics. Detailed microscopic analyses of these deposits show that the makeup, characteristics, and even the chemistry vary from layer to layer.

While occasionally some refractory materials are found in the expander deposits, a majority of the deposits are accumulations of

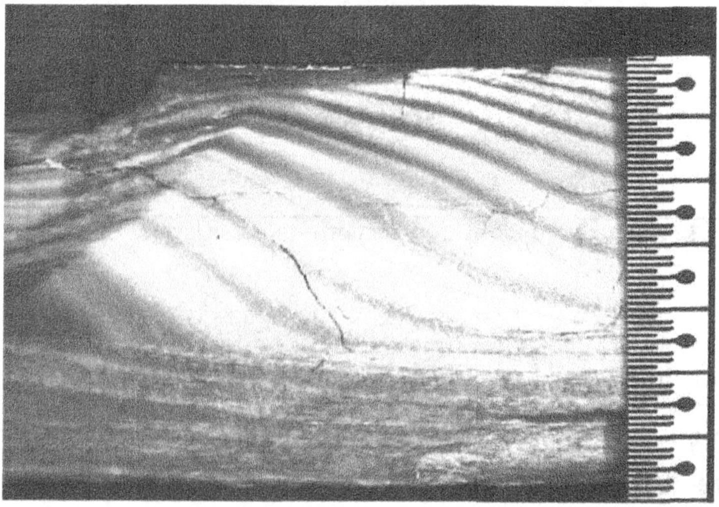

Figure 3.32 Cross section of Type C catalyst deposit.

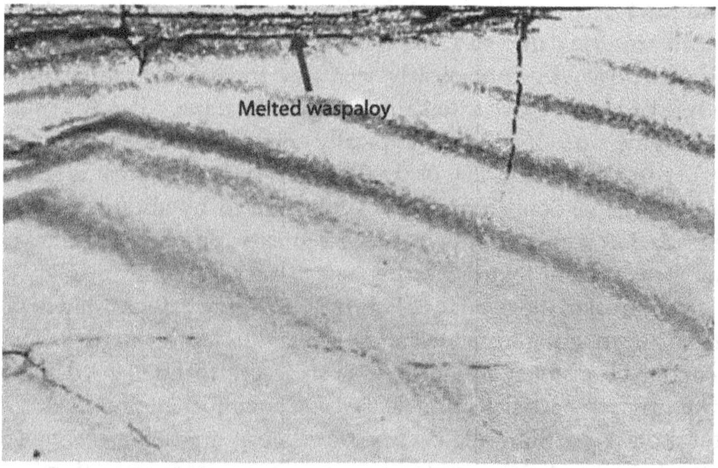

Figure 3.33 Catalyst deposit cross section with melted blade material.

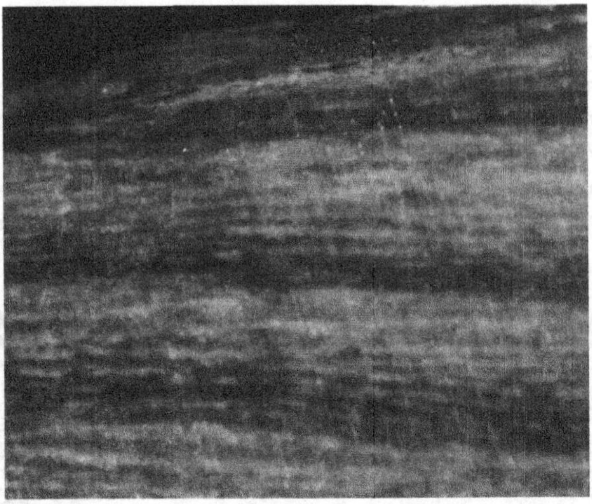

FIGURE 3.34 Catalyst deposit cross section—multiple layers.

CCU catalyst fines. Over time and at the normal operating temperatures, the catalyst deposits form very hard ceramic deposits.

3.4.5 Microscopic Characteristics of Expander Flow-Path Deposits

Surface Observations

Microscopic examinations of the surfaces of the catalyst deposits can be seen in Figures 3.35 and 3.36. Figure 3.35 shows a typical surface condition for a catalyst deposit that is relatively new and has not experienced sintering due to time, temperature, or mechanical contact. In this deposit, it can be clearly seen that the deposit is made up of discrete catalyst particles that are spherical in shape. All of the catalyst particles are very fine in size. Typically being less than 2 micrometers (μm) (>0.00001 in) in size or smaller. This type of deposit is referred to as a Type A deposit and is easily removed by abrasive cleaning methods. On a unit shutdown, these deposits can often be remove by hand, brushing, or hydro-blasting.

In contrast to the deposit of Figure 3.35, Figure 3.36 shows a deposit that has been sintered. Sintering is when the deposit particles begin to combine and form an amorphous structure without the definition of the original particles. The catalyst is actually changing back to its natural clay base. Sintering occurs over time at normal operating temperatures or can be accelerated by over-temperature operation, compaction, or frictional heating from rubbing of the rotor blades against

FIGURE 3.35 Microscopic view of catalyst surface layer—gas side.

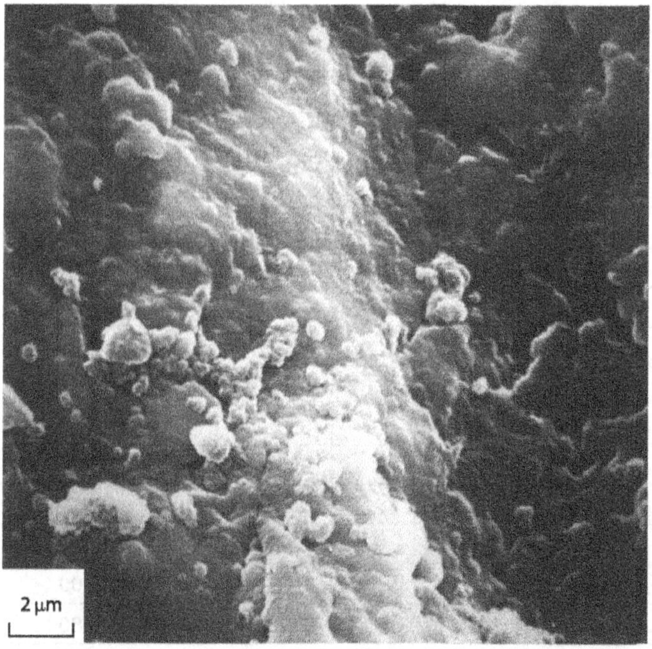

FIGURE 3.36 Microscopic view of catalyst surface layer—shroud side.

the deposit or chemically. These deposits tenaciously adhere to the expander parts (usually the shroud inside diameter) and are very difficult to remove. Online abrasive cleaning is totally ineffective in removing these Type C or D deposits. The only method for online removal is an extensive thermal cleaning or a unit shutdown.

Deposit Chemistry

Chemical analyses of expander deposits indicated that the bulk of the deposits are made up of CCU catalyst and not refractory. In comparison to the amount of catalyst passing through the expander, the amount of material depositing is actually quite low (less than few pounds). Even with a proper functioning TSS, a typical CCU expander can be passing approximately 100 lb (45 kg) of catalyst fines per hour. That translates to more than 1 ton of catalyst in a single day or more than several hundred tons of catalyst in a year.

Table 3.4 compares the chemistries of equilibrium catalyst versus expander flow-path deposits.

Element	Symbol	Measurement	Equilibrium Catalyst Level	Catalyst Deposit Level
Sulfur	S	wt %	0.1–0.5	0.2–8.0
Aluminum	Al	wt %	27.0–55.0	10.0–55.0
Carbon	C full burn	wt %	0.05–0.15	0.05–0.60
	C partial burn	wt %	0.10–0.40	
Calcium	Ca	ppm	<1000	500–3000
Potassium	K	ppm	<5000	1000–6000
Silicon	Si	wt %	35–70	10–70
Iron	Fe	wt %	<0.75	0.60–4.40
Tin	Sn	wt %	0.02–0.2%	0.60–7.50
Sodium	Na	wt %	0.10–0.50	0.30–1.60
Copper	Cu	ppm	90–4500	
Nickel	Ni	wt %	0.01–0.5	0.01–0.5
Chromium	Cr	ppm	<500	0–500
Titanium	Ti	ppm	<4000	0–2000
Vanadium	Vn	ppm	90–5000	500–8000

TABLE 3.4 Comparison of Equilibrium versus Catalyst Deposit Chemistries

Chemical analysis of catalyst samples has shown a number of interesting findings:

- Most catalyst deposits show abnormally high levels of sulfur.
- Many catalyst deposits also show abnormally high levels of iron, which is often traced back to corrosion of the steam distribution systems.
- High levels of calcium and potassium are often traced back to poor-quality boiler feed water treatment and/or steam production.
- High concentrations of sulfur, calcium, potassium, iron, tin, and sodium.

Above normal metal concentrations such as aluminum, nickel, and vanadium are generally carried over in the flue gas. Other contaminants such as calcium, iron, potassium, and sodium are often carried over from the steam systems. The presence of sulfur oxides in the flue gas provides an additional sintering mechanism. The SO_3 reacts with the alumina in the catalyst to form aluminum sulfate, which acts as a binder to hold the particles together.

The presence of water in the flue gas has been found to be a major contributor to the rate and quantity of catalyst deposition. The water droplets absorb SO_3 from the flue gas and form sulfuric acid in the droplet. The acid both captures catalyst particles and dissolves some of the metals or alumina in the catalyst and forms metallic oxide binders.

3.4.6 Online Cleaning

Two successful online cleaning methods have been developed over the years for CCU expanders. The first method uses the injection of an abrasive media into the expander for cleaning. The impaction and scraping action of the cleaning media against the catalyst deposits dislodges them. The second cleaning method involves thermal spalling of the catalyst. In both cases, the dislodged catalyst is carried out the exhaust of the expander.

3.4.7 Abrasive Cleaning

Abrasive cleaning has been found to be effective in controlling the amount and rate of catalyst deposition. Abrasive cleaning involves the injection of an abrasive into the inlet of the expander as shown in Figure 3.37.

A variety of abrasive media has been used including ground-up walnut shells, peach pits, grape seeds, whole grain rice, and even

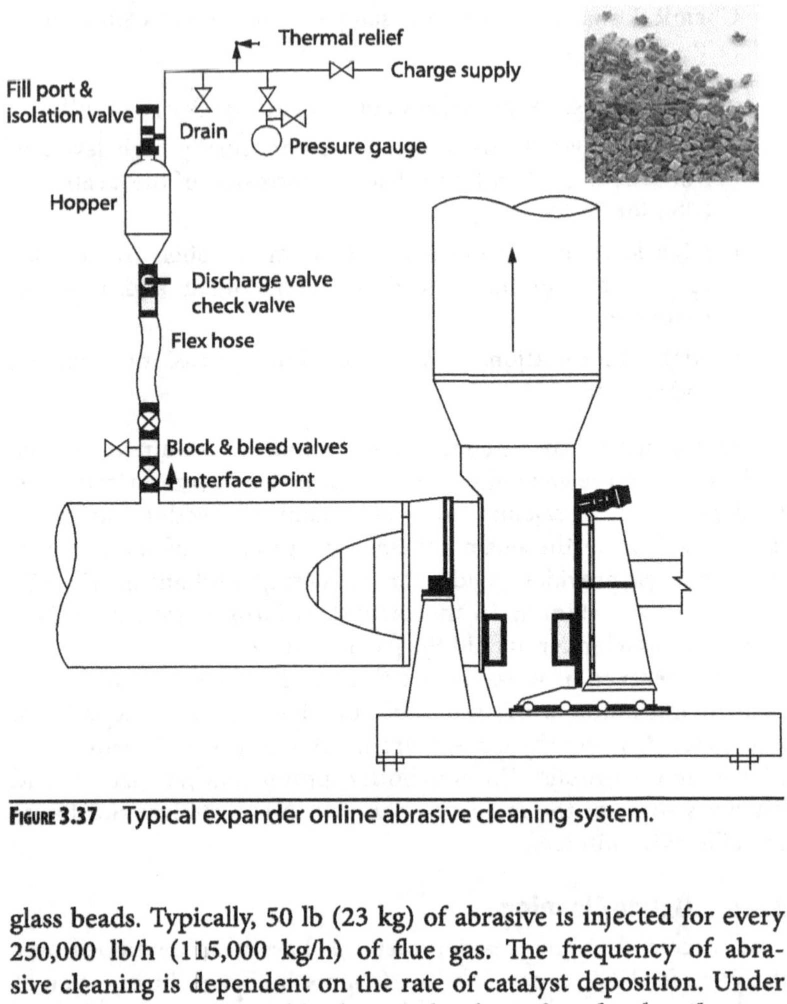

FIGURE 3.37 Typical expander online abrasive cleaning system.

glass beads. Typically, 50 lb (23 kg) of abrasive is injected for every 250,000 lb/h (115,000 kg/h) of flue gas. The frequency of abrasive cleaning is dependent on the rate of catalyst deposition. Under normal operation, a weekly cleaning has been found to be adequate. In units experiencing high catalyst deposition, abrasive cleaning is performed as frequently as every operational shift.

It is important to emphasize that abrasive cleaning is a preventative maintenance activity used to limit the buildup of catalyst in the expander. Abrasive cleaning is effective in cleaning off new lighter catalyst deposits (Type A). Abrasive cleaning is not very effective in removing thick, heavily sintered catalyst deposits (Types C and D).

3.4.8 Thermal Cleaning

An alternative online cleaning procedure is referred to as a thermal cleaning. As the name implies, this cleaning procedure involves changing the operating temperatures in the expander. It has been found that the differential expansion between the catalyst and the expander flowpath components causes the catalyst to spall off the expander parts. The coefficient of expansion of the expander parts is between 8.5 and 10.3 E-6 in/in/°F. The coefficient of expansion of the ceramic catalyst deposit is between 4 and 5 E-6 in/in/°F. As the temperatures of the parts are changed, the parts and deposits grow at different rates and the bond between the parts and the deposit fractures. With the high rotational speeds, the catalyst deposit is thrown off or spalls off the expander parts.

Experience has shown that reducing the expander inlet temperature between 400 and 600°F (220 to 330°C) is required to cause the catalyst deposits to spall off. The actual amount of temperature change is very dependent on the type of catalyst deposit, the thickness of the deposit, the degree of sintering in the deposit, and the geometry of the expander. The temperature reduction into the expander may require the process throughput to be reduced or even shut down.

3.4.9 Controlling Catalyst Deposits in CCU Expanders

Although the elimination of catalyst deposits in a CCU flue-gas system is very difficult, the deposits can be controlled and their effect on unit operation minimized. The following courses of action have been effective in controlling catalyst deposition:

1. Proper sizing, operation, and maintenance of the particle separation equipment are probably the most influential items in controlling expander erosion and deposition.
2. Upgrading older Third Stage Separation equipment. The technology available today is much more efficient and will greatly reduce the amount of catalyst entering the expander.
3. Operate the TSS with higher amounts (>3%) of underflow. Higher underflow rates improve the particle separation efficiency of the TSS. This is particularly true for the separation of fine catalyst particles.
4. Minimize the length and number of turns and elbows in the TSS underflow line. Install online cleaning/rodding provisions to clear a blocked underflow.
5. Instrument and monitor the operation of the underflow line to detect flow reductions and blockages.
6. Minimize or eliminate the use of water quenching. High-pressure atomizing nozzles should be used if water quenching is required.

7. Minimize or eliminate the use of steam quenching. High-quality dry steam should be used if required.
8. Minimize the use of steam purging, particularly at low temperature (less than 400°F (205°C).
9. Minimize the use of catalyst additives that contain high levels of salts or metals.
10. Minimize catalyst contaminants such as heavy metals and salts.
11. Design for aerodynamically smooth flow passages to eliminate flow interruptions and stagnation areas.
12. Design the expander flow path with slow and smooth transitions and polished surface finishes.
13. Monitor catalyst deposits while in service through devices like the expander viewports.
14. Clean out deposits before the more detrimental catalyst deposit types develop.
15. Develop site-specific online cleaning procedures.

3.5 CCU Expander Corrosion

3.5.1 Unique Environment and Corrosion

The CCU Hot Gas Expander is subjected to a gas environment that is high in temperature, erosive in nature, and also highly corrosive. Corrosion is generally defined as a degradation of a material owing to corrosive elements in the environment. The corrosive nature of expander flue gas is caused by significant levels of sulfur (H_2S or sulfur oxides [SO_x]), alkali salts (Ca, Na, K) and metal oxides (Fe, Ni, Vn, Sn). The source of these corrosive agents is the feedstocks being cracked in the CCU.

The corrosion potential in the gas expander is highly dependent on the operating temperature of the various expander components. The expander inlet temperature varies between 1150°F and 1425°F (620°C and 774°C). These inlet temperatures translate to rotor disc and blade temperatures in the 1000°F to 1300°F (620°C and 774°C) range. At these temperatures and in an oxidizing environment, the metallic elements in the base materials will form metallic oxides. These oxides can be stable and offer significant corrosion protection if undisturbed.

Unfortunately, the CCU flue-gas composition has very low oxygen levels and the formation of protective oxide scales is limited and occurs at very slow rates. Oxygen levels in the CCU flue gas range from

Gas Component	Incomplete Combustion	Complete Combustion
Nitrogen	70.0%	73.0%
Carbon monoxide	7.0%	0.0%
Carbon dioxide	10.4%	14.0%
Water	12.5%	11.0%
Oxygen	0.0%	2.0%
Sulfur (SOx)	50–1800 ppm	50–1800 ppm

TABLE 3.5 Typical FCC Flue Gas Compositions

virtually none to maximums of only a few percent. In comparison, atmospheric air or even combustion air in a gas turbine or boiler, has oxygen levels above 20%.

Table 3.5 shows the flue-gas composition of the two main operating modes of a CCU: incomplete and complete combustion. Units operating in an incomplete combustion mode have significant levels of carbon monoxide in the flue gas that has not burned in the CCU regenerator. Complete combustion operation does burn up the carbon monoxide in the gas and has a small amount of excess oxygen remaining. Even minor levels (0.5% to 1.0%) of excess oxygen in the flue gas can have a major effect on the corrosion potential of the expander.

The corrosion potential of an expander can be evaluated using an analytical tool called a phase equilibrium diagram. Two typical phase equilibrium diagrams can be seen as Figures 3.38 and 3.39. These diagrams show the combined nickel and chromium diagrams for two different flue gas compositions. Figure 3.38 illustrates a flue-gas composition with only 0.2% oxygen and 0.5% sulfur. Even with this low level of oxygen, nickel, and chromium sulfides are likely to form on the base material surfaces. Figure 3.39 shows a different flue-gas composition that has 1.5% excess oxygen and 0.5% sulfur. The nickel and chromium elements will have a tendency to form protective nickel and chromium oxides.

In addition to the low levels of oxygen, the CCU flue gas contains significant levels of sulfur. Under normal operation, the sulfur tends to carry over as a sulfur oxide (SO_x). The metallic elements in the base materials will attach to sulfur in a manner similar to its affinity to attach to oxygen. Metallic sulfides are formed in this reaction. Iron, nickel, and chromium sulfides are common corrosion products.

Unfortunately, the FCC environment also contains significant levels of metallic salts (sodium, calcium, vanadium, etc.). At the

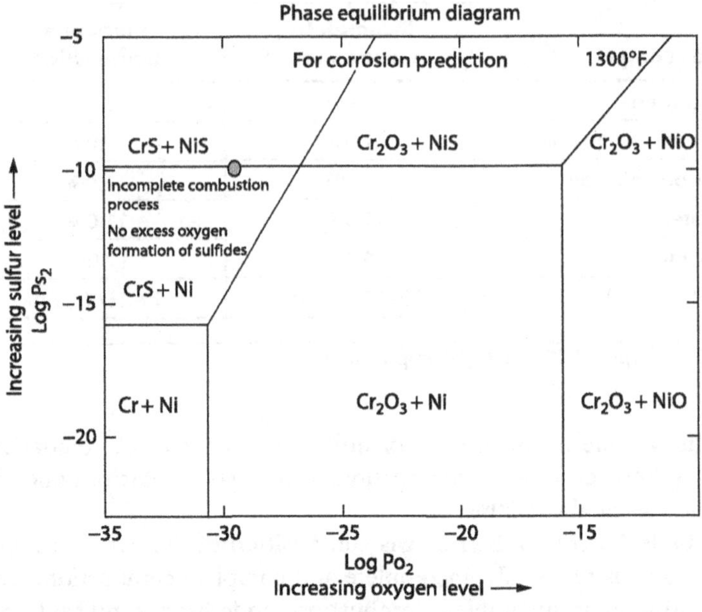

FIGURE 3.38 Phase equilibrium diagram—incomplete combustion.

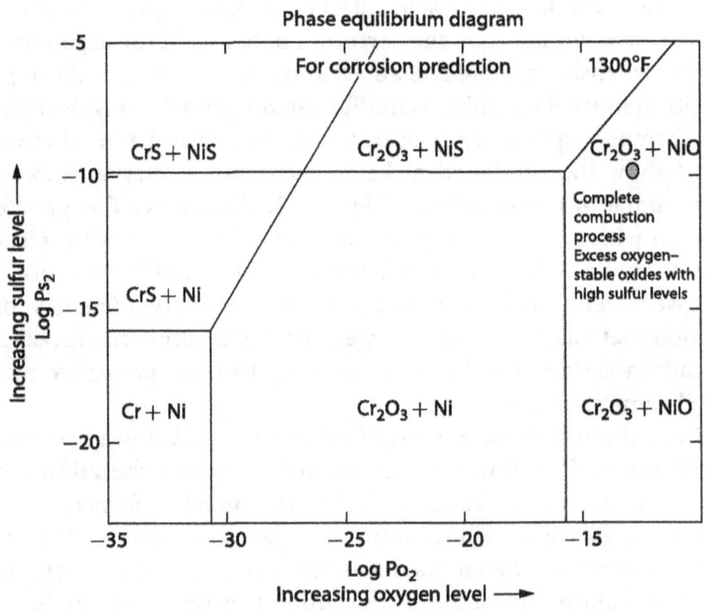

FIGURE 3.39 Phase equilibrium diagram—complete combustion.

expander normal operating temperatures, these salts act as a flux and accelerate the formation of the damaging metal sulfides.

It can be clearly seen that the FCC environment has the potential to be highly corrosive. The corrosion potential varies from unit to unit and even within a given unit depending on the operating conditions and the feedstocks being run at any given time. Different types of corrosion have been noted and are described below.

3.5.2 Corrosion Types Noted in CCU Expanders

A number of types of corrosion have been noted in the CCU expander and are often referenced in the evaluation of expander components.

- *Uniform surface corrosion* (Figure 3.40 (A)) is characterized by a corrosive attack proceeding evenly over the entire surface area or a large fraction of the total area. Surface corrosion can be a general discoloration of the part or an actual thinning of a part. If the corrosion is severe, the integrity of the part may be compromised when the operating stresses exceed the materials capabilities. A common example of the CCU expander surface corrosion is the black discoloration noted on many parts. This discoloration is the formation of a chromium sulfide. Another example is the surface oxidation that occurs in low-alloy-steel expander casings.
- *Pitting corrosion* (Figure 3.40 (B)) is a localized form of corrosion in which cavities or holes are produced in the base material. Pitting corrosion occurs when areas of a material undergo rapid attack while the adjacent surface remains virtually unaffected. Pitting corrosion is considered to be more dangerous than uniform corrosion damage because it is more difficult to detect, predict, and design against. CCU expander pitting is often the result of small salts deposits or surface condensation.
- *Crevice corrosion* (Figure 3.40 (C)) is a localized form of corrosion usually associated with a stagnant solution on the micro-environmental level. These local environments tend to occur in crevices or shielded areas. The local micro-envirnoment is often very different than the bulk area environment. Threaded regions, clamped joints, and even the expander blade attachement region are typical examples of crevices that do experience localized crevice corrosive attack.
- *Integranular corrosion* (Figure 3.40 (D)) is a localized corrosive attack along the grain boundaries, or immediately adjacent to the grain boundaries in the base material. This form of corrosion is

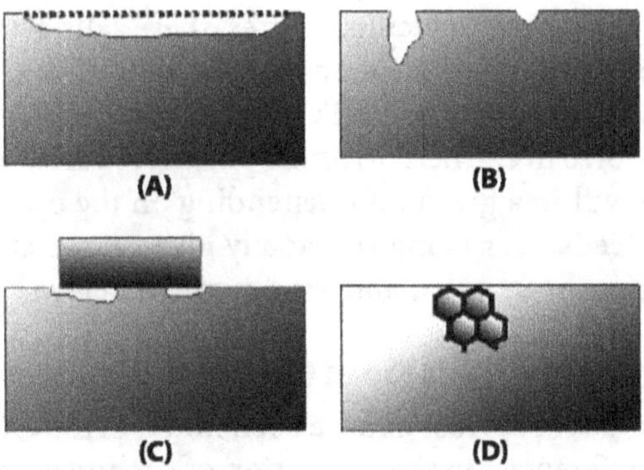

FIGURE 3.40(A–D) Different corrosion types.

usually associated with chemical segregation effects or specific phases precipitating in the grain boundaries. Intergranular base material oxidation and sulfidation is commonly noted in CCU expander blades and discs.

- *Stress corrosion cracking* (SCC) (Figure 3.40 (E)) is the cracking induced from the combined influence of tensile stress and a corrosive envirnoment. Three main factors are required for SCC—stress, environment, and temperature. Often most of the main surface remains unattacked, and only fine cracks penetrating into the material are found. The cracks can have an intergranular or transgranular morphology. Macroscopically, SCC fratures have a brittle appearance. SCC is classified as a catastropic form of corrosion, as the detection of such fine cracks can be very difficult and the damage not easily predicted. Most of the CCU expander base material exhibit a susceptibility to SCC.
- *Erosion corrosion* (Figure 3.40 (F)) is associated with a flow-induced mechanical removal of the protective surface film that results in a subsequent corrosion rate increase. The erosive nature of the FCC catalyst enhances the potential for an erosive-corrosive phenomina.
- *Corrosion fatique* (Figure 3.40 (G)) is the result of the combined action of an alternating or cycling stresses and a corrosive environment. The fatique process can break down the protective passive films on the base material and make the fretting region highly susceptible to corrosive attack. Much lower failure stresses and failure times occur in a corrosive environment compared to the situration where the alternating stress is in a noncorrosive environment. The fatigue fracture is brittle and

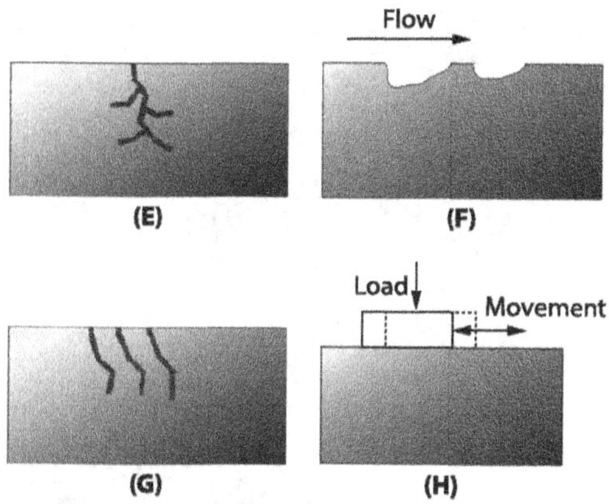

FIGURE 3.40(E–H) Different corrosion types.

the cracks are most often transgranular, as in stress corrosion cracking, but not branched.

- *Fretting corrosion* (Figure 3.40 (H)) refers to corrosion damage at the asperities of contact surfaces. This damage is induced under load and in the presence of repeated relative surface motion. Damage occurs at the interface of two highly loaded surfaces that are not designed to move against each other. The rubbing action tends to break down the protective oxide films and exposes fresh active metal to the corrosive action of the atmosphere. The interface areas between the rotor-blade and disc attachment are a common location to find fretting corrosion. Once the material surfaces are degraded, more aggressive corrosion damage often occurs.

3.5.3 Importance to Control

Corrosion control in FCC expanders is very important since corrosion failures are the number-one cause of expander rotor-blade and disc failures. Some FCC expander failure databases show that corrosion has been a contributor in more than 45% of all rotor-blade and disc failures (see Figure 3.41).

Corrosion problems have become even more of a concern over time. The typical expander operating time between turnarounds and inspections has increased over time. In the 1960s it was common to operate an FCC unit for only 2 years between shutdowns and periodic maintenance. Since 2000, it has become common for FCC units to operate 4 to 6 years between the normal schedule shutdowns and maintenance. These longer run times obviously allow

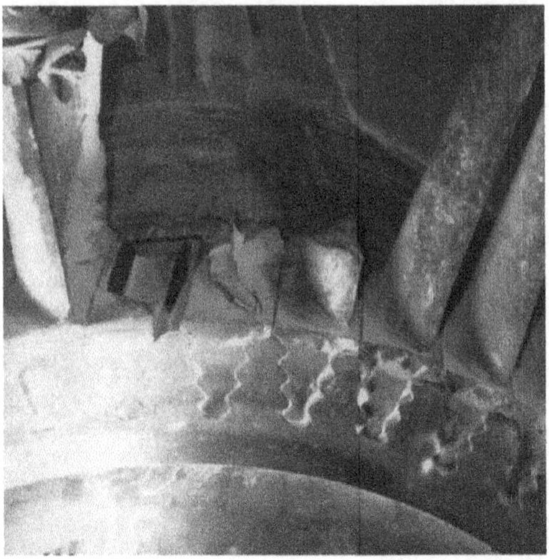

FIGURE 3.41 Expander rotor-blade failure due to corrosion fatigue.

more corrosion damage to occur and prevents the periodic cleaning and mitigation of the corrosion.

Today's fluid catalytic and residual oil Cracking Units operate with both heavier and more sour feedstocks. This obviously increases the amount of corrosion-causing contaminants (sulfur, salts, metals, etc.). The heavier feedstocks also produce higher coke makes that require greater amounts of combustion air and result in higher regeneration temperatures.

3.5.4 Types of FCC Expander Corrosion

FCC expanders experience two primary types of corrosion—in-service hot corrosion and ambient temperature or shutdown.

In-Service Hot Corrosion

The most damaging corrosion to the expander is a high-temperature corrosion that occurs continually during normal expander operations. The corrosion is generally characterized by relatively thick corrosion products occurring in the blade and disc attachment regions. Figure 3.42 shows corrosion pitting damage in the fillet of the attachment region of a rotor blade. Figure 3.43 shows typical fretting and crevice corrosion damage in the interface region of a blade and disc attachment. These corrosion products are primarily composed of Cr_2O_3, CrS, and Ni_3S_2. The corrosion products tend to be very brittle in nature and are prone to cracking. The cracking often carries into the base material and exposes the internal base material to the corrosive

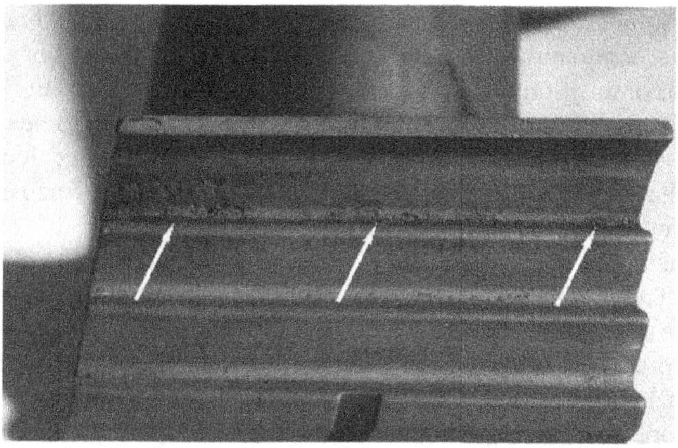

FIGURE 3.42 Pitting-type corrosion in the outer neck of a rotor-blade attachment.

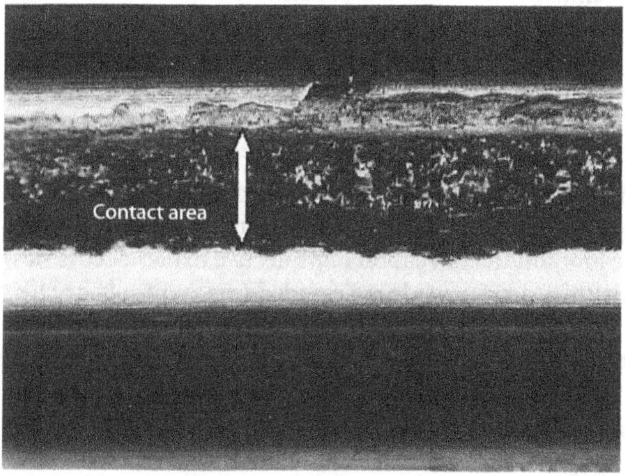

FIGURE 3.43 Localized corrosion owing to attachment fretting.

environment, thereby promoting additional corrosion and crack propagation. The corrosion penetration is almost always intergranular with sharp-faceted features that suggest the failure mode is related to a stress-corrosion mechanism.

Hot corrosion noted in FCC expanders is quite unique when compared to other high-temperature industrial turbines. The FCC hot corrosion is similar to Type I corrosion that occurs in industrial gas turbines. Type I hot corrosion is often defined in combustion turbines as corrosion that occurs above 1380°F (749°C) where molten metal sulfates form. In the FCC environment, molten metal sulfates occur at temperatures *as low as* 1150°F (621°C). Type II hot corrosion in combustion turbines is the type of corrosion that occurs below the melting point of the sulfates and is generally defined to be corrosion that occurs in the 1110 to 1380°F (599 to 749°C) range. It can be seen that the FCC expander is experiencing Type I hot corrosion in the temperature range of Type II corrosion.

The noted corrosion is often concentrated locally in the blade or disc attachment regions versus a general corrosion over the entire surface of the part. This local corrosion is related to the degradation of the protective oxide scales on the part surface. The protective scales can be broken down by mechanical cracking and spalling, erosion, abrasion, and fretting, similar to that seen in Figure 3.43. Another cause of localized corrosion and preferential attachment corrosion is this region tends to collect corrosive component-laden catalyst and the creation of crevice corrosion.

The fundamental corrosion problem is related to the local flue-gas environment in the expander. At the elevated operating temperatures (>1000°F [538°C]), the metallic elements (iron, nickel, chromium, etc.) that make up the rotor disk and blades alloys tend to form either an oxide or sulfide. The formation of stable surface oxides is desirable and provides corrosion protection against further corrosion. This is what makes many of the expander rotor alloys very corrosion resistant at the operating temperatures. In the FCC environment, the oxygen levels are quite low and the formation of protective oxides occurs slowly. This means that if a surface oxide is degraded (mechanically or chemically), the surface may not be self-healing and a new protective oxide layer may not form. Without the protective oxide film, the metallic elements are exposed to the sulfur in the flue gas and the metallic sulfides will form.

Often the corrosion of FCC expander components occurs in stages. The first type of corrosion is a sulfidation of the nickel in the alloy. A nickel sulfide corrosion tend to form on the surface of the base material. The corrosion tends to pull/diffuse the nickel out of the base alloy. The new base material chemistry then allows the sulfur compounds to attack the chromium in the base material and form chromium sulfide corrosion products in the base material (see Figure 3.44). The resulting chromium sulfide corrosion product is brittle in nature and tends to both penetrate intergranularly into the base material. The form of base material cracking is common. An example of base material intergranular cracking can be seen in Figure 3.45. Whereas the corrosion shown in Figure 3.45 was directly in the region where the blade contacted the disc, the corrosion noted in Figure 3.46 is not in the area of contact, but next to the contact region in an area that

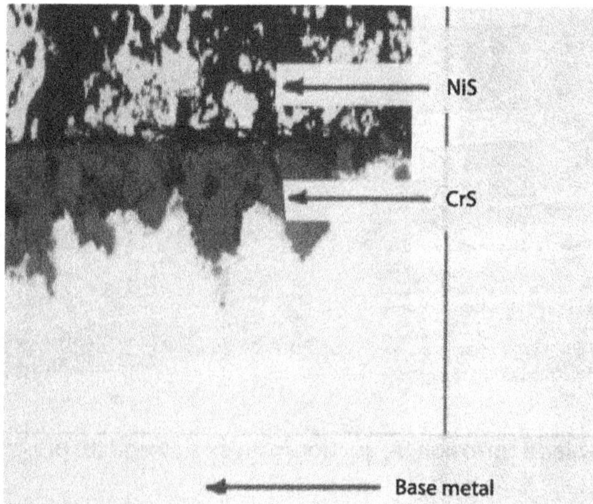

Figure 3.44 Formation of corrosion products in nickel-based alloys.

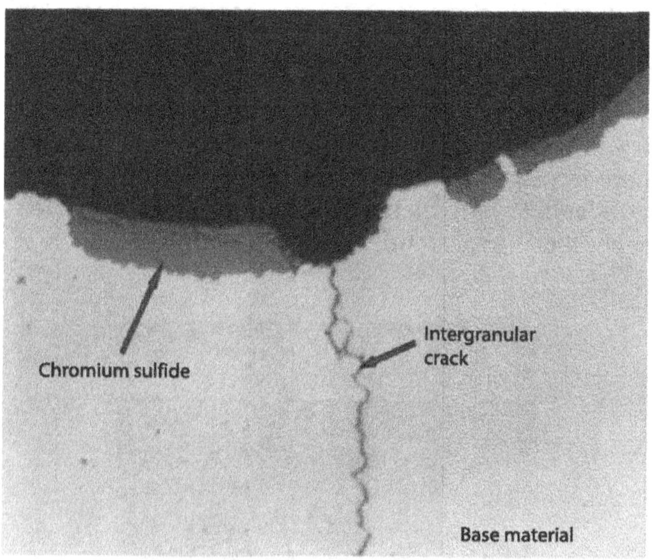

Figure 3.45 Attachment fillet radius with corrosion and cracking.

has approximately 0.005 in (0.13 mm) of clearance. This region was found to be heavily packed with FCC catalyst and is a good example of crevice corrosion where an aggressive micro-environment is formed.

The local environment in the attachment region has also been found to be detrimental to the base material's high-temperature strength and fracture toughness. Creep and fatigue strengths of the base material in

FIGURE 3.46 Rotor-blade corrosion (A) suction/convex side and (B) pressure/concave side.

the FCC environment can be reduced to a third of the level the material exhibits in air. This stress corrosion mechanism has been duplicated in the laboratory under various flue-gas conditions and when catalyst is packed around the test speciments. The phenomena is seen in the field in that some blade or disc materials experience corrosion-assisted failures with very little corrosion damage and other materials do not fail while experiencing extensive corrosion damage.

Expander hot corrosion has been found to be very sensitive to the component metal temperature. Figures 3.47 and 3.48 is an example

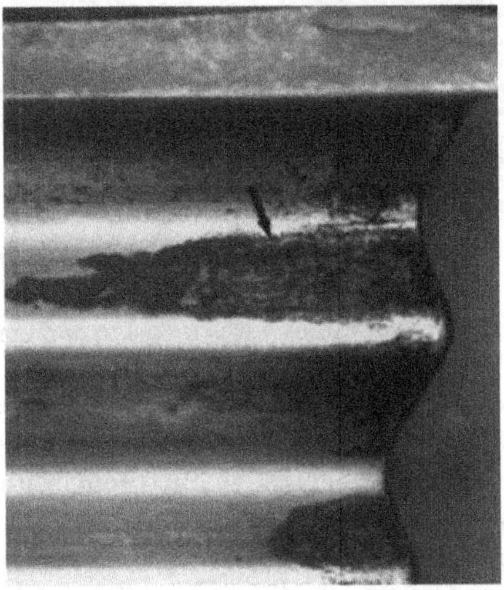

FIGURE 3.47 Cracked blade attachment.

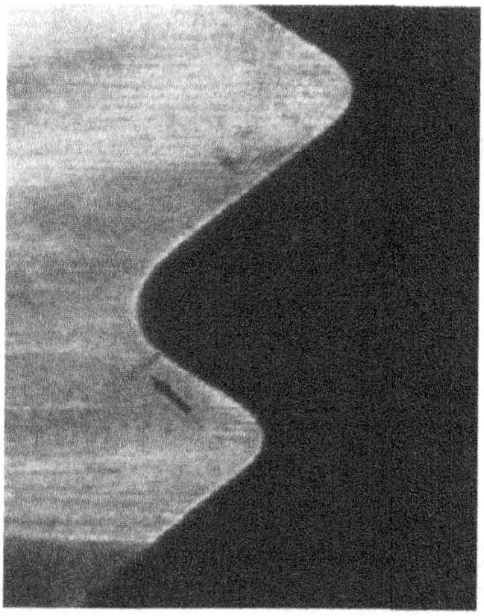

FIGURE 3.48 Cracked blade attachment.

of this preferential attack and show common characteristics noted in many FCC expanders:

- Corrosion tends to be concentrated toward the discharge side of the blade.
- Corrosion tends to be concentrated toward the outer necks and decreases with radial diameter.
- The corrosion is generally heavier on the suction (convex) side of the blade.
- The corrosion is often associated with local contact or fretting damage.

Another unique aspect to FCC expander corrosion is erosion. The flow-path components are subject to catalyst erosion. The erosion tends to wear away or degrade the protective oxide scales on the base materials and makes them more susceptible to the corrosive species in the flue gas.

3.5.5 Material Selection

FCC expander rotor discs and blades are manufactured from high-temperature superalloys typical of those used in combustion turbines or flight engines. Both iron-base and nickel-base alloys have been used. The most commonly used materials include the following:

Rotor Blades

A-286—Fe, 25% Ni, 15% Cr

Inconel X-750—Ni, 15% Cr

Inconel 738—Ni, 16% Cr, 8.5% Co

Nimonic 90—Ni, 19% Cr, 18% Co

Waspaloy—Ni, 20% Cr, 14% Co

Rotor Discs

A-286—Fe, 25% Ni, 15% Cr

Incoloy 90—FE, 43% Ni, 13% Cr

Waspaloy—Ni, 20% Cr, 14% Co

These materials were originally selected for use because of their high-temperature strength (tensile, creep, and rupture) and corrosion resistance. Wrought, cast, and forged alloys have been used. Over time, operating experience has shown that the operating experience in another industry is not transferable to the FCC industry and the materials do not necessarily respond in a similar manner. Some of the interesting findings have been:

- The material tensile, creep, rupture, and fatigue strengths are significantly lower in the FCC environments than those in air or a combustion turbine.
- The protective oxide scales formed in combustion turbines do not form as readily in the FCC environment.
- The oxide scales are often not self-healing in the FCC environment.
- Some of the iron base materials have better corrosion resistance because of the thermal stability of the iron sulfide. The iron based corrosion products generally do not become molten until the metal temperatures exceed 1300°F (704°C).
- The nickel-base alloys readily form nickel and chromium sulfides that can become molten at temperatures as low as 1100°F (593°C). Molten corrosion products can be washed or abraded off the blade or disc to expose fresh material to the corrosive environment, which, in turn, allows additional corrosion to occur.
- Local micro-environmental conditions have a major effect on the corrosion potential of the alloys in use.
- The presence of catalyst deposits in contact with the base materials has a major effect on the corrosion potential of the alloys in use.
- Major differences in corrosion rate occur between the different alloys in use. Figure 3.49 shows the condition of corrosion coupons removed from the ducting of an FCC expander.

FIGURE 3.49 Corrosion test coupons after long-term field exposure.

Note the wide amount of corrosion on the various material samples. The corrosion rate is often controlled by some of the minor alloying elements (aluminum, titanium, molybdenum, etc.) in base material.

Over time, the various expander manufacturers have developed metallurgical upgrades to improve the corrosion resistance of the rotor materials including:

- *Modified chemistry*. Special derivatives of the standard base materials have been developed by some manufacturers.
- *Material pre-oxidation*. Some expander manufacturers are increasing the corrosion resistance of the expander rotor components by pre-oxidizing the components. Proprietary controlled environment heat treatments are used to generate protective chromium and nickel oxide surface scales.
- *Coatings*. In recent years, the use of expander blade and disc corrosion coatings have become more common. A variety of coatings

have been used include chromizing, silica alumina, titanium nitride, and zirconium nitride. These coatings can be difficult to apply due to the complex geometry of the blade and disc attachment regions. The coatings can be applied by a variety of methods including high-velocity plasma spray, detonation gun, plasma vacuum deposition (PVD), plating, and pack diffusion methods.

3.5.6 Corrosion Prevention

Corrosion protection of expander rotors is generally a multi-action approach. It generally includes metallurgical upgrades as previously discussed, material upgrades, increased rotor cooling, and operational changes.

Rotor Cooling

Rotor cooling has been found to be an effective deterrent to blade and disc corrosion. Lowering the actual operating temperatures of the expander components has numerous positive effects including:

- Reduced corrosion potential.
- Reduced corrosion kinetics (lower temperature means slower corrosive attack).
- The cooling medium helps to increase the local partial pressure of oxygen and thereby decreases the corrosion potential.
- The formation of more stable protective metal oxides.
- The formation of stable and solid nickel and chromium sulfides versus the formation of molten sulfides that can be carried away with the flue gas and exposes fresh parent material.
- Increased material capabilities, thereby increasing structural margins and decreasing the potential of corrosion induced cracks.

Rotor cooling is generally accomplished by bathing the components in either steam or air. The lower-temperature cooling medium can reduce the part components by more than 200°F (111°C). Steam is often used for cooling because it has higher heat capacity and provides increased levels of cooling.

Figure 3.50 shows several of the commonly used expander internal cooling designs. Each of these designs can be used either individually or in combination. The designs may include the following:

- *Normal cooling*. Injection of cooling media into the rotor disc cavity. Cooling may be introduced on either side (upstream or downstream) of the rotor disc.
- *Impingement cooling*. Injection of the cooling media directly at and onto the rotor blade and disc.

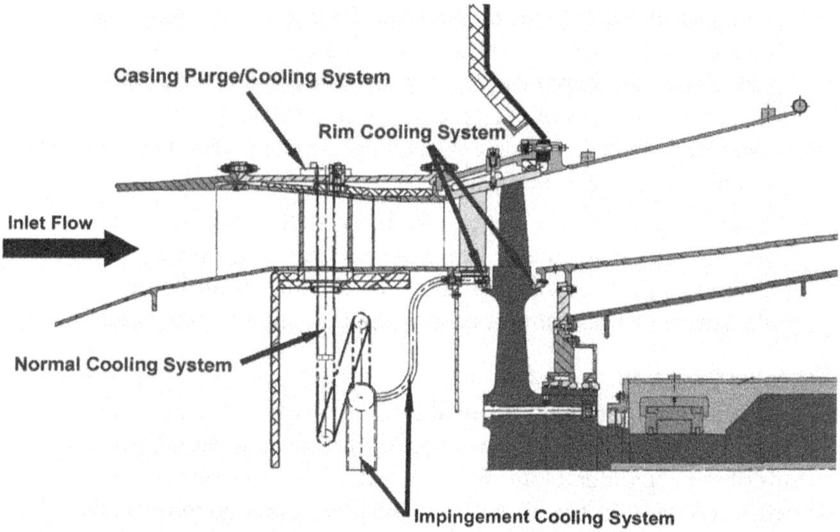

FIGURE 3.50 Typical expander rotor cooling arrangements.

- *Rim cooling.* Uses rotor disc rim seals that direct cooling media at the blade and disc attachment regions and help to minimize the amount of corrosive gases in the rotor disc region.

Operational

Changes in the FCC operation can greatly affect the corrosion potential of a specific expander. The following process operational changes have been found to significantly reduce the corrosion potential in the Hot Gas Expander.

- *Flue-gas oxygen level.* The flue-gas composition has the greatest potential for changing the corrosion potential. It has been found the corrosion potential is greatly eliminated by operating the regenerator at higher oxygen levels, which greatly reduces the corrosion potential in the expander. Expander corrosion has been greatly reduced with oxygen levels as low as 0.5% and virtually completely eliminated at oxygen levels above 1.0%.
- *Flue-gas sulfur level.* Although less significant than oxygen levels, the reduction of sulfur in the flue gas also greatly decreases the corrosion potential in the expander.
- *Flue-gas temperature.* As previously noted, the flue-gas temperature is critical in determining the type of corrosion products, the corrosion kinetics and the state of the corrosion products. The lower the flue-gas temperature, the lower the corrosion potential.

- *Flue-gas and catalyst salts loadings.* Flue-gas and catalyst salt levels can also have a significant effect on the corrosion potential in the FCC expander. Salts in the flue gas can act as both a flux and accelerant of the hot corrosion. Salts can be introduced from both the FCC feedstocks and from system steam or water quenches or injection points.
- *Flue-gas quenching.* Flue-gas quench systems have been found to be a major source of flue-gas contaminates, particularly a primary source of metallic salts. The cleanliness and quality of the steam or water sources should be thoroughly controlled.

Shutdown Corrosion

Inasmuch as high-temperature oxidation can occur during normal operation, a second type of corrosion occurs during the shutdown periods of the expander. Shutdown corrosion is an aqueous corrosion related to moisture in the air and moisture originating from the flue gas or steam sources. The use of steam purges throughout the CCU system is a major source of moisture in the flue-gas systems. Residual amounts of sulfur from the flue gas combine with any moisture or water present and form strong sulfur-bearing acids. Shutdown corrosion can occur on both rotating and stationary parts of the expander. Figure 3.51 shows a general surface attack of an expander casing. In more than one application, the expander flow-path erosion coatings

Figure 3.51 Low-alloy steel discharge casing with surface corrosion.

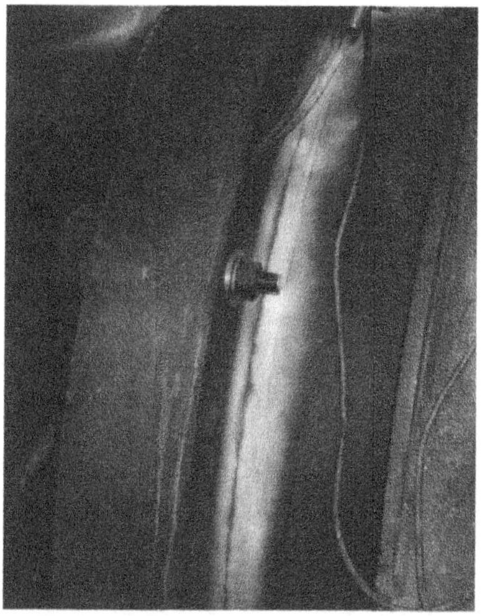

FIGURE 3.52 Catalyst deposit in the critical flow orifice.

have been damaged from shutdown corrosion attacking the nickel binder in the coating.

One of the most common acid attacks found in CCU systems is from polytheonic acid. Polytheonic acids are formed in services that involve mixtures of hydrogen sulfides, metallic sulfide scales, water, and air or oxygen. The CCU industry has shown that polytheonic acid stress corrosion cracking is most common in condensation-prone areas of the system. Intergranular stress corrosion cracking of the flue-gas ducting or expander casings can be quite severe as can be seen in Figure 3.52. SCC is a common mode of failure in stainless-steel components and is generally associated with material sensitization and chromium carbide precipitation caused by heating in the material into the 797°F to 1517°F (425°C to 825°C) temperature range. A number of the 18-8 stainless steels are very susceptible to SCC from long-term operation.

The available prevention or remedial actions to prevent SCC are as follows:

- Lower the applied or residual tensile stresses.
- Modify or eliminate the specific SCC agents.
- Change the alloy or increase alloy content.
- Soda ash neutralization during a unit shutdown per NACE RP01-70.

CHAPTER 4

CCU Expander Auxiliary Systems

A number of auxiliary systems are required for the CCU expander. Although many of these systems are common to most rotating machinery, there are a number of unique systems specific to the CCU expander that includes:

- Bearing lubrication
- Expander cooling system
- Expander flue gas seal system
- Expander online cleaning system(s)
- Online blade photography
- Instrumentation

Proper design, manufacture, installation, and operation of the auxiliary systems are critical to the power recovery train operation. Many reliability studies have shown that the number-one cause of PRT shutdowns is not caused by the PRT machinery itself, but the auxiliary systems that support the machinery.

4.1 Expander Piping

CCU expanders use large complex ducting to direct the flue gas to and from the expander. Figure 4.1 shows a typical layout for CCU ductwork. The ductwork often dwarfs the actual expander assembly in physical size. The smaller expanders might use a 48-in (1.2-m) inlet duct and 60-in (1.5-m) discharge duct, whereas the largest expanders use a 72-in (1.8-m) inlet duct and an 84-in (2.1-m) discharge duct. The ducting is sized based on the flue-gas volume flow. A common design basis is to have the flue-gas velocity in the range of 125 to 150 ft/s (38 to 46 m/s).

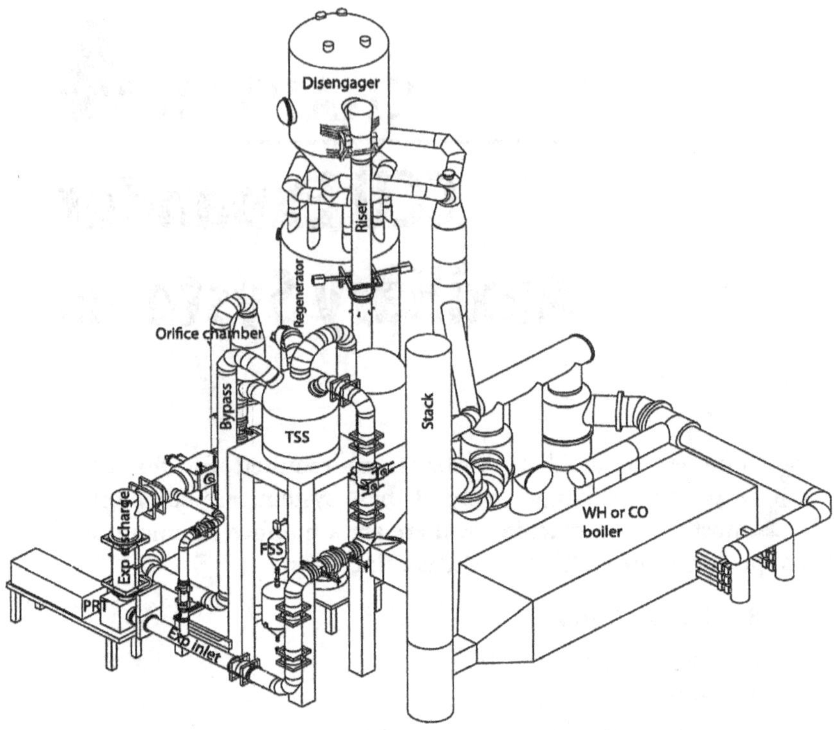

FIGURE 4.1 Typical CCU complex with high-temperature ducting.

This large ducting experiences large thermal growths and deflections, is very heavy, and can generate very large loadings on the anchor points of the piping system. Unfortunately, virtually all expander installations use the expander as an anchor point. The expander manufacturers provide limits to the loads that can be applied to the expander flanges. The limitations are intended to prevent expander operational problems or machine damage. The manufacturer limitations tend to be a small fraction of even the weight of the expander ducting.

Although a wide variety of piping systems have been applied to the CCU expander applications, the most successful designs include supporting the ducting on its centerline via overhead spring supports. The designs include strategically located guide keys and supports to control the directions of movement of the ducting.

4.2 Structural Design Considerations

4.2.1 Piping Wall Thickness and Weight

Expander piping is generally designed in accordance with the ASME Code Section 8. The most commonly used specification is ASME/ASI

B31.1 for pressure piping, API-605 and ANSI B16.47. Expander inlet duct piping typically is 0.75 to 1.25 in (1.8 to 3.2 cm) in thickness. Discharge piping typically is 0.375 to 0.75 in (1 to 1.9 cm) in thickness:

$$t_m = \frac{PDo}{2(SE + Py)} + A$$

t_m—Minimum wall thickness (in)
P—Maximum allowable internal working pressure (psi)
Do—Pipe outer diameter (in)
SE—Maximum allowable stress (psi)
A—Minimum thickness to compensate for mechanical strength and/or corrosion (usually 0.125 in)

A typical 60-ft (18-m), 60-in diameter expander inlet duct, weighing 340 lb/ft (47 kg/m), can end up weighing over 20,000 lbs (9,250 kg). With an inlet valve and an inlet transition piece with flanges, the total weight of the horizontal inlet run into the expander can easily exceed 45,000 lbs (20,400 kg), which exceeds the total weight of the entire expander. Applying this type of load to the expander would obviously create high loadings on the expander casings.

4.3 Piping Pressure/"Blow-Off" Load

The expander casings are also subject to a steady piping pressure blow-off load that pulls axially on the expander casing (see Table 4.1). This load is equal to the duct area times the operating pressure. These loads can be very large as can be seen in the following example:

A typical 60-in expander inlet duct operating with a 30 psig (2.11 kg/cm^2) inlet pressure and a 2 psig (0.14 kg/cm^2) discharge pressure:

$$PF = P\left(\frac{\pi d^2}{4}\right)$$

PF—Pressure force (lbs)
P—Pressure (psig)
d—Duct diameter (in)

	Duct Size (diameter)		Pipe Area		Pressure Load	
	Inches	Meters	Inches^2	Meters^2	Pounds	Kilograms
Inlet	60	1.52	2,827	1.82	85,000	38,500
Discharge	72	1.83	4,072	2.63	8,100	3,700

Table 4.1 Piping Blow-Off Load Imposed on the Expander Inlet

The piping loads are resolved with equal and opposite loadings in the expander structural members. The inlet piping blow-off load is reacted by the pressure drop across the stator vanes and rotor. The stator vane pressure drop (approximately 60% of the total pressure drop) is transmitted into the inlet casing and subtracts from the piping blow-off load at the inlet casing supports. The rotor thrust resolves the remaining inlet pressure loading and is transmitted into the bearing support and then into the baseplate. The discharge duct blow-off load is in the vertical direction and tends to try to lift up the expander casings.

4.4 Allowable Force and Moments

From the above examples, it can be seen that large forces can easily be applied to the casings. These forces need to be remediated to prevent the overloading of the expander casings. The allowable flange piping loads has always been a contentious area between the expander manufacturers and piping system designers. The expander manufacturers like to see virtually no expander loadings being applied to the expander flanges. The piping designers like to use the expander as a rigid piping support and anchor.

Process designers, architect engineers, and even the expander manufacturers continue to use industry standards that do not address the requirements of the CCU expander. The most commonly used industry standard used to calculate allowable piping loads is NEMA SM-23 issued in 1969. This standard was developed for steam piping systems connected to mechanical-drive steam turbines. The calculation procedure has proven satisfactory for that application that uses much smaller piping and rigid turbine casings. API has also adopted this calculation procedure for application to industrial centrifugal compressors. API-617 specifies that the compressor/expander allowable loadings should be increased to 1.85 times the NEMA-allowable values.

For many years, the expander manufacturers specify that the maximum expander loads could only be 60% of the values calculated in accordance with the NEMA SM-23 specification. These values result in very low allowable loads that are typically less than a few thousand pounds. Over time, some of the manufacturers have increased these allowable forces and moments to be equal to 100% of the NEMA values. Increasing the values to 100% NEMA still results in very low allowable loadings.

The biggest problem with the NEMA SM-23 specification is that the standard is only geometry-based and applicable to rigid casings. CCU expander casings are not rigid, and significant deflections and stress can result from the piping loads. The specification does not address certain conditions including:

- The potential for casing lifting
- Contact conditions
- Casing misalignment (internal and external)
- Excessive casing deflection
- Excessive casing stresses
- Casing flexibility

Using the industry-standard NEMA-SM23, the following equations are used to determine the maximum flange loadings of the expander. The calculation procedure accounts for both the expander inlet and discharge nozzles:

1. The total resultant force and total resultant moment imposed on the expander must not exceed the following:

$$F = \frac{500D - M}{3}$$

   F—Resultant force in pounds including pressure forces where unrestrained expansion joints are used
   M—Resultant moment in ft lbs
   D—(16 + Inside pipe size)/3 in

2. The combed resultants of the forces and moments of the inlet and discharge connections, resolved at the centerlines of the discharge connection, must not exceed the following two conditions:

   a. These resultants must not exceed:

$$Fc = \frac{250Dc - Mc}{2}$$

   Fc—Combined forces of the inlet and discharge in lbs
   Mc—Combined resultant of the inlet and discharge moments and moments resulting from forces in ft lbs
   Dc—(18 + Equivalent diameter)/3 in

3. The components of these resultants shall not exceed the following:

$$Fy = 125\text{ Dc} \qquad My = 125\text{ Dc}$$
$$Fz = 100\text{ Dc} \qquad Mz = 125\text{ Dc}$$
$$Fx = 50\text{ Dc} \qquad Mx = 250\text{ Dc}$$

   Fy—Vertical components of Fc

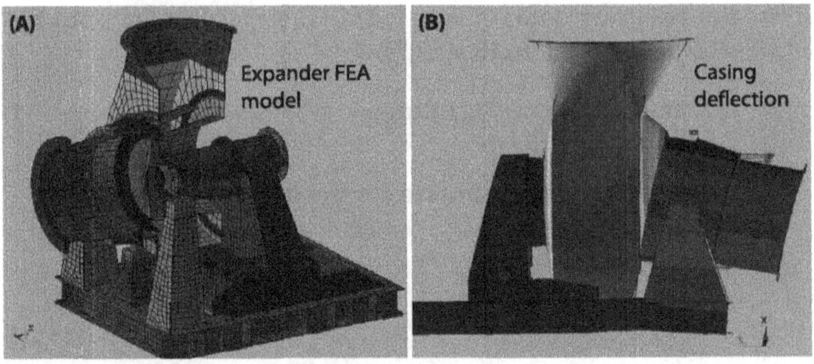

Figure 4.2 FEA model (A) and deflection results (B) for evaluating expander piping loads.

Fz—Horizontal components of *Fc* at right angles to the expander shaft
Fy—Horizontal components of *Fc* parallel to the expander shaft
Mx—Component of *Mc* in a vertical plane at right angles to the expander shaft
My—Component of *Mc* in a horizontal plane
Mz—Component of *Mc* in a vertical plane parallel to the expander shaft

Although the industry standards for allowable piping loads have been used for decades, more recently, an engineering approach has been implemented by some expander manufacturers. This new approach was made possible by using upgrades in the finite element analysis programs and improved computational capabilities. Complete FEA models of the expander have been developed as can be seen in Figure 4.2(A). These models allow the assessment of any load combination on the expander. Casing and support stresses and deflections can be accurately evaluated. In addition, the loading effects on the rotor clearances can also be predicted and evaluated. Figure 4.2(B) also shows a typical output from an FEA of a problematic expander. The casing deflections can be seen for a large axial force being applied to the expander inlet flange. The use of FEA for the evaluation of piping loads has greatly reduced the number of expander problems caused by expander piping systems.

4.5 Flange Design

The piping flanges are also designed in accordance with the ASME B31.1 or B31.3—Standards for pressure piping. Flange geometries in accordance ANSI B16.1, B16.5 and B16.47 (API-605) are commonly used. The most common CCU expander duct design uses weld neck

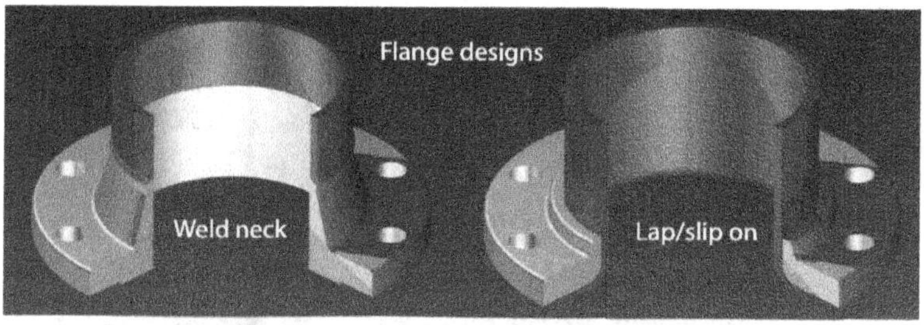

FIGURE 4.3 Typical weld neck and lap joint flanges.

flanges, although some installations have had success with lap joint flanges. Figure 4.3 shows these two flange designs.

4.5.1 Weld Neck Flanges

Weld neck pipe flanges (Figure 4.3) attach to the pipe by welding the pipe to the neck of the pipe flange. This allows for the transfer of stress from the weld neck portion of the flange to the pipe itself. The weld neck design reduces high-stress concentrations associated with the attachment of the thinner wall pipe to the relatively large and massive flange. Weld neck flanges are made integral to the piping with full penetration welds. The inside diameter of the weld neck is machined to match the inside diameter of the piping. The weld neck flanges are generally the raised face design.

4.5.2 Lap Flanges

Lap joint pipe flanges are loose and have no attachment to the pipe as can be seen in Figure 4.3. The hubs slide over the pipe and are most commonly used with stub-end fittings. A pipe is typically welded to the stub end and the lap joint pipe flange is free to rotate around the stub end. The benefit of this is that the flange can be rotated around the circumference of the piping during assembly to accommodate bolt hole misalignment.

Flanges designed in accordance with the ASME Code tend to be quite large in size as the code attempts to address the problems associated with high-temperature operation. At the expander operating temperatures, the typical flange material is 300 series stainless steel. While these material provide good corrosion resistance, their allowable stresses are quite low. For example, a 304 stainless-steel material operating at 1,200°F (649°C) would have an allowable stress of only 6,100 psi (429 kg/cm^2). At 1,350°F (732°C), the allowable stress falls off to 2,900 psi (204 kg/cm^2). It is very difficult to meet these low allowable stress values without making the flanges very large. In addition, when temperatures exceed 1,000°F (538°C), the ASME Code requires the use of a flange with a 300-lb rating. This makes the flanges even

FIGURE 4.4 Typical expander inlet duct with uninsulated flanges.

larger in size. For example, a 60-in, 150-lb class flange would have an outer diameter of 67.9 in (172.5 cm) and a flange thickness of 2.94 in (7.47 cm). A 60-in, 300-lb class flange would have an outer diameter of 74 in (18.80 cm) and a flange thickness of 5.88 in (14.9 cm). That makes the 300-lb flange more than 3 in (7.6 cm) larger radially and twice as thick. The heavy flanges can be clearly seen in Figure 4.4, which shows a typical expander inlet system. Thick flanges actually cause significantly higher thermal stresses in the flanges and often lead to excessive distortion of the ducting. The use of uninsulated flanges is a common practice and greatly increases the flange thermal stresses further since the flange average operating temperature is typically 250 to 350°F (139 to 194°C) cooler than the duct wall temperature. The cooler heavy flange does not grow as much radially as the hot duct and acts as a belt that limits the duct radial thermal growth.

4.6 Fasteners

To maintain the effectiveness of the gasket seal throughout an operating campaign, sufficient stress must remain on the gasket surface for years of operation. This means that the residual bolt load on the gasket should at all times be greater than the internal pressure forces acting on it. This implies that the fastener preload must be maintained above a minimum value throughout the operating run that could exceed five years of operation. Careful material selection and loading are required as many of the commonly used fastener materials will experience significant creep relaxation during the operating campaign.

The problem is further accentuated by the thermal compatibility of the fasteners to the duct materials. Unfortunately, the materials that are thermally compatible with the flange materials have the lowest creep resistance. The materials with the highest creep resistance have significantly different thermal coefficients of expansion. The lower operating temperature of the fasteners, combined with the thermal coefficient of expansion, causes the fastener stresses to increase in operation. When using high-strength fasteners, lower preloads may be applied to account for the increased loadings that occur in-service at the operating temperatures. Obviously, the preload must be sufficient to prevent flange leakage both at ambient and operating conditions.

The piping flange fasteners are designed in accordance with ASME Boiler and Pressure Vessel Code Section VIII Division I. The various design requirements and considerations are discussed further in the section on CCU expander casing design (section 2.3.8).

One area in which it has been difficult to get piping designers to agree is whether or not it is best to insulate the ducting flanges and fasteners. Most duct flanges are intentionally left uninsulated to reduce the operating temperatures. Lower temperatures have higher allowable stress values. The use of insulation significantly increases the operating temperatures of both the flange and the fasteners. As the temperatures increase, the relaxation of the fasteners increases. Alternatively, the use of insulation makes the flange and fastener operating temperatures more uniform and consistent. The uniform temperatures reduce the thermal stresses in both the flanges and fasteners.

Some expander manufacturers use Belleville (conically shaped) washers/springs on the expander piping flanges to both accommodate the increased thermal loadings and to maintain joint preload throughout the operating campaign. This practice has been quite successful, although there is a significant cost associated with the use of high-temperature Belleville springs.

Overall, each installation must carefully design and select the fasteners for the specific operating conditions.

4.7 Flange Gaskets

Whereas the numbers of flanges used in CCU piping systems are minimized, flanges need to be used to allow assembly and disassembly of major components such as the expander, control valving, expansion

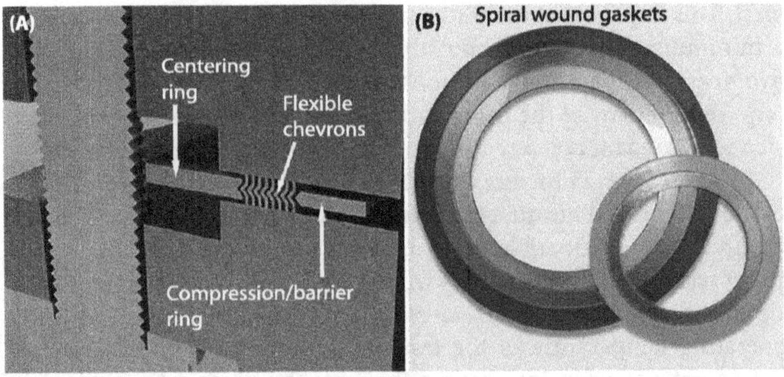

FIGURE 4.5 Typical FCC expander piping gasket (A); spiral wound (B).

joints, and so forth. The flanges typically use gaskets to prevent leakage of the flue gas to the atmosphere. The most commonly used gaskets include the spiral-wound gasket shown in Figure 4.5, and the Flexpro (formerly known as a Kammprofile) gasket shown in Figure 4.6. Both designs have had good operating experience.

These gaskets provide the following benefits:

- Good for high-temperature and pressure applications
- Good for cyclic loadings
- Have high blowout strengths
- Use a compression ring to prevent gasket overload
- Help to reduce the potential for flange erosion
- Are resistant to inward buckling

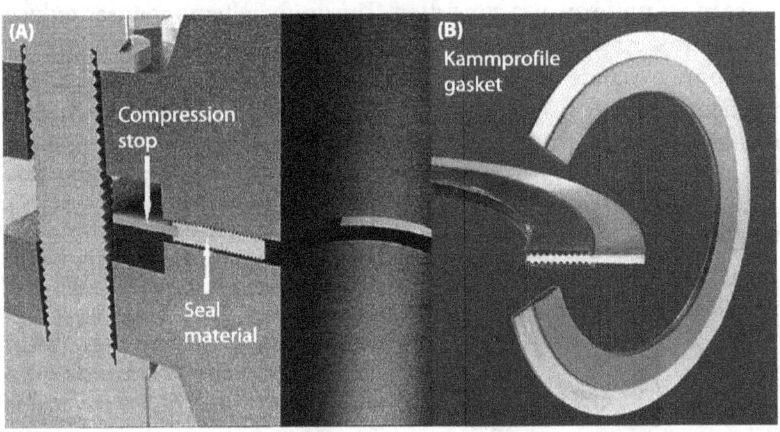

FIGURE 4.6 Typical FCC expander piping gasket (A); Flexpro (B).

- Are self-centering
- Reduce turbulent flow

An important aspect in the selection of the gasket is the use of a compression stop. Even with excessive fastener preload, the compression stop will prevent overloading of the gasket seal region and "crushing" of the gasket. The use of an internal compression/barrier ring is also commonly used to prevent the gasket from extruding or unwinding into the piping system.

The stresses acting on a gasket in a pressurized joint include the compressive stress produced by the fasteners, the side loads owing to internal pressure trying to extrude the gasket out of the flange, and the internal working pressure trying to separate the flanges. For a gasket to operate successfully, the residual gasket stress after the application of the loads needs to be significantly higher than the fluid pressure.

Most CCU gaskets are manufactured from a 300 series stainless steel. The filler material in the spiral-wound gaskets and the surface seal material in the Flexpro gaskets are a graphite-based specialty filler.

The gasket design is often a customized gasket with reduced seating loads. Some of the standard gaskets require seating loads of more than 20,000 psi (1,406 kg/cm^2). To apply this load, the stresses in the fasteners may be excessive. Generally the seating load for the expander applications is limited to 10,000 psi (700 kg/cm^2) to reduce the preload requirements of the fasteners.

Some expander installation ducting use a third flange gasket design commonly called a "lip seal." The lip seal involves installing two flat metallic gasket rings between the flanges, and welding the inside diameter of the seal rings to each flange face. The outside diameters of the two seal rings are also welded thereby making up a fully welded assembly with no leakage paths. One disadvantage of this design is the need to grind away the outer weld whenever the flange needs to be taken apart. The operating success of this design has been very good.

Care needs to be taken during the installation of expander duct gaskets as they are physically large in size and are quite sensitive to handling damage. It is also very important that the flanges be uniformly tightened in steps to prevent crushing the gaskets and assuring uniform loading around the circumference. A typical tightening procedure would include the following steps:

- Lubricate the fasteners with an appropriate anti-seize compound. High-purity nonsulfur-based lubricants must be used to prevent fastener corrosion.
- Center the gasket with all fasteners in place.

- After final flange assembly and alignment, snug all fasteners against the flange.
- As the flange fasteners are tightened, continuously measure the flange gap to assure uniform pull up around the flange circumference. Fasteners should be tightened in a sequence that assures uniform pull up (i.e., –0°, 180°, 270°, 90°, etc.).
- Torque all fasteners per the above sequence to 25% of the specified torque value.
- Torque all fasteners per the above sequence to 50% of the specified torque value.
- Torque all fasteners per the above sequence to 75% of the specified torque value.
- Torque all fasteners per the above sequence to 100% of the specified torque value.
- Recheck all fasteners to assure 100% loading.
- Hot retorque after start-up and temperature equalization.

4.8 Expansion Joints

FCC expander installations are quite complicated with long runs of large ducting. With the high temperatures of the flue gas, the ducting experiences large thermal growths. This is particularly true for the expander inlet line. The expander inlet line is of a hot wall construction and manufactured in a 300 series stainless steel that has a high coefficient of thermal expansion. Thermal expansions of several inches are common. This growth must be allowed to occur freely or excessive forces and moments can be imposed on the attachment of the duct to the vessels and the expander.

The large thermal movements of the expander ducting cannot be readily accommodated with a single expansion joint accommodating the axial, lateral, and angular movements. These movements are best accommodated by using multiple expansion joints that absorb the movements in rotations similar to that shown in Figure 4.7.

Universal, hinged, or Gimbel expansion joints are typically used in FCC piping systems. Pressure-balanced expansion joints are commonly used for expander discharge lines. Three or more expansion joints are typically used in both the expander inlet and discharge lines.

FCC expansion joints can be built in a hot or a cold wall configuration. Hot wall designs are used between the Third Stage Separator and the expander to ensure the refractory does not enter the expander. Many expander discharge lines are of a hot wall design to the first

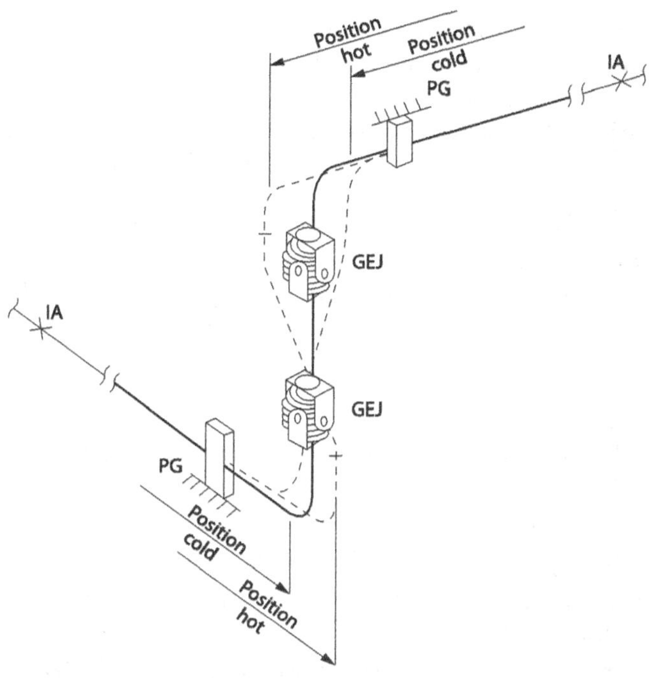

FIGURE 4.7 Multidirectional thermal expansion.

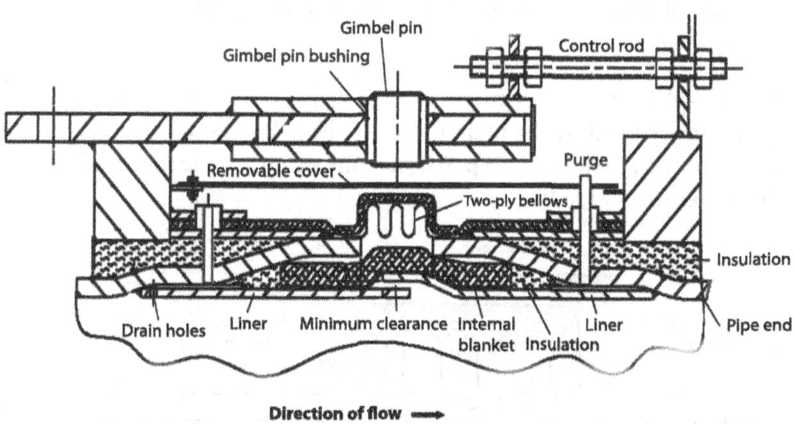

FIGURE 4.8 Typical FCC hot wall expansion joint design.

bend downstream of the expander. The duct then transitions to a cold wall design. Cold wall piping is used throughout the remaining sections of the flue-gas system. The cold wall design uses carbon-steel piping walls and is refractory lined. The refractory acts as an insulator and is highly erosion resistant.

4.8.1 Materials of Construction

FCC expander expansion joints have been manufactured from 300 series stainless steels conforming to ASTM A-240 Types 304, 321, and 347, Incoloy 800 per ASTM B-409 and Inconel 600 per ASTM B-168. The best operating experience (although the most costly) has been with Inconel 600 LCF.

4.8.2 FCC Expansion Joint Special Design Features

FCC expansion joints require a number of special features to assure proper operation and long-term life. Typical special-design features of FCC expansion joints include the following:

- Flow liners to protect the bellows from the environment degradation
- Internal insulation to reduce the operating temperature of the bellows
- Two-ply bellows to provide backup protection should a bellow fail
- Inter-ply bellows leakage indicators
- Drain provisions to prevent the accumulation of liquids during a unit shutdown
- Purge provisions to prevent the accumulation of catalyst in the expansion joint
- Low-friction guide pins and bushings
- Control rods to limit bellows movement or squirm
- Removable weather cover

Figure 4.8 shows the construction of a hot wall expansion joint design typically used in the expander inlet line. Figure 4.9 shows the construction of a cold wall expansion joint design typically used in the expander discharge line.

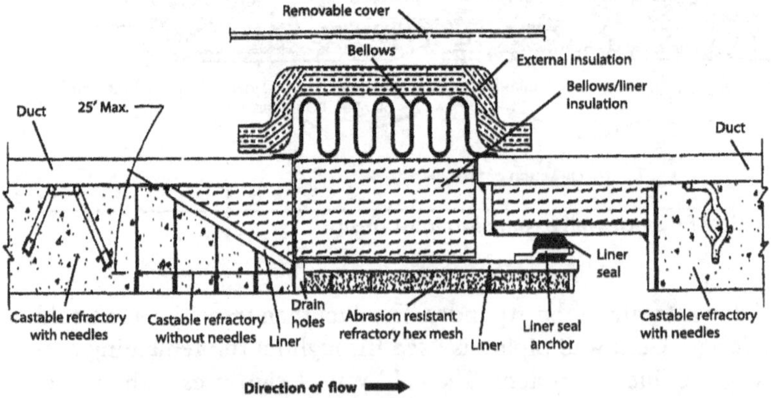

FIGURE 4.9 Typical FCC cold wall expansion joint design.

4.9 Lube Oil System

Power recovery trains typically use API-614 lube oil systems. Figure 4.10 shows a typical lube oil console schematic for a system using centrifugal pumps. Oil pumps can be either centrifugal or positive displacement pumps.

Typical features of the lubrication system include:

- Main and spare oil pumps and drivers with different energy sources
- Oil reservoir
  - Heaters
  - Purge
  - Level indicator(s)
- Main and spare full-capacity oil coolers with an online transfer valve
  - Instrumentation to monitor oil and water side performance
- Main and spare full-capacity oil filters with an online transfer valve
  - Instrumentation to monitor filter differential pressure
- Oil regulation valves
- Low-pressure sensing, alarms, and trip systems
- Stainless-steel supply and drain headers

Additional components in the lube oils system often include:

- Emergency rundown tank
- Pump transition tank
- Oil conditioner/purifier
- Reservoir vacuum system
- Generator jacking/lift oil system

Commonly monitored lube oil parameters:

- Reservoir oil temperature
- Pump discharge pressures
- Oil temperature into and out of coolers
- Filter differential pressure
- Lube oil supply pressure
- Jacking oil pressure

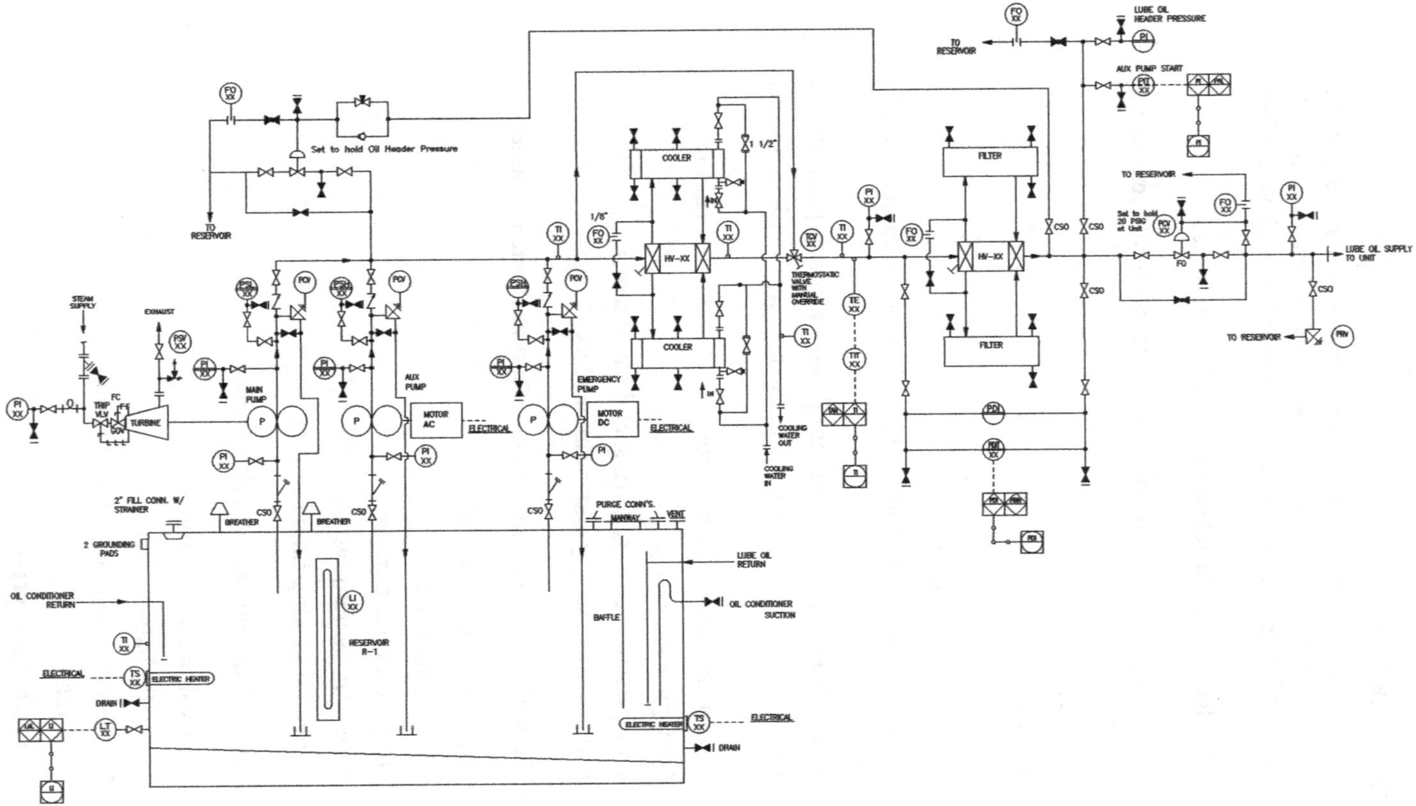

FIGURE 4.10 Typical FCC power recovery train lube oil system.

Common lube oil system alarms:

- Low lube oil supply pressure
- High-cooling water supply temperature
- High oil filter differential pressure
- Cooling water high and low levels in expansion tanks
- Console flame detection
- Unexpected auxiliary pump operation
- Jacking oil low pressure

Common lube oil system shutdowns:

- Low-low lube oil supply pressure (automatic)
- Flame detection (operator initiated)

4.10 Expander Cooling System

One of the most critical auxiliary systems for the FCC expander is the expander cooling system (see Figure 4.11). Expander cooling is critical in controlling the operating temperatures of various expander components including the casings, nose cone, and, most importantly, the rotor blades, discs, and tie bolts. Insufficient cooling of many components can greatly reduce the component lives and cause premature failures. Similarly, overcooling of many components can cause excessive thermal stresses that also can reduce the component lives, increase the amount of distortion, and cause premature failures. The various expander manufacturers use different cooling schemes as appropriate for the individual expander design. The preferred cooling scheme has always been contentious.

Cooling is accomplished by directing steam or air to critical locations of the cooled components. The internal cooling flow paths and purposes are further described in the component design section of this book. The expander cooling steam has been controlled in three different methods that include external critical flow orifices, supply pressure control with external and/or internal orifices, and via flow control. More recently built expanders use flow control because it assures the expander is getting the proper amount of cooling despite changes in steam supply conditions, pressure losses, orifice sizing or wear, and internal clearance changes.

Expander cooling systems were mounted on or around the expander baseplate in early expander applications. To make the area around the expander more accessible, newer expander applications usually have separate skid-mounted controls and instrumentation that are located remotely from the expander. The expander cooling

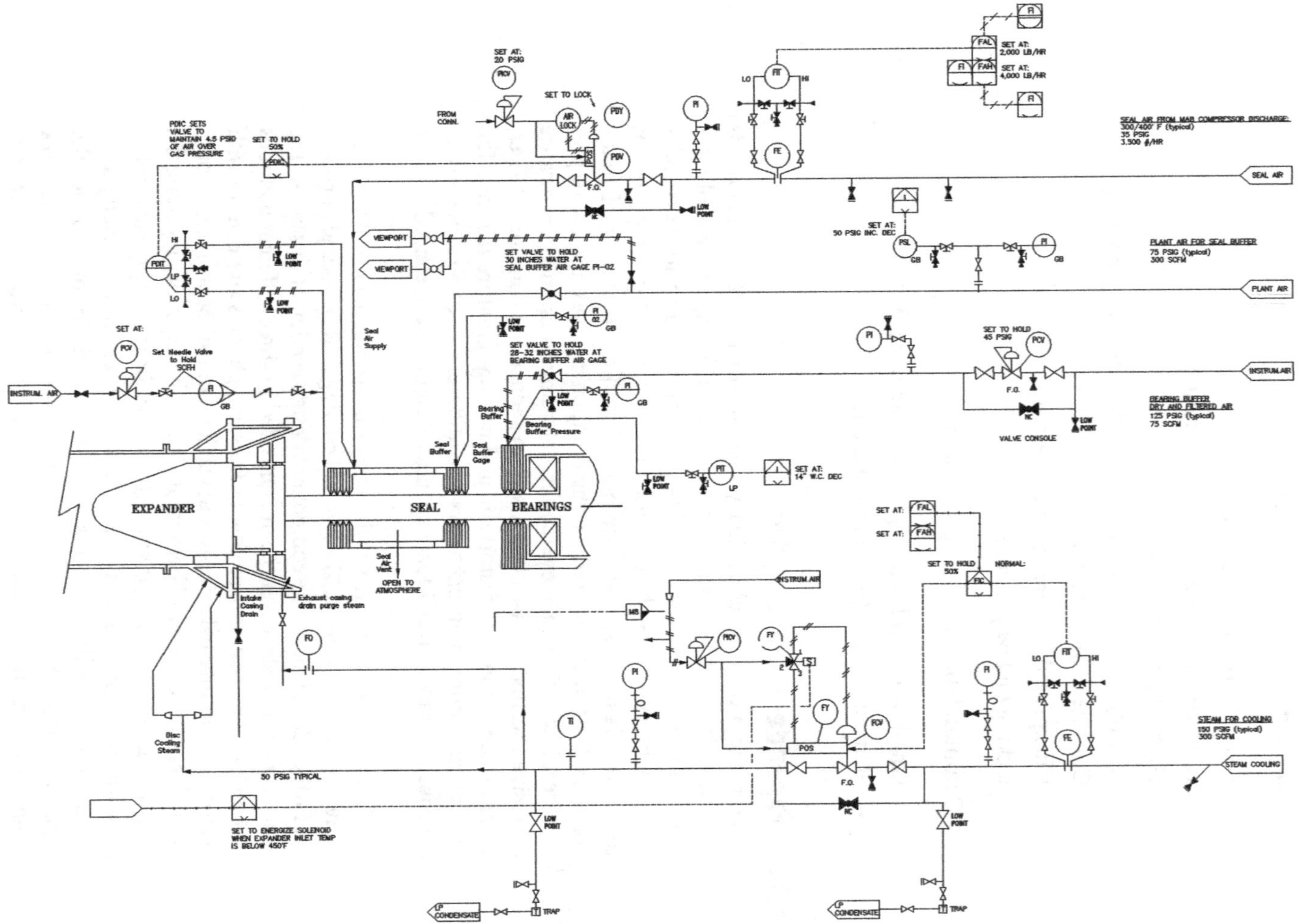

FIGURE 4.11 Typical CCU expander cooling and seal schematic.

system is generally fairly simple and usually includes a number of control and instrumentation components including:

- Isolation valves(s)
- Supply pressure and temperatures
- Flow control valve or orifices that control the cooling flow rate
- Flow controller that regulates and enables the cooling media when the expander inlet or outlet duct temperature matches that of the cooling steam
- Steam traps at all low point drains to prevent condensation carryover into the expander
- Blow down provisions to remove condensate from the steam lines prior to commissioning

A number of early expanders were also built with an afterburn provision as part of the expander cooling steam system. The system increased the amount of cooling steam provided to the expander when the expander inlet temperature exceeded a prescribed value. While this cooling steam did provide some protection to the expander rotor, it was found that the afterburn steam actually caused a thermal shocking of both the expander casings and rotors and could produce transient thermal stresses that were more detrimental than the higher temperature operations alone. In addition, the afterburn lines were often the source of condensate carryover and system shocking since the afterburn provision were dead legs and not properly blown down prior to activation. Most recent expanders have been built without an afterburn steam-cooling provision. Today, the afterburn protection is usually located upstream of the expander at the regenerator discharge and is intended to reduce the flue-gas temperature to the CCU downstream equipment.

Commonly Monitored Expander Cooling System Parameters

- Cooling media supply pressure
- Cooling media supply temperature
- Flow

Cooling System Alarms

- Low-cooling steam flow
- High-cooling steam flow
- Low-differential steam pressure

4.11 Expander Sealing System

The number-one purpose of the expander seal system is to seal the flue gas in the expander casings (see Figure 4.11). Flue-gas seal leakages are significant and potentially dangerous. The gases are hot (generally above 1,000°F (538°C) and pose both a fire and burn hazard to personnel. In addition, the gases are laden with catalyst particulate that is both an irritant and sight hazard. Most significant, the flue gas is potentially lethal since it contains both hazardous sulfur-bearing gases, salts, metals, and is almost void of oxygen.

A second function of the expander seal system is to prevent lubrication oil leakage from the bearing housing. With both high-ambient and component temperatures, a strong fire hazard exists if the lube oil were to leak from the bearing housing onto the expander casings. Even minor leakages (slow oil drip) or oil mists can soak the casing insulation and ignite. In addition, the bearing housing seal is also intended to prevent the entry of flue gas and catalyst into the bearing housing.

A third function of the expander seal system can be to provide downstream rotor disc cooling.

Several different flue-gas seal designs have been used over time and by various expander manufacturers. Figure 4.11 shows the typical flue-gas seal system used on many CCU expanders. Virtually all flue-gas seals are of the buffered labyrinth design. Both steam and air-buffer media commonly have been used. The most common methods of controlling the seal flows are a differential pressure control, a differential pressure control with flow measurement, or strictly flow control. The advantage of measuring the seal flow is that it assists in monitoring and assessing the condition of the expander seals.

The high side of the differential pressure control is the supply pressure of the buffer media. This pressure should be measured at the seal as significant pressure drops can occur between the controller and the seal housing. The low side of the differential pressure control is the expander exhaust pressure. A small air purge is often used on the low-side pressure tap because it has a tendency to plug up with catalyst.

Most early FCC expander installations used steam for the flue-gas seal media. With the use of steam, the atmospheric side of the gas seal is educted to prevent the discharge of steam or flue gas to the environment. The vacuum is created via a steam eductor. The educted gases are condensed in a gland condenser and discharged to a dirty water drain. These systems have been notoriously unreliable and experience significant corrosion due to flue-gas mixing with the condensables and create strong sulfur-bearing acids. The use of educted seal housings has been greatly in decline in more recent times.

The use of air as the buffer media is most common today. The air cleanliness is less of a concern in the flue-gas seal, and main air-blower discharge air has been a cost-effective source.

Many seal systems use a backup air-buffered seal. The backup seal is required in an educted seal-housing design to prevent the induction of large quantities of air. The backup seal can also be used to seal the flue gas if the primary seal is leaking.

Flue-Gas Seal Instrumentation

- Seal media supply pressure(s)
- Flue-gas differential pressure
- Seal media flow
- Flue-gas seal differential pressure
- Instrument purge pressure

Flue-Gas Seal Alarms

- Low-differential pressure
- Low or high seal flow(s)

4.12 Bearing Housing Seal System

The bearing housing shaft end seal is also sealed with a buffered labyrinth seal. A dry, filtered air source is required for the buffer media. Instrument air is commonly used for the bearing buffer air. Any water or dirt that enters with the buffer media will enter into the bearing housing and lubrication system. The bearing buffer air flow is controlled with a pressure control valve.

The bearing housing and coupling guards in the PRT are vented to prevent the buildup of pressure inside the housing that may result in bearing housing leakages. The sizing and design of the vents and the physical location are important in preventing the emissions of both nuisance and dangerous oil mists. Occasionally, the lube oil systems are operated with a slight vacuum to prevent oil mist discharge.

Bearing Housing Seal System Instrumentation

- Seal media supply pressure
- Seal differential pressure

Bearing Housing Seal System Alarms

- Low seal supply air pressure

4.13 Expander Start-Up Steam

Some expanders require the use of start-up steam to assist in controlling the expander heat-up rate and to maintain the expander exhaust temperature at acceptable levels under low loading conditions. Start-up steam is applicable to those expanders that are not started up with the CCU process. Although commonly used in the expander/generator applications, conventional PRT expanders may also require start-up steam. On occasion, start steam may also be used to thermally clean an expander of catalyst deposits. Start-up steam is injected into the flue-gas ducting upstream of the expander, as can be seen in Figure 4.12. The commissioning of the start-up steam must be done very carefully. Because the line is normally closed, the formation of condensation in the supply line is highly likely. The start-up steam line needs to be self-draining away from the duct and have suitable steam traps and drains. The start-up steam should not be commissioned until the duct temperature equals or exceeds the steam supply temperature. This is done to minimize the formation of condensation and prevent thermal shocking of the ducting and expander. A temperature controller is normally used to control the start-up steam flow rate.

FIGURE 4.12 Typical start-up steam-duct connection.

4.14 Expander Casing and Duct Drains

One of the most noncomplex, although very important, auxiliary systems is the expander drain system. Many expanders have been damaged owing to the collection of condensate in the expander duct and casings. The condensation occurs during normal start-up and shutdown operations and even while the PRT may be stationary. When the duct temperature drops below 350°F (177°C), liquid condensates can form. The condensation can be the result of water in the flue gas, process quenches, expansion joint purges, control valve purges, expander cooling, sealing media, and other factors. The amount of condensation can be quite large, and some expander inlet ducts and casings have literally filled up with several feet of water. Extensive duct and expander casing damage has occurred because of the large thermal gradients that develop during the unit start-up.

Liquid condensation is most problematic from a corrosion perspective. Sulfur from the flue gas and on the catalyst can combine with the liquids and form very aggressive sulfur-bearing acids. Any polytheonic acid attack of the stainless-steel ducting or the expander casings can be quite damaging and result in major in-service cracking and failures. Numerous small bore piping and instrument connection failures result from this type of corrosion. Many PRT installations have had to protect the stainless-steel components during unit shutdowns by neutralizing the acids with soda ash washes.

Condensate drains are critical during shutdown and transient operation to minimize the amount of condensation collection. Unfortunately, many duct and expander manufacturers do not provide adequate drain provisions. They are often undersized, difficult to access, and manufactured in corrosion-susceptible materials. It must also be remembered that drain lines are the low points in the system and susceptible to collecting catalyst and plugging.

Experience has shown that the system drains should include the following features:

- The drains need to be sized for all sources of system condensation.
- Drain line sizes should be substantially large enough to prevent plugging and mechanical damage.
- Rodding/cleanout provisions need to be included in areas susceptible to catalyst plugging.
- Flanged drain connections should be used to allow access and cleaning.
- The drain lines should be manufactured in a suitable stainless-steel material.

The use of the system drains during unit start-up and shutdown is critical and needs to be a priority item on the operator's operational instructions and checklists. The use of the drains needs to be done purposely as leaving them open can pose significant safety issues to the operators. Significant high-temperature, catalyst-laden flue gas will discharge from the drains once the unit is put online. The operators clearly need to know when and how to use the drains and when not to use the drain provisions.

4.15 Online Expander Flow-Path Cleaning

Expander flow-path deposits can cause rotor vibrations and if left to continue, rotor-rubbing and component failures can result. Fortunately, two online cleaning methods have been developed to help control the build of catalyst in the expander. The first method is an abrasive cleaning procedure that involves the injection a cleaning media to remove the deposits. The second procedure involves thermally cycling the expander to break off the deposits.

Blade viewport photography, rotor and casing vibration data, bearing temperatures, rotor axial position and expander pressures, temperatures, and flow characteristics are used to determine if an expander online cleaning is required. Blade photography obviously gives a direct look at the condition of the blades. Expander vibration characteristics signal the buildup of catalyst on the rotor and the state of unbalance. The vibration characteristics often change in magnitude and change from a steady vibration signal to a wandering or "hashy" vibration signal. Expander performance monitoring also highlights the degree of catalyst buildup. As the deposits build up in the expander, the expander often gets more efficient as the blade-tip clearances are reduced. At the same time, the flow capacity of the expander decreases as the deposits become a blockage in the flow path. Blockages can occur over the rotor blades, in the stator or rotor blade throat regions, and as a general deposit over the entire airfoil.

4.16 Abrasive Cleaning System

Successful cleaning of the expander flow path has been achieved in many expanders by injecting abrasives into the inlet duct of the expander. This method of cleaning has been found to be effective in the removal of the lighter catalyst deposits. The abrasive cleaning is not as effective on hard sintered catalyst deposits that are tenaciously adhered to the expander flow-path components.

Various abrasives have been used to clean the expander flow path including walnut shells, ground-up peach pits, grape pits, rice, sandblast media, and even glass beads. The amount and type of abrasive must be chosen carefully as many abrasives can be quite erosive to the expander blading and can cause damage. Most of the organic materials are not excessively abrasive and do provide effective cleaning. Each expander installation needs to determine which cleaning media and what size media is most effective for their type of deposits.

Fine abrasive particles 0.039 to 0.059 in (1.0 to 1.5 mm) in diameter tend to clean more areas of the flow path. Larger abrasive particles 0.079 to 0.138 in (2 to 3.5 mm) in diameter tends to be more aggressive and clean the more stubborn deposits, although they clean a smaller area of the flow path.

The cleaning media is generally injected at a quantity of 50 lbs (23 kg) for every 250,000 lbs/h (115,000 kg/h) of flue gas. A typical FCC expander passing 750,000 lb/h (340,000 kg/h) would require 150 lbs of abrasive at each cleaning. Excessive amounts of abrasive should be avoided because it can cause flow-path wear and affect the expander performance.

The abrasive is injected into the pressurized flue-gas inlet duct upstream of the expander from specially built cleaning hoppers as shown in Figure 4.13. The hoppers are usually designed to hold at

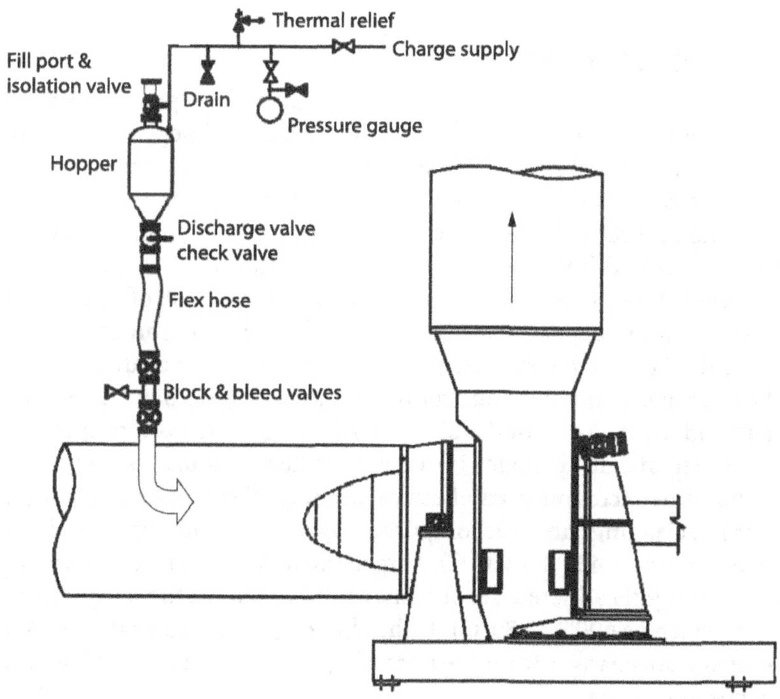

Figure 4.13 Expander abrasive cleaning system.

least a single load of abrasive media. The injection hopper is usually located between 10 and 30 ft (3 and 9 m) upstream of the expander. If the injection point is too far away from the expander, the abrasive will burn up in the inlet duct and will not strike the flow-path components. If the injection point is too close to the expander, the abrasive will not be uniformly distributed in the duct and uniform flow-path cleaning will not occur. The hopper uses either air or nitrogen as a motive fluid to push the abrasive into the duct. Nitrogen is used as an injection media if the abrasive particles are burning too quickly or if moisture is a problem and causes abrasive sticking. The abrasive media needs to be injected very quickly with the full discharge occurring in just a few seconds.

The expander flow-path design has a significant influence on the effectiveness of the cleaning. The abrasive particles have significant momentum and do not readily follow rapid changes or steps in the expander flow path. Abrasive cleaning is most effective in a smooth, gradually changing flow path.

It needs to be emphasized that abrasive cleaning is a preventative maintenance practice used to remove light and newly deposited catalyst. It is not effective in removing thick, sintered catalyst deposits that are tightly adhered to the expander flow-path components.

4.17 Thermal Cleaning

The most effective method of cleaning expander flow-path deposits is via thermal cleaning. Thermal cleaning can clean off deposits in many areas where the abrasive media does not reach areas such as both the upstream and downstream faces of the rotor disc. The thermal-cleaning procedure involves a cooling down of the expander at a controlled rate and holding at a reduced temperature for a given time. The cool down takes advantage of the low coefficient of expansion of the ceramic deposit and the high coefficient of expansion of the expander flow-path components. When a temperature change occurs, the deposit expands or contracts differently than the expander component and the deposit bond is sheared, allowing the deposit to spall off.

A significant change in operating temperature is generally required to accomplish an effective cleaning. For most expanders, a thermal cleaning can be accomplished by reducing the expander inlet temperature by 400°F (220°C). Larger expanders or those experiencing very tightly adhered deposits may need to reduce the temperature by as much as 600°F (330°C). It should be emphasized that these are temperature changes from the normal operating temperature and not an absolute value.

Once the target temperature drop has been achieved, the reduced temperature is held for a couple of hours to allow the expander components to stabilize and come to a uniform temperature. This is particularly important if the expander is experiencing rotor disc deposits as the rotor disc response to a change in temperature is quite slow.

In addition to the reduction in temperature, the rate of temperature change is also important. If the temperature is reduced quickly, the shroud and its deposits will close down the blade-tip clearances. Blade-tip rubbing can occur and result in rotor vibrations or even mechanical damage. In the extreme, a tripped expander rotor can actually seize and bind up from catalyst deposits being caught between the blade tips and the shroud. An expander disassembly is usually required to remove the stuck deposits. A safe cooldown rate has been found to be 200°F (111°C) per hour. Similarly, the same rate is usually used to heat the expander up after the thermal cleaning. In some expanders, the full effects of the thermal cleaning are not realized until the expander operating temperature is restored to the original level. In addition, the expander may experience an increase in rotor vibrations during the thermal cleaning process.

Cooling down the flue gas is generally accomplished by pulling feed from the FCC unit. As the coke make is reduced, the regenerator discharge temperature slowly decreases. In expander/generator applications or when there is enough auxiliary power to drive the blower in a conventional train, the cooldown may be accomplished by shutting off or reducing the flow to the expander. Experience has shown that it is not practical to cool down the full flue-gas flow. Flue-gas quenching is then accomplished by injecting cooling media such as air or steam. Water is generally not used as experience has shown that water quenching can cause thermal shocking of the system and can actually greatly increase the rate of catalyst deposition. Extra care needs to be taken to assure that the quench lines are fully blown down to remove any condensation.

Properly operating FCC units require minimal online cleaning. A monthly abrasive cleaning and a biannual thermal cleaning would be common. In more problematic installations with above normal catalyst carryover, weekly abrasive cleaning and quarterly thermal cleaning might be expected. In the extreme, daily abrasive cleaning and monthly thermal cleaning might be required.

4.18 Online Blade Photography

4.18.1 Photography Equipment

Online expander blade photography has proven to be a valuable tool in monitoring the condition of expander flow-path components. The

Figure 4.14 Typical operational blade photograph showing flow-path catalyst deposits.

two greatest concerns, blade erosion and flow-path deposition, can be seen visually and closely monitored. Blade photography has been helpful in determining expander turnaround intervals and workscopes, monitoring normal wear and/or deposition, monitoring the effects of process upsets, identifying spare part requirements, and monitoring expander performance. Blade photography is an integral part of most FCC expander preventative maintenance programs (Figure 4.14).

Viewport photography is a special high-speed photo technique that stops the action or motion of the blading. This is a difficult task when one considers the blading is traveling at a high rate of speed. The blade tips are typically traveling at speeds between 900 and 1,400 ft/s (275 and 425 m/s). Even using a high-speed camera with shutter speeds of 1/1,000 of a second, it is not possible to stop the action and get a clear photograph because the blade is actually moving 12 in (305 mm) in that time period.

A special high-speed flash unit is required to "stop the action." This is best accomplished with the use of a high-speed stroboscope. While the stroboscope can be used in a strobe light mode where the light flashes are synchronized with the rotational speed of the rotor, the best photographic quality is not achieved in this mode. Stroboscope sensitivity and minor variations in PRT speed makes it difficult to fully stop the blade motion. In addition, stroboscope flash intensity is lessened when used in a strobe-light mode and the photographs are often quite dark.

Higher quality photographs are achieved when the stroboscope is used in a single-flash mode. For high-quality photographs, the stroboscope must have a flash duration of less than 8 microseconds. It is this short flash duration that stops the movement of the blades. In addition to the short-duration flash, the flash intensity needs to be greater than

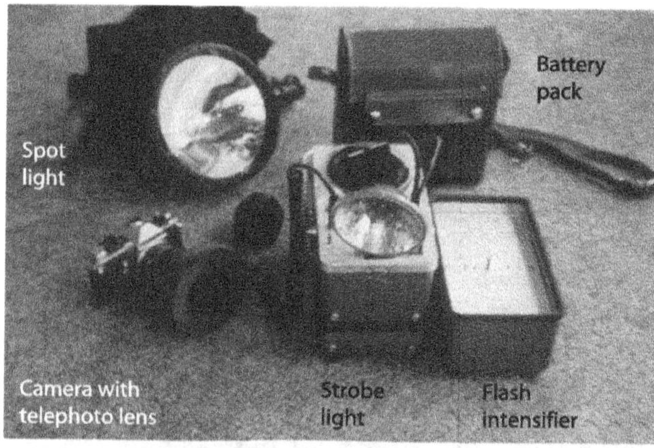

FIGURE 4.15 Blade photography equipment.

15,000,000 candelas. Some larger expanders require more light and the use of a flash intensifier that can increase the light intensity three times the base unit intensity. Figure 4.15 shows some of the common photography equipment used to take expander blade photographs.

The stroboscope is connected to a standard 35-mm camera via a special patch chord. The patch chord allows the camera to trigger the stroboscope with the camera shutter. Most users today use digital cameras for taking the photographs. The camera needs to have manual focus and exposure provisions, a synchronized flash, and hot shoe provisions. In addition, the use of interchangeable lenses is helpful. A fixed-length telephoto or zoom lens helps to fully fill up the field of view and to zoom in to specific regions of the blades. The camera lenses are usually fitted with a fiber backlight screen to prevent light reflections in the viewport glass. Alternatively, the use of a hood or nighttime photography can be used to prevent backlighting from reducing the quality of the photographs.

Handheld continuous spotlights are generally included in the photograph kit. The spotlight allows visualization of the rotating blades and most importantly assist in the focusing of the camera lens. A number of optional accessories have been used by some expander users that include a battery pack, key phasor module to tie the flash into a specific rotor position, and a time delay to be used with the key phasor to allow the viewing of selected blades. Many expander users have also included videography in their blade evaluation programs.

4.18.2 Viewports

The blade photographs are taken through two viewport assemblies. One viewport is used for the camera and one viewport is used

FIGURE 4.16 Expander viewports mounting.

for the light source (stroboscope). The viewports are located on the exhaust side of the expander at any location that is convenient for the photographers. Many users have found side-mounting viewports to be most convenient so the photographers do not have to climb up on the hot expander (Figure 4.16). The most important features are to make sure they are large enough to allow full viewing of the expander flow path, and they are properly aligned to the blades and to each other. The minimum viewport size is 3 in (75 mm), although 4 in (100 mm) is preferred and allows easier alignment and greater viewing area. The sight glasses must be a high-temperature ceramic glass capable of withstanding the expander exhaust temperature of approximately 1,000°F (535°C). The sight glasses are removable in service and generally need to be cleaned prior to each photography session. Extra care must be exercised in handling and installing the sight glasses because they are very prone to handling damage. The sight glasses are purged with dry air or nitrogen to reduce the amount of catalyst depositing on the sight glasses.

The isolation valves allow the removal of the sight glass assemblies while the expander is online. Full-port ball valves have been the most reliable valve design. Although gate valves are commonly used and take up less axial space, they do tend to stick open or closed from catalyst packing the seat or stuffing box regions.

Blade photographs are typically taken on a monthly basis for normal FCC operation. More frequent photographs may be necessary if the expander has experienced adverse operating conditions or upsets such as after burns, catalyst upsets, catalyst deposition, mechanical damage, or other problems. Many expander installations take blade photographs before and after major process operational changes or even expander online cleaning. A round of operating and supervisory instrument data should also be taken whenever a set of photographs is taken.

Each set of blade photographs should be evaluated by someone experienced in evaluating expander blade photographs. The evaluations usually include:

- Rotor blade coloration
- Rotor blade deposits—type, location and magnitude
- Rotor blade erosion—primary and secondary
- Rotor blade-tip clearance
- Rotor blade condition—signs of rubbing, wear or erosion
- Rotor disc condition
- Shroud geometry
- Shroud deposition
- Shroud erosion

One disadvantage of the blade photography is the inability to see the stator vanes or upstream side of the rotor blades.

4.19 Supervisory Instrumentation

Expanders are equipped with a full complement of instrumentation to air in the condition monitoring (Figure 4.17). Typical supervisory instrument includes:

- Radial vibration proximity probes (two per bear; four total)
- Housing accelerometers (two bi- or tri-directional)
- Axial position (two or three probes)
- Radial-bearing temperature (minimum of two per bearing)
- Thrust-bearing temperature (every other shoe; three minimum)
- Key phasor probe (one)
- Rotational speed (three)

FIGURE 4.17 Expander coupling end bearing with numerous supervisory instrument connections.

The proper selection of the instrumentation is very critical to assure long-term reliable operation of the instruments. The instruments can be subjected to ambient temperatures in excess of 300°F (148°C) during normal operation which can cause rapid deterioration and failure of the probes and connection cables. Casing flue-gas leakages can cause even higher local ambient temperatures.

In addition, the routing of the probe leads is important to minimize their exposure to high-temperature operation. Conduit purging and the routing of the probe leads inside the bearing housing have been effective methods for reducing instrumentation failures. Unfortunately, the high local temperatures and restrictive access often make online probe change out impossible. Alternatively, many expander operators have chosen to install multiple probes inside the bearing housings that can have their leads swapped out in a local junction or conduit box.

4.20 Expander Alarm and Shutdowns

The expander supervisory and condition monitoring instrumentation is used to both monitor the condition of the expander as well as shut down the power recovery train should certain limitations exceed safe operating levels. While alarm and shutdown philosophies vary from operating company to operating company and even within a company, some of the more common alarms and shutdowns include:

Instrumentation Alarms

- High inlet temperature
- High discharge pressure
- High radial vibration
- High axial position
- High radial-bearing temperature
- High thrust-bearing temperature
- Low flue gas-seal differential pressure
- Low cooling flow
- Low lube oil pressure
- Backup lube oil pump running
- Low oil level
- High reservoir temperature
- Low-pressure HPU emergency accumulator
- Low HPU pump pressure

Automatic Shutdown Parameters

- High-high inlet temperature
- High-high rotational speed
- High-high radial vibration
- High-high axial position
- Low-low lube oil pressure

Operator-Initiated Shutdown Parameters

- High-high discharge pressure
- High-high discharge temperature
- High-high radial-bearing temperature
- High-high thrust-bearing temperature
- Low-low lube oil level

Condition Monitoring Parameters

- Inlet temperature
- Inlet pressure
- Exhaust temperature
- Exhaust pressure

CHAPTER 5

Common Operational Problems and Failure Case Studies

5.1 Plugging of Critical Flow Orifice

Essential to the reliable operation of the Third Stage Separator (TSS) and the Fourth Stage Separator (FSS) is the maintenance of the critical flow orifice. The TSS underflow line is used to remove the separated catalyst from the TSS or FSS systems (if applicable). Typically the underflow rate is 1% to 3% of the regenerator overflow. Some Catalyst Cracking Units operate with underflow rates as high as 5%.

A reduction or loss of TSS underflow will greatly reduce the efficiency of the TSS. Both larger amounts of catalyst and higher amounts of larger catalyst particles will be carried over into the gas expander and cause increased expander flow-path wear and deposition. A complete plugging of the underflow line will fill the TSS with separated catalyst and cause all of the catalyst in the flue gas to be carried over into the expander. Expander blade life is reduced from more than five years with a properly operating TSS system to less than four months.

The most common causes for underflow line plugging are catalyst deposits plugging the CFN. In the unit shown in Figure 5.1, the underflow rate was reduced to less than 30% of the original design value. Other causes of underflow line plugging can be refractory failures in the Third Stage Separator or excessive catalyst carryover such as a catalyst "dump."

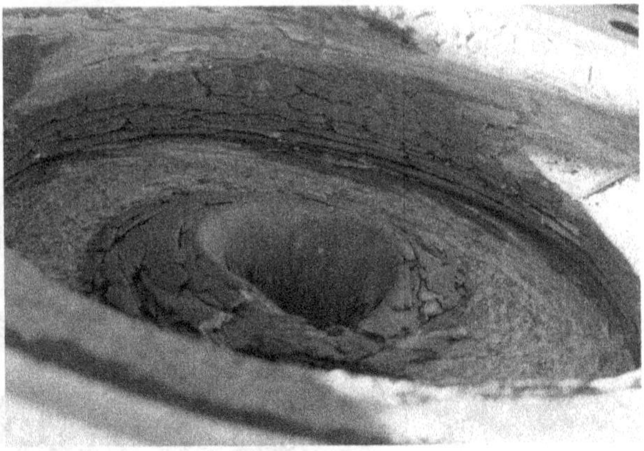

FIGURE 5.1 Catalyst deposit in the critical flow orifice.

Proper design is critical to reliable operation of the underflow system. The underflow system design should include isolation valves, an underflow inlet debris screen, long radius elbows in the piping, elbow rodding provisions, underflow line blow down and purging provisions, as well as continuous temperature monitoring.

5.2 Inlet and Piping Flange Failures

FCC expander piping systems are quite complex. As can be seen in Figure 5.2, the ducting is physically very large and the system design needs to accommodate a wide range of operating conditions including:

- Wide range of ambient pressure and temperature conditions
- Weather variations including rain and snow
- Wind loadings under extreme conditions such as major storms or hurricanes
- Earthquake loadings
- Large thermal growths
- Process normal operating pressures and temperatures
- Process upset conditions such as afterburns
- Variety of materials of construction including both hot and cold wall designs

Improperly designed, installed, or maintained piping systems have resulted in excessive piping and expander casing loadings. These loads have resulted in expander casing distortion, blade-tip rubbing, casing

FIGURE 5.2 Typical FCCU piping system.

cracks, duct cracks, control valve distortions, and body cracking, expansion joint failures, duct support failures, and other problems.

The two major causes of expander duct failures are excessive piping forces and moments and stress corrosion cracking. Figure 5.3 shows an expander inlet duct flange that has experienced major cracking. Failures of this type can cause unit shutdowns, pose significant safety issues, and are generally preventable.

Careful duct system design is critical in controlling the duct thermal expansions and loadings. The ducting must be free to expand in all directions. This is best accomplished by designing a free-floating ducting system. The duct is fully supported by overhead spring hangers attached to the centerlines of the duct. The duct growths are completely free to occur unrestricted, but are fully guided and controlled. Specially designed expansion joints are judiciously located to allow the larger thermal growths.

Stress corrosion cracking can be a unit shutdown problem. Sulfur-bearing acids such as polytheonic acid are formed and can attack the grain boundaries in stainless-steel ducting. Keeping the ducting and expander dry during a unit shutdown is critical in the prevention of stress corrosion cracking. See Chapter 3 for further discussions on stress corrosion cracking and polytheonic acid attack.

Chapter 4 discusses the piping design requirements in further detail.

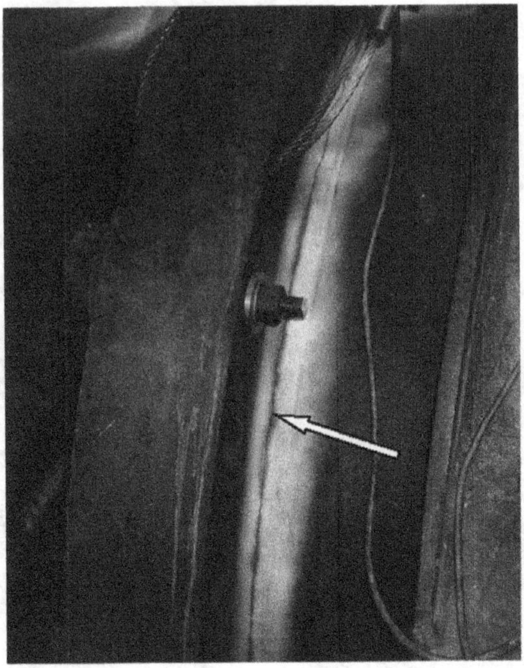

FIGURE 5.3 Expander inlet duct flange failure.

5.3 Nose Cone Failures

A number of expanders have experienced a variety of nose cone failures including:

- Inlet strut cracking
- Inlet strut creep, resulting in a rotor rub
- Nose cone bullet failures
- Baffle plate failures

Nose cones are subject to significant steady state and transient loadings. Under normal operation, the nose cone assembly is subjected to the full pressure drop across the stator vanes, which are typically 50% to 60% of the total expander pressure drop. In addition, the nose cone is subjected to the full inlet temperature of the flue gas. These high operating temperatures reduce the material capabilities significantly and creep damage and distortions will occur over time.

The most common nose cone problem is in the regions of the inlet struts. The struts transmit the bullet and stator vane pressure loadings and can be subjected to larger transient thermal stresses. Designing

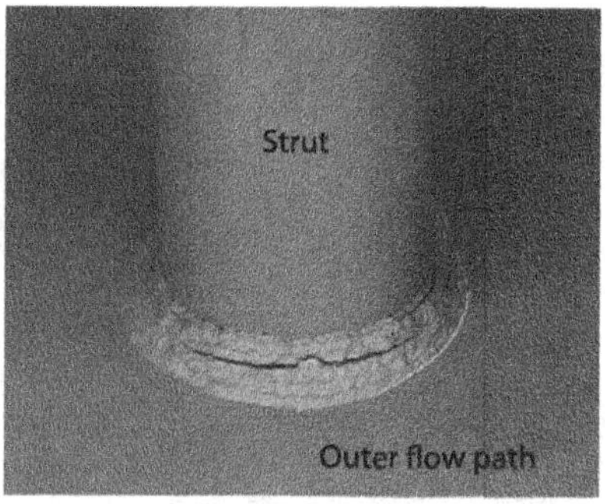

Figure 5.4 Nose cone inlet strut cracking.

of the inlet struts is a challenge as the struts need to be fairly thin and aerodynamically shaped to prevent flow pressure losses, to reduce rotor blade excitations, and to prevent nonuniform catalyst entry into the stator vanes. Figure 5.4 shows extensive nose cone inlet strut cracking. Inlet strut cracking has been significantly reduced over time by applying improved designs, upgraded weld designs and procedures, changes in heat treatment, and improved quality assurance practices.

Several expanders have experienced major rotor failures as the result of nose cone strut creep and distortion. Over time, the nose cone bullet has moved downstream toward the rotor.

Eventually the nose cone distortion has been sufficient to allow it or an attached part contact the rotor. This rubbing has caused an overheating of both the rubbed component and the rotor, often resulting in surface melting. These rubbed regions often end up developing cracking that propagates through the parts to a point of material separation.

Some of the more significant nose cone failures have resulted from high-thermal stresses often caused by thermal shocking of the nose cone from upstream quenches or internal cooling system failures. The nose cone bullet buckling noted in Figure 5.5 is the result of excessive temperature gradients caused by upstream liquid quenching of the flue gas. While the quenching per se did not cause the collapse, the quenching caused radial distortions/indentations in the nose cone bullet. Under normal long-term operation, the imperfection weakens the shell structure and the bullet collapsed under the normal pressure and temperature loads.

Thermal gradients are the major cause of high nose cone stress levels. In addition to the quenching issues previously noted, other potential causes of thermal gradients are misdirected cooling steam flows impinging on the inside of the nose cone bullet, the collection of condensate in the nose cone, expander casings and inlet ducting,

FIGURE 5.5 Collapsed nose cone bullet.

and poorly supported nose cone (internal and external) insulation. Many nose cone designs line the nose cone bullet or outer liner/flow path with insulation to reduce in-service thermal gradients. These insulations work well until they deteriorate, shift in position, or are sucked out of their location via the turbulent flows around the rotor discs. One of the most common failures that occurs in the expander is the shutdown issue, in which the nose cone internal insulation gets soaked with condensate and "slumps" down into the bottom of the casing. When the expander is returned to service, the region with the insulation remains at its normal operating temperature, but the noninsulated area develops large radial, axial, and circumferential temperature gradients.

5.4 Expander Rotor Disc Failure—Corrosion

The number-one cause of CCU expander blade and disc failures is high-temperature corrosion (Figure 5.6). More than 40% of all expander rotor failures are caused directly by or have corrosion as a major contributor to the failure. CCU blade and disc corrosion is a very complex phenomenon that was discussed in greater detail in Chapter 4. Factors that contribute to corrosion include the following:

- Flue-gas operating temperature
- Flue-gas chemistry
  - Oxygen level
  - Sulfur level

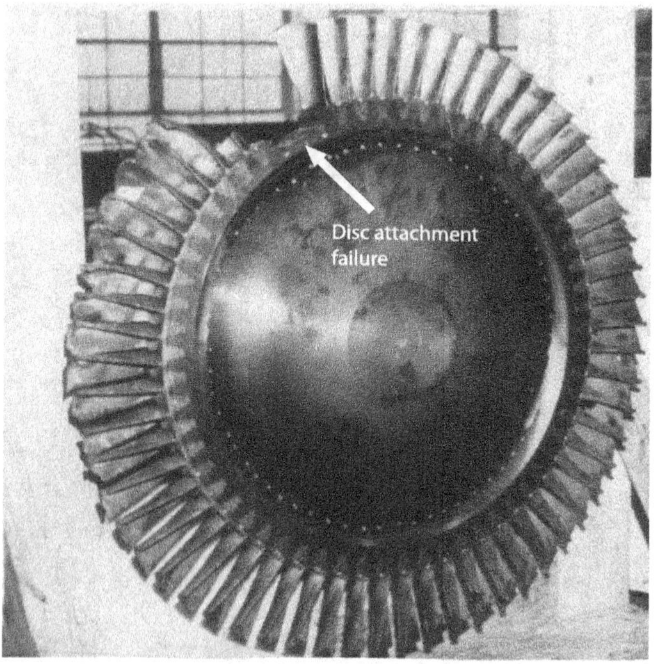

FIGURE 5.6 Expander rotor-blade failures.

- Base material chemistry
- Corrosion accelerators including:
    - Type of catalyst in use
    - Use of catalyst additives
    - Localized fretting
    - Steam purity

Figure 5.7 shows the rotor disc failure close up and the complexity of many expander failures. A significant number of attachments were found to be cracked. The cracks initiated from local corrosion damage in the fillet of the fir tree attachment is highlighted by the arrow. Cracking emanated out from the local corrosion and progressed in a creep fatigue mode. Eventually the attachment failed in a tensile overload when the remaining uncracked area could no longer sustain the remaining stresses. The attachment on the left in Figure 5.7 was the first attachment to completely fail. The adjacent blade (right side) was found to be cracked to a lesser extent, and a majority of the failed region is tensile overload from impaction of the first blade separation.

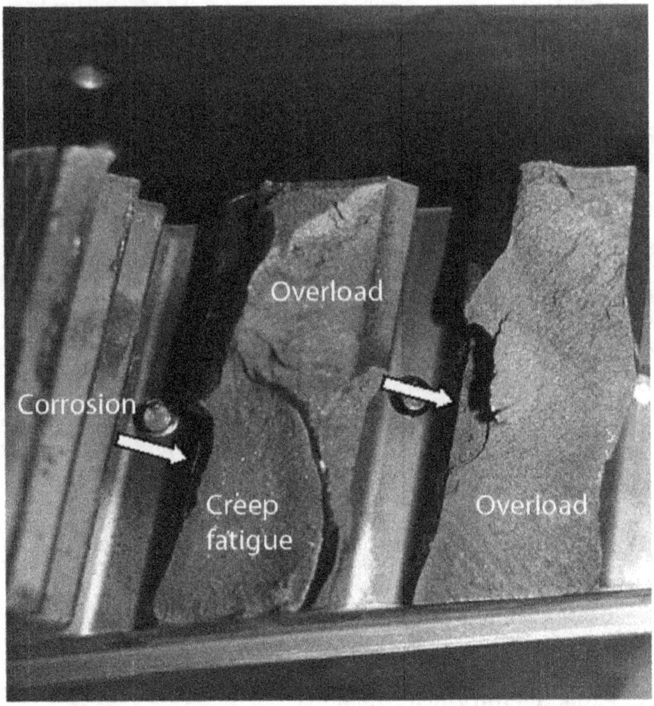

FIGURE 5.7 Failed disc attachments.

In the subject example, all of the rotor blades and disc attachment regions showed evidence of corrosion in the fillet region. The corrosion damage can be seen at a high magnification in Figure 5.8. The micrograph shows an attachment that didn't fail, which was sectioned and metallurgically examined. Classic hot corrosion was found and can be seen as the gray-colored scale on the surface and penetrating into the attachment. The scale penetrates into the base material approximately 0.008 in (0.20 mm).

The corrosion product was found to be mixture of chromium oxide and chromium sulfide. Emanating from the corrosion product are three small crack initiation points approximately 0.002 to 0.005 in (0.05 to 0.13 mm) in length.

The subject disc failure occurred after only 10 months of operation in a Waspaloy rotor disc. The cause of the corrosion was found to be a combination of low (<0.25%) oxygen levels, significant sulfur levels (1.4% typical), and relatively high metal operating temperatures of 1,100°F (593°C). Increasing the cooling steam flow amount was found to be effective in preventing a repeat failure.

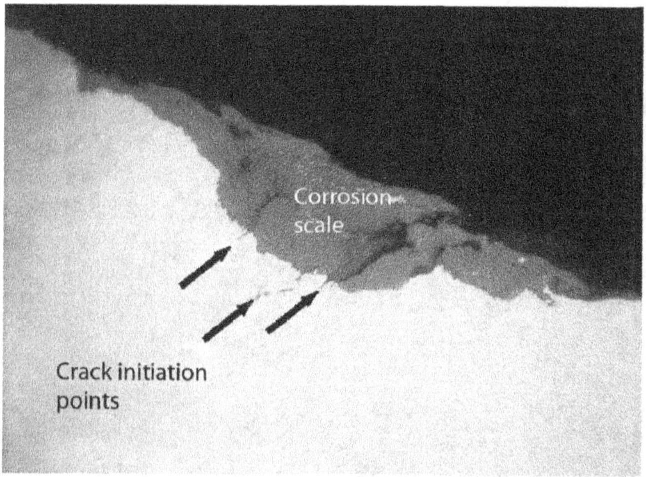

FIGURE 5.8 Micrograph showing fillet surface corrosion and crack initiations.

Case Study 5-1 Stator Vane Failures

CCU expander stator vanes are generally low stressed, capable of sustaining significant erosion damage, are generally not subject to the corrosive effects of the flue gas, experience only minor in-service aging, and are overall trouble-free. That being said, operational problems can develop from improper design and/or unit operation.

Vane Tuning

It is very important that the stator vanes (both freestanding and shrouded) be properly tuned to assure that the vanes are not operating in a resonant condition. The vane natural frequencies need to be determined analytically and verified via testing. The frequencies need to be checked against the rotor blade passing excitation to assure that the vane natural frequencies are not excited. Figure 5.9 shows fatigue cracking in a set of cantilevered stator vanes that operated in a resonant condition.

Operational Problems

Operational problems that have caused stator vane distress include the following:

- *Excessive catalyst carryover*. Above-normal catalyst loadings can lead to above-normal erosive wear similar to the vanes shown in

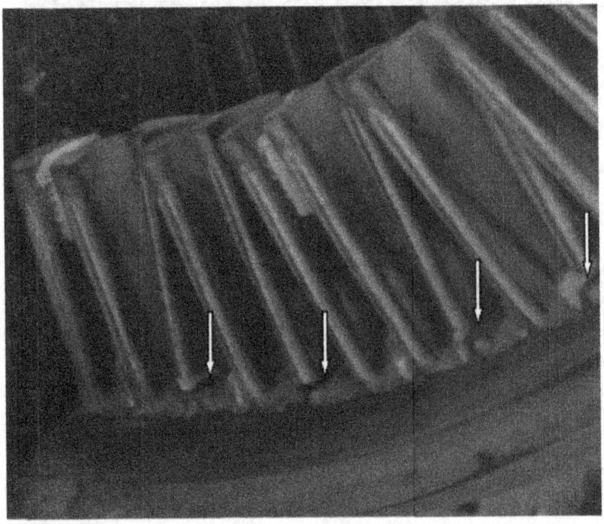

FIGURE 5.9 Cracked stator vanes.

Figure 5.10. Excessive vane wear can result in above-normal operating stresses and changes the vane natural frequencies.

- *Foreign object damage.* Pieces of stator vanes have broken out when foreign objects have been ingested into the expander. Extensive expander damage can occur if stator pieces or the foreign object lodges in the stator and strikes the rotor blading. Figure 5.10 shows just some of the very interesting items that have been found in expander inlets. Discharge screens on the exit of the

FIGURE 5.10 Scaffolding clips found in expander stator vanes.

Third Stage Separator have been effective in preventing foreign material carryover into the expander. In addition, a thorough inlet line inspection prior to closing up the ducting is critical.

- *Over temperature from CCU afterburning.* Excessive flue-gas operating temperatures can cause stator vane damage. The most common over-temperature consequence is thermal cracking of the vanes. Some expanders have experienced stator vane creep damage. Creep damage is the result of above-normal temperatures over a period of time. Some vanes have deflected downstream toward the rotor blading from creep distortion.
- *Thermal shocking.* The most common cause of cracking in stator vanes is from rapid temperature changes. Figure 5.11 shows a number of cracked stator vanes from rapid temperature changes. Flue-gas quenching is a common cause of stator vane thermal shocking. Quenching can result from process quenches at the discharge of the process vessels, expander inlet line quenches, or even the expander cooling system. Flue-gas quenching must be done carefully to assure that uncontrolled cooling does not occur or that liquids are not carried over. When quenching is required, steam quenching has been a safer practice. When liquid quenches are required, the use of high-pressure atomizing nozzles has helped to assure controlled cooling.

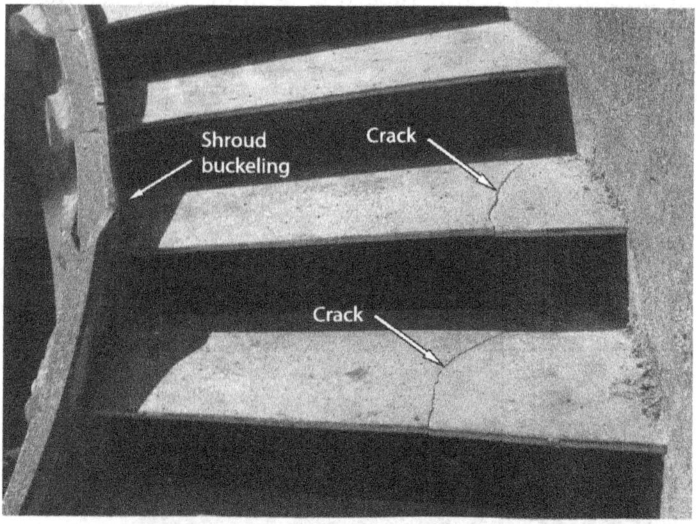

FIGURE 5.11 Cracked stator vanes.

CCU expanders that start up quickly (such as an expander generator set) have a high potential for rapid temperature changes. The power recovery train is usually not started up until the process is up to full operating conditions. If the full temperature flue gas were introduced to a cold expander, the expander inlet temperature would go from ambient to more than 1,100°F (593°C) in less than a minute. Extremely high-transient thermal gradients and stresses would result under these conditions. To prevent these excessive temperature gradients, the expander inlet temperature is controlled by preheating the expander with small amounts of flue gas and start-up steam. It is very important that the start-up and heat-up procedures be strictly adhered to. A comfortable heat-up and cool-down rate would be 200°F per hour (111°C). A fast heat-up or cool-down rate would be above 400°F per hour (222°C).

5.5 Rotor Rubbing

Rotor rubbing can be very damaging to both the rotor and the rubbed component. Virtually all of the materials in use in the expander are either high in nickel content or nickel based. Rubbing of these materials generates a large amount of heat that can severely damage the base materials. These materials also have a strong tendency to gall when rubbed together.

Figure 5.12 shows a heavily rubbed rotor in the blade attachment region. The rubbing was sufficient to cause base material melting, annealing, and thermally induced blade and disc cracking. A rotor-blade separation resulted from the rubbing shown in the above example.

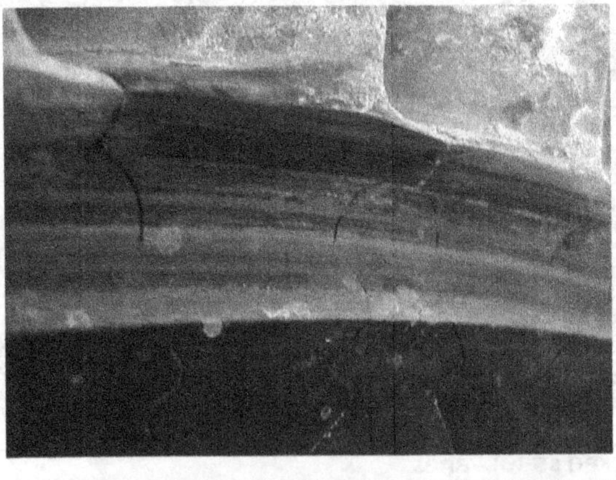

FIGURE 5.12 Severe rub on the face of a rotor disc resulting in blade failure.

Because catalyst deposition is the most common operational problem of the CCU expander, blade-tip rubbing is not uncommon. Although most expanders experience only minor tip rubs, occasionally the tip rubbing can be quite severe as shown in Figure 5.13. The catalyst buildup on the rotor blade airfoil and shroud is extensive with deposit thickness in excess of 0.5 in (13 mm). The shroud deposit has caused blade-tip rubbing and wear. It is common for the maximum wear to occur in the mid-chord region of the blade. Rubbing of this type should be avoided because the rubbing action can cause an overheating of the blade tips and result in both blade-tip wear and cracking. Figure 5.14 shows three different installations that have experienced blade-tip rubbing against a shroud deposit.

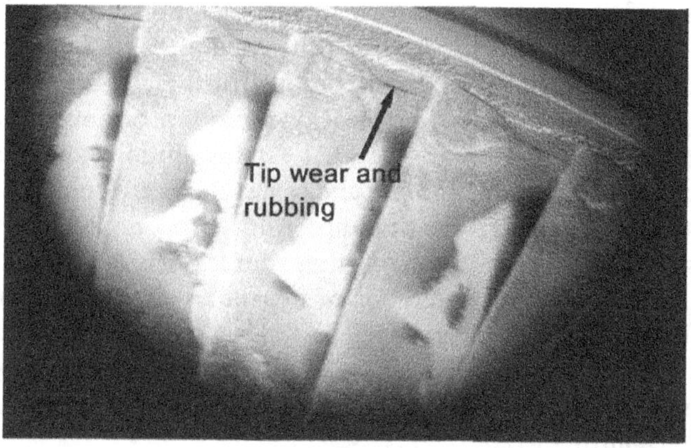

Figure 5.13 Expander viewport photograph of expander blade-tip rubbing.

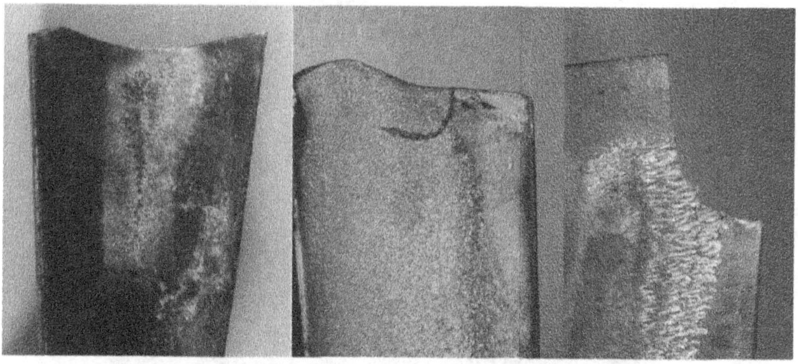

Figure 5.14 Three examples of blade-tip damage that can result from blade-tip rubbing.

The first blade shows just tip rubbing and wear. The second blade tip has cracked as a result of the tip rub. The rubbing action has resulted in the blade-tip overheating and a cracking of the tip. The crack often travels radially inward and turns axial toward the leading or trailing edge. The final photograph shows a blade that has lost a large piece of the trailing edge. Obviously the best prevention for damage of these types is to monitor the condition of the rotor blade via the view-port photographs and to clean the expander on a regular basis before blade-tip rubbing occurs.

5.6 Expander Casing Drains

One of the most neglected parts of the expander is the casing drains. All expanders have casings drains to allow the drainage of large amounts of condensate that forms during unit start-up and shut-downs. It must be remembered that the expander is usually the low point in the flue-gas system between the Third Stage Separator and ducting into the waste heat boiler. Condensation can occur at flue-gas temperatures as high as 340°F (171°C). In addition, condensation can form if the expander cooling and seal systems are commissioned prior to the expander heating up, from process steam purges, expansion joint, valve purges, or other sources. The quantity of condensation is high and a number of expanders and expander inlet ducts have collected several feet of water during a unit start-up. As the casings and ducting heat up, very large temperature gradients can form around the circumference and large thermal stresses develop (Figure 5.15).

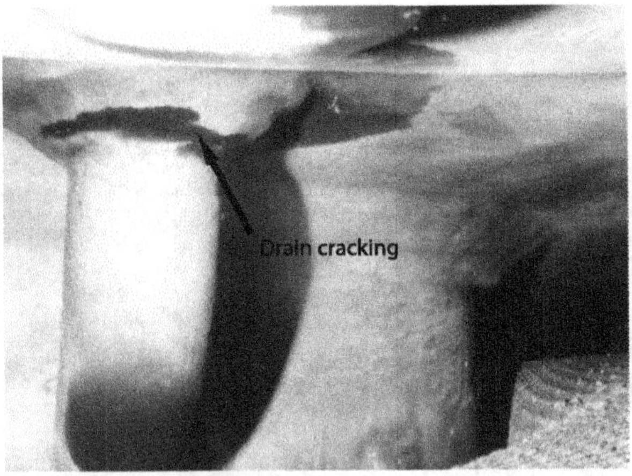

Figure 5.15 Expander inlet drain cracking.

Casing drains are problematic for a number of reasons including:

- Improper material selection—the drain piping and valve internals are often made from carbon steel that quickly corrodes from the acids that form in the lines.
- Poor location of drain valves makes it difficult for operators to access.
- The lack of understanding of the condensation potential.
- Plugging from the collection of catalyst.
- Corrosion of the piping and connections to the expander casings.

The most common failure of a drain line is stress corrosion cracking. Aggressive acids form from the presences of sulfur and water in the drain line. Online failures of a drain line can be dangerous because high-pressure and high-temperature toxic flue gas blows out of the failed line.

To prevent in-service failures of the drain lines, the following practices should be implemented:

- Drains should be open whenever the expander internal temperature falls below 340°F (171°C).
- Operators should be suspicious if no condensate is draining from the unit during start-up and shutdown conditions.
- Commissioning of condensation sources should be delayed until the expander internal temperature exceeds 212°C (100°C).
- The drain line terminations should have appropriated personnel protection to prevent exposure to the high-temperature toxic gases.
- All drain lines should have online "rodding" or "blow-out" provisions to allow plugs to be safely cleared. Flanged connections also allow partial disassembly should a blockage occur.
- All drains should be manufactured in a 300 series stainless steel.

5.7 Expander Online Cleaning

Online cleaning has become commonplace in virtually all CCU expander installations. Flow-path deposition is the most common operational problem with today's expanders. Two online cleaning procedures are commonly used: abrasive cleaning and thermal cleaning.

5.7.1 Abrasive Cleaning

Abrasive cleaning involves the injection of an abrasive into the flue-gas line just upstream of the expander. The abrasive removes the flow-path deposits by impacting the surfaces of the flow path. The abrasive cleaning is done frequently (from once per week to as frequent as once per shift). It is used to remove light catalyst deposits from the flow path and will not remove the heavier more tightly adhered deposits. Abrasive cleaning is a preventative maintenance procedure that is used to prevent significant accumulations of catalyst.

5.7.2 Thermal Cleaning

Heavier and more tightly adhered deposits can only be removed by thermally cleaning the expander. Thermal cleaning involves cooling the expander down a minimum of 400°F (222°C) and allowing the deposits to spall off the flow-path components. Many expanders required larger temperature drops to fully remove all deposits. Figure 5.16 shows the results of a typical online thermal cleaning. It can be seen that online cleaning can remove a significant amount of the deposits but cannot be guaranteed to remove them all.

Typical operating data through an expander thermal cleaning is depicted in Figure 5.17. In this case, the expander was operating at low vibration levels. Over a period of several days, the vibration began to climb from 1 mL (25 microns) to 5 mL (125 microns). Viewport inspection confirmed blade-tip rubbing.

Based on the operating data, a decision was made to thermally clean the expander. The expander was unloaded to allow a lowering of the inlet temperature. This is reflected in the rise in exhaust

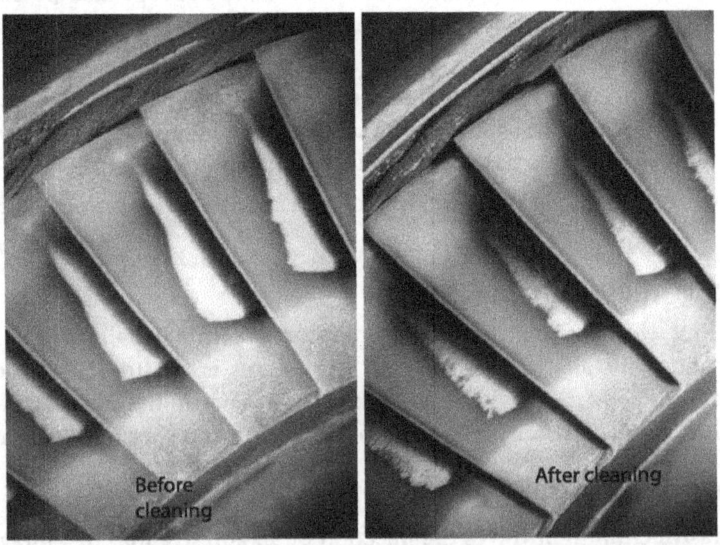

FIGURE 5.16 Expander blades before and after thermal cleaning.

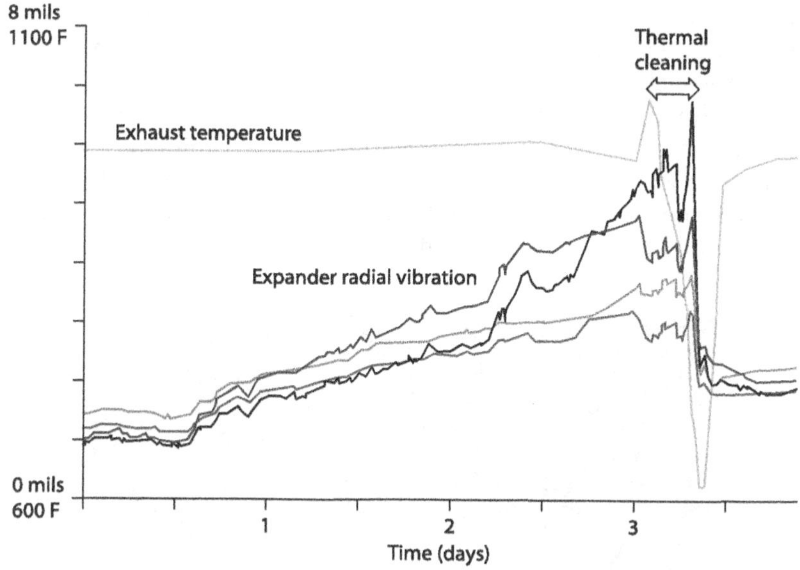

FIGURE 5.17 Expander exhaust temperature and vibration data.

temperature prior to cleaning. The expander inlet temperature was decreased approximately 600°F (316°C) over the course of a shift. Upon restart of the expander, the expander vibration levels decreased from the precleaning level of 5 mL (125 microns) to 2 mL (50 microns).

It is not unusual for the expander to experience some increase in vibration during the cleaning process. In this case, the vibration levels peaked around 6.5 mL (165 microns).

More details on the cleaning of the expander flow path can be found in Chapter 2.

5.8 Catalyst Deposition and Plugging

Expander flow-path catalyst deposition is a complex problem that has many causes. The presence of liquids is one of the greatest contributors to catalyst deposition. Liquids can form during unit start-up and shutdown conditions as well as during normal operation. The formation of liquids during the start-up and shutdown conditions is generally related to liquid condensation that occurs from water sources such as the flue gas or external sources such as valve stuffing box or expansion joint purges and the expander cooling and sealing systems. The most common online source is flue-gas quenches located on the discharge side of the regenerator. Heavy catalyst deposits similar to those shown in Figures 5.18 through 5.20 have also occurred when liquids have been injected upstream of the expander for thermal cleaning purposes.

Catalyst deposits of the type shown in this example have a significant effect on the expander performance. The additional surface roughness and flow-path blockage causes a dramatic reduction in both expander flow capacity and efficiency. Although most expander deposits do not reduce the expander performance by more than a few

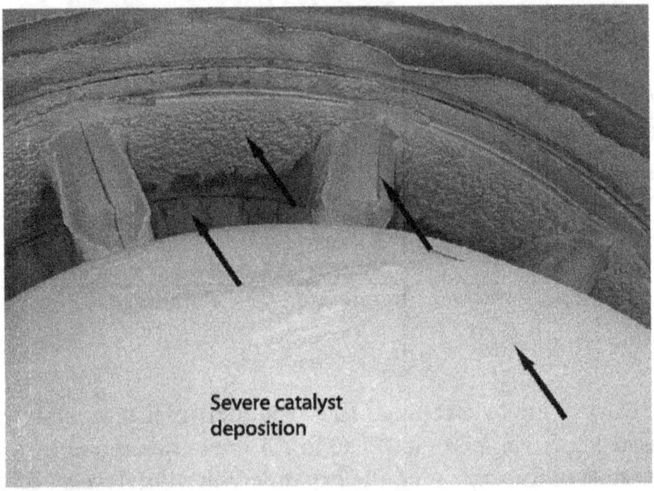

FIGURE 5.18 Expander nose cone and stator vanes—heavy catalyst coating.

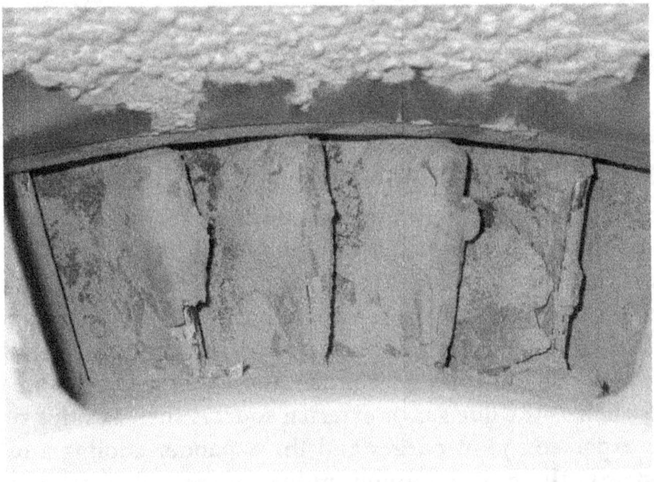

FIGURE 5.19 Severe catalyst deposits on stator vanes.

FIGURE 5.20 Severe catalyst deposits on rotor blades.

percent, excessive catalyst like those shown in this example can result in a reduction in power in excess of 10% of the normal power.

Wetted catalyst deposits have a different consistency and texture than the normal catalyst deposits that occur under normal operation. They generally have the appearance of dried mud. These deposits are removable via thermal cleaning or abrasively during a unit shutdown.

5.9 Exhaust Diffuser Failure

Whereas the expander exhaust diffuser is thought to be a low-stressed component subjected to small pressure drops and relatively low velocities, the diffuser design is critical to prevent in-service failures as shown in Figure 5.21. Several ft (1 m) of this exhaust diffuser is missing from an in-service failure. The failed diffuser fell down into

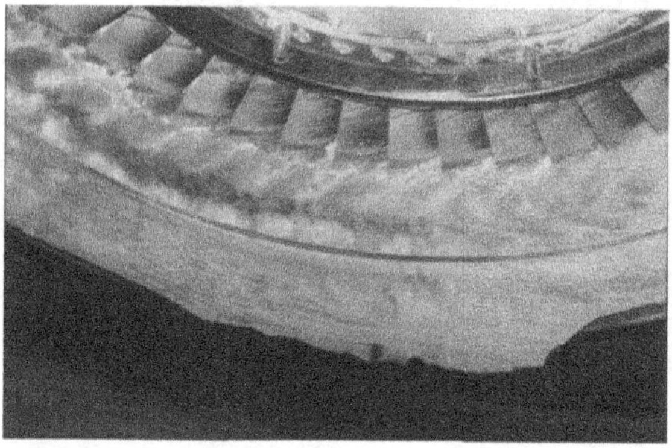

FIGURE 5.21 Failed exhaust diffuser.

the bottom of the expander exhaust casing. Failures of this type are almost always the result of turbulent flows through the diffuser that excite the diffuser natural frequencies.

Failures of the type shown in the figure are preventable by performing a detailed vibration analysis of the diffuser and adding sufficient thickness and stiffeners to increase the natural frequencies.

Another common problem with exhaust diffusers is the attachment of the diffuser to the exhaust casing. Diffusers are usually attached by bolting or welding them onto the exhaust casing. High-thermal stresses often occur in the attachment region owing to larger thermal mass differences between the attachment flange, the casing to which the diffuser is being attached, and the thin diffuser cone. Large thermal deflections generally occur in this area and result in both fastener and weld failures.

5.10 Expander Support: Unit Vibration

A number of expanders have experienced unusual vibration characteristics (Figure 5.22). In this example, the vibration characteristics developed after a major unit turnaround. The expander vibration amplitudes were acceptable on start-up. As the expander was loaded, the synchronous vibrations increased dramatically along with the sub-synchronous vibrations. The vibration characteristics were unsteady and not repeatable.

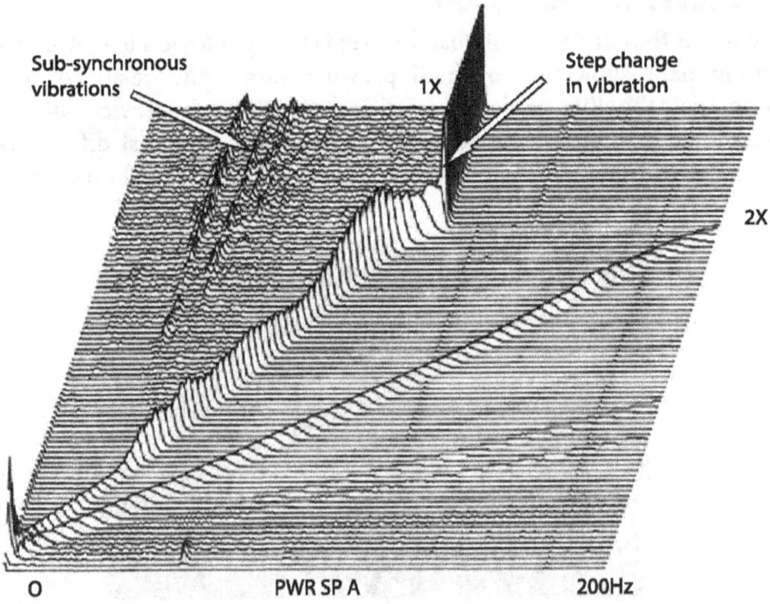

Figure 5.22 Expander disc end vibration characteristics.

Extensive operational reviews showed the expander vibration responding to a number of operating parameters including the expander inlet temperature, power, and even the site-ambient conditions.

Detailed reviews of the unit overhaul showed the expander to be properly assembled with all assembly requirements (rotor blade tip, bearing and seal clearances, etc.) being within the manufacturer's specifications. A large number of analytical analyses were conduct to determine the cause of the expander vibrations. The analytical work included the following:

- Vibration analysis
  - Rotor operating characteristics
  - Casing and support vibration analysis
- Operational analysis
- Rotor dynamic analysis
  - Critical speed
  - Forced response analysis
  - Stability
  - Bearing design
  - Unbalance sensitivity
  - Support stiffness influences
- Piping system design review and field inspection
- Modal analysis

During a foundation inspection and evaluation, numerous grouting and foundation cracks were uncovered.

The modal analysis work revealed large structural vibrations that correlated with the expander radial vibrations (Figure 5.23). Eventually, after many operational problems, the unit was scheduled to be shut down and the issues investigated further. The expander baseplate was removed and a number of installation problems were found including:

- Poor baseplate grouting
- Cracks in the grouting and foundation
- Soft foot conditions
- Corrosion of alignment shims
- Loose hold-down bolting
- Improperly installed anchor bolts—anchor bolts were fully grouted into the top of foundation

Figure 5.23 Grout cracking.

Figure 5.24 shows the top of the expander foundation after removal of the expander baseplate. The grout was found to be oil soaked and literally coming off in layers. Obviously the expander was being supported on a very soft support system that caused the unusual vibration characteristics.

Foundation repairs were made along with the clearing of the anchor bolt sleeves of grout. The anchor bolts were requalified at

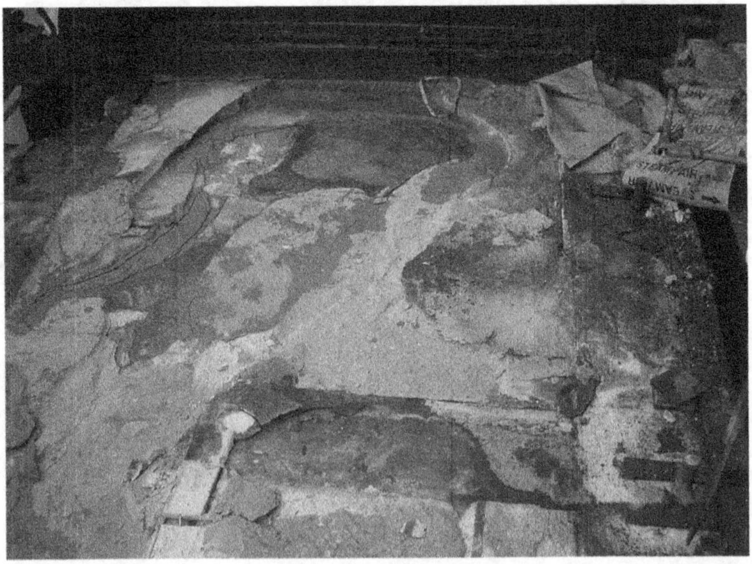

Figure 5.24 Grouting issues under expander baseplate.

FIGURE 5.25 Reinstalled baseplate with new grout.

the proof load. All of the old grout was removed and the top of the foundation was chipped down to good material. The baseplate was reprepared for installation, leveled with stainless-steel shims, and reinstalled. Finally the baseplate was grouted into position as can be seen in Figure 5.25.

After the foundation repairs, the expander was successfully restarted and operated trouble free. The expander vibrations were steady and at the lowest levels ever recorded.

5.11 Air-Blower Failures

An operational problem that has been experienced in a large number of operating Hot Gas Expander trains involves the CCU air blower. Operation of the blower in undesirable regions of the performance map has repeatedly caused major blower failures.

Early CCUs used centrifugal compressors for the main air blower. Over time, as CCUs increased in size and with the desire for increased efficiency, axial compressors became the compressor of choice. Virtually all new main air blowers today use axial compressors as the main air blower. This design is most often used because it is the best selection for a high-flow volume, high efficiency, and modest discharge head application. These compressors are capable of very high air-flow rates. To permit optimum performance and maximum range, the air blowers are operated in a variable speed mode, variable inlet guide vanes and stators, or a combination of both variable speed and variable geometry. In some applications, where there is a need for higher pressure air, for example, for transport flow, a last-stage centrifugal compressor impeller may be used in the compressor as can be seen in Figure 5.26.

The potential for problems is, in part, a result of the basic design which uses freestanding, unshrouded rotor blades and, in many cases, unshrouded stator vanes. Freestanding blading has much less inherent damping than shrouded blading and has a greater potential for failure from resonance or aero excitation. The first few rows of the inlet of the compressor are more likely to experience a failure due to the inlet stages being longer with the associated higher stresses. In addition, the inlet stage blades and vanes are more susceptible to corrosion damage. Unfortunately, a single blade failure will often fail or damage the remaining blades throughout the rotor as can be seen in Figure 5.27.

FIGURE 5.26 CCU main air blower.

FIGURE 5.27 Failed compressor rotor—looking from the discharge toward the inlet.

Compressor problems can occur under a wide variety of operating conditions. The compressor start-up is often a problematic period because the compressor often operates in or near the various operational limits. Figure 5.28 shows a typical performance map for a typical axial compressor and defines the compressor limitations. These limitations can be due to little flow (rotating stall), too much flow (stonewall), too high a discharge pressure (surge), and too low of a discharge pressure (choke). The compressor operating point is limited by controlling the compressor discharge pressure, operating speeds, or variable geometry positioning.

Operations under the conditions of surge and choke are the most common cause of axial compressor blade failures. Surge can be a very violent and audible condition in which there can be complete flow reversal within the compressor. Inasmuch as surge quickly reveals itself, unit operators tend to more quickly respond to the problem. In addition, modern-day control systems are quite effective in preventing the condition of surge. Because surge is caused by excessive compressor discharge pressure, the solution is to reduce the back pressure on the compressor.

In contrast, choke is difficult to detect, and unit operators are not likely to know it exists. For this reason, choke flutter is called the "silent killer" of axial-flow compressors. Choke flutter results when the compressor discharge is too low. This produces high internal air velocities within the compressor, and aerodynamic stall can occur. Aerodynamic

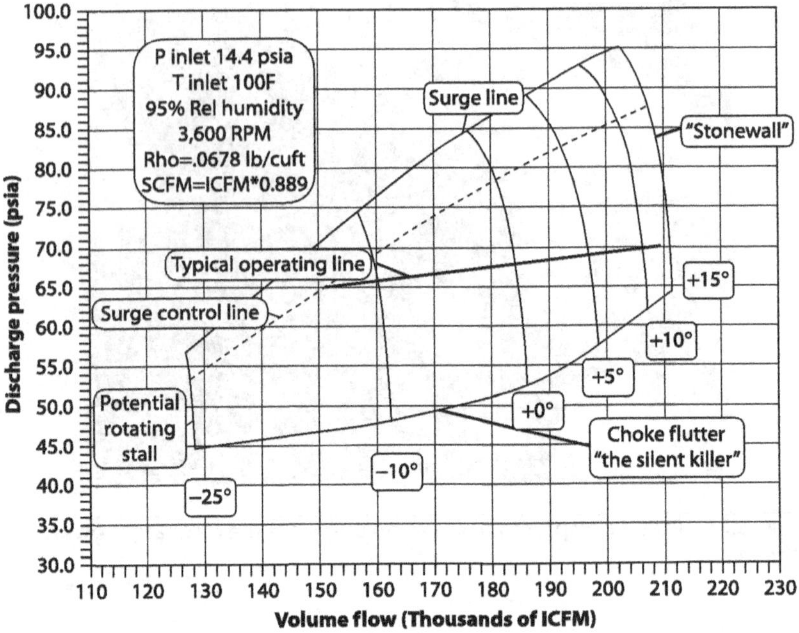

Figure 5.28 Typical compressor operating limits.

stall can cause rotor blade excitations and excessive stress that result in blade failures. The solution to prevent choke flutter is to make sure that the compressor operates with an appropriate discharge pressure. This is particularly true during unit start-up and shutdown conditions. Unfortunately, there is no antichoke control system to prevent choke operation, and it requires operator interaction.

Another common cause of air-blower blade failures involves the variable inlet guide vanes and stators. Viable inlet guide vanes and stators must be precisely set to assure proper operation and compressor performance. The inlet guide vanes and stators use a variety of drive rings, linkages, gimbals, and bearings to control the position. If the inlet guide vanes and stators are not properly set up or wear, loosen in-service, or get out of synchronization, the flow in the compressor will not be uniform and flow excitation of the vanes, stators, and rotor blades can occur. If these disturbances have a frequency content that coincides with a resonance of the vanes or blading, a fatigue failure can occur.

This example has been provided to emphasize that attention must be paid to all the turbomachinery components within the power recovery train.

CHAPTER 6

Nitric Acid Expanders for Chemical Applications

6.1 Introduction

Nitric acid (HNO_3) is an important component in the manufacture of nitrogen fertilizers, synthetic fibers, and explosives. Hundreds of nitric acid plants can be found throughout the world. Early (pre-1950s) nitric acid plants were fairly small and produced generally less than 110 tons (100 tonnes) per day. Today, there are plants that produce as many as 2,200 tons (2,000 tonnes) nitric acid per day. Because the process reactions occur under pressure, a turbomachinery compression train is an integral part of the nitric acid plant. The compression train typically includes an air compressor, often a nitrous gas compressor, a gas expander, and a steam turbine. The power recovery expander has been used in the nitric acid plant to improve the overall process efficiency for more than 60 years. The production of nitric acid is an exothermic process and a significant amount of heat energy can and needs to be removed. This is best accomplished via a series of heat exchangers and a Hot Gas Expander.

Power recovered by the expander is generally used to drive the main process compressor(s) and/or an electric generator. The typical nitric acid process train has the expander driving two compressor bodies with four stages of compression, and an auxiliary steam turbine or gear motor. The first body compressor is an air compressor with interstage cooling. The second body may be either a second air compressor for high-pressure applications or a tail gas compressor for many of the lower-pressure applications. The steam turbine or motor is used for unit start-up and in a makeup mode during normal operation. The power levels of the expander range from 2,000 to over 40,000 horsepower (1,500 to 30,000 kW). Over the last 50 years, more than 1,200 pieces of turbomachinery have been installed in nitric acid plants around the world.

6.2 Nitric Acid Process

The principal product of the nitric acid process is ammonium nitrate. About 75% of the nitric acid produced worldwide is consumed as an intermediate in the manufacture of fertilizers. Understandably, a majority of the nitric acid plants are located in agricultural regions because of the high demand for fertilizer in these areas (see Figure 6.1). Another 5% to 10% of the nitric acid produced is used for organic oxidation in adipic acid manufacturing. Other end uses for nitric acid are gold and silver separation, military munitions, steel and brass pickling, photoengraving, and acidulation of phosphate rock.

6.2.1 Simplified Nitric Acid Plant

Nitric acid is produced by two methods. The first method uses oxidation, condensation, and absorption to produce a weak nitric

FIGURE 6.1 Typical nitric acid plant.

acid. Weak nitric acid has concentrations ranging from 30% to 70% nitric acid. The second method combines dehydrating, bleaching, condensing, and absorption to produce a high-strength nitric acid from a weak nitric acid. High-strength nitric acid contains more than 90% nitric acid. The air is mixed with ammonia and burned over platinum catalyst to form H_2O and NO. An exothermic reaction occurs between ammonia and oxygen, and nitric oxide and water vapor are produced. The combustion products are cooled in a cooling tower. During the cooling process, NO in the presence of O_2 forms NO_2, which reacts with condensed H_2O to form a weak HNO_3 (Figure 6.2). This weak product is absorbed in a bubble cap water

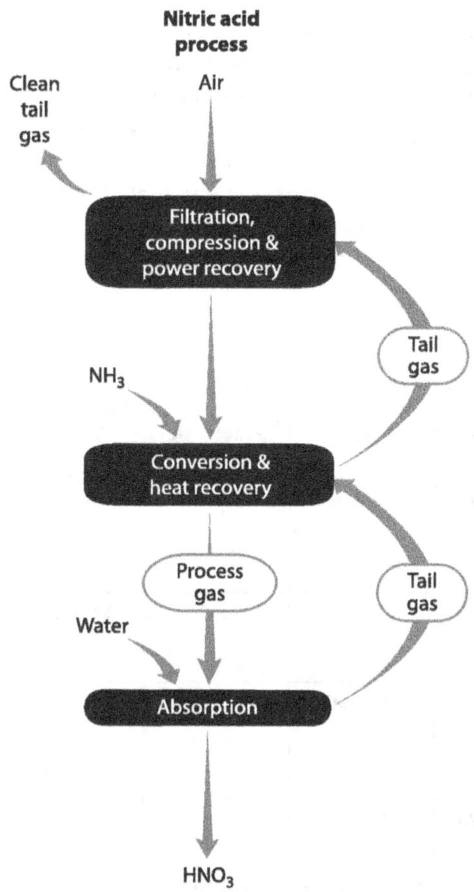

Figure 6.2 Typical nitric acid process flow diagram.

tower to form 60% HNO_3. The nitric acid is separated from the remaining process gas in the absorption system. In the final stage, nitric oxide, nitrogen dioxide, oxygen, and water are combined in an absorber column, forming nitric acid of the desired strength. A portion of the reaction energy is recovered in a heat exchanger train and is used to reheat the tail gas to provide power for the air compressor by driving the Hot Gas Expander. The tail gas (mostly N_2) leaves the top of the tower and is heated by burning H_2 or methane to achieve a tail gas temperature of 1,000°F to 1,200°F (538°C to 649°C), which enters the gas expander.

The efficiency of the air and ammonia mixing in the first production step is accomplished best at low pressures, whereas those of the second/third production steps are best accomplished at higher pressures. These considerations give rise to two basic types of nitric acid plants known as single-pressure plants and dual-pressure plants. Each of these two types of plants can be operated at either high or low pressures. This results in four basic process types used for the production of nitric acid (see Table 6.1). They are distinguished by their pressure levels for combustion and absorption.

In the single-pressure plant, such as that shown schematically in Figure 6.3, the oxidation and absorption steps take place at essentially the same pressure. The ammonia combustion and absorption occurs

Process Type	Combustion Pressure Psig	Absorption Pressure Psig	Schematic Figure	Combustion Pressure Kg/cm^2—g	Absorption Pressure Kg/cm^2—g
Low/medium single pressure	40 to 90	40 to 90	6–4	2.8 to 6.3	2.8 to 6.3
High single pressure	115 to 190	115 to 190	6–5	8.1 to 13.4	8.1 to 13.4
Low/medium two pressure	Atmospheric	55 to 90	6–5	Atmospheric	3.9 to 6.3
Medium/high two pressure	100 to 220	100 to 220	6–6	7.0 to 15.5	7.0 to 15.5

Table 6.1 Typical Nitric Acid Plant Operating Pressures

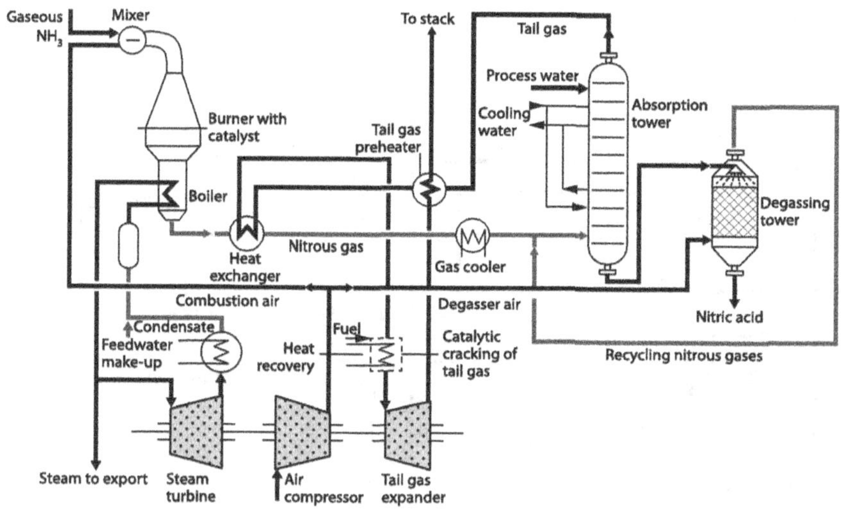

Figure 6.3 Flow schematic: Typical single-pressure process with intercooling.

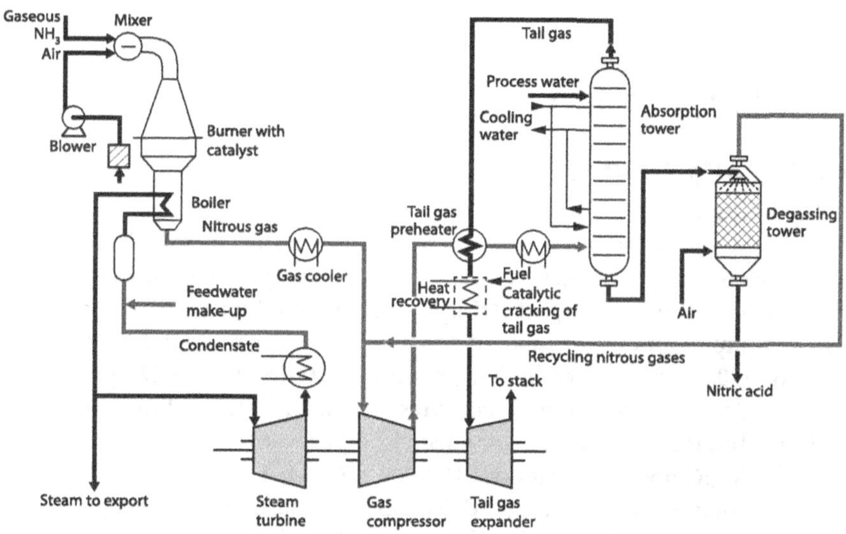

Figure 6.4 Flow schematic: Typical two-pressure process with low-pressure combustion.

at the air-compressor discharge pressure. In the dual-pressure plants, absorption takes place at a higher pressure than the oxidation stage. As can be seen in Figures 6.4 and 6.5, the dual-pressure plants use both an air compressor and a nitrous gas compressor.

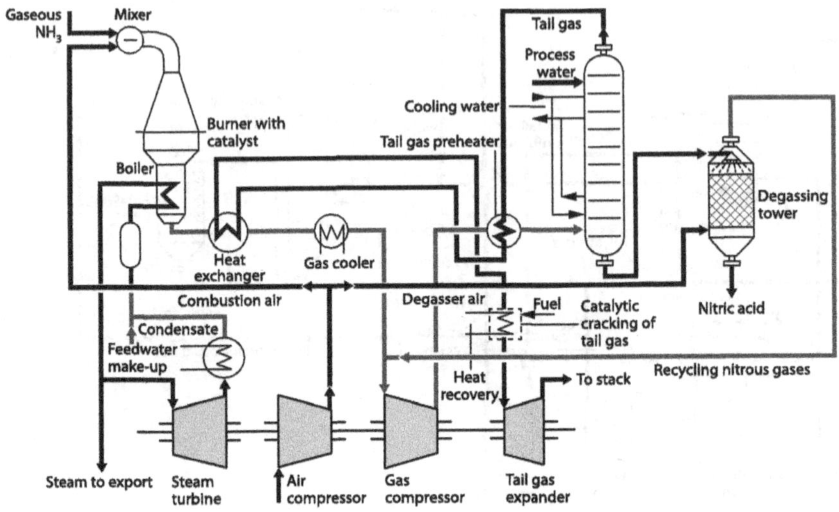

FIGURE 6.5 Medium/high two-pressure process cycle.

6.3 Machinery Selection

6.3.1 Compressors

Air Compressor

Obviously, the compression requirements are related to the size of the process and its oxygen needs. For the small plants with lower flow rates, a single-casing centrifugal compressor typically is used. These compressors are capable of producing process pressures up to 85 psig (6.0 kg/cm^{2}g). For higher pressure operation (up to 225 psig [15.8 kg/cm^{2}g]), centrifugal compressors with intercooling and multiple bodies are used as can be seen in Figure 6.6.

For high-flow applications 100,000 SCFM (170,000 m^3/hr), axial compressors have been used. The axial compressor offers high-volume flow capability, high efficiency, and wide operating range capabilities. At a given pressure ratio, the axial flow compressor polytropic efficiency can be between 3% and 7% higher than that of a centrifugal compressor. Variable stator geometry is applied to fixed-speed operating compressor trains. A combination axial/centrifugal compressor may also be used for the higher pressure applications.

In recent years, integrally geared compressors have been applied to small plants using the medium and/or high-pressure processes with pressures above 85 psig (6.0 kg/cm^{2}g). These compressors can be direct drive, include interstage cooling, and integrate the tail gas expander into a single compressor assembly.

Nitrous Gas Compressor

A nitrous gas compressor is used in the two pressure level processes. Multistage centrifugal compressors are used to compress the nitrous gas. The NO_x compressor has very specific design requirements because the nitrous gas forms ammonium nitrite, nitrate, and HNO_3. The gases are poisonous, highly corrosive, and allow chemical reactions to take place during the compression. Specific design features included in the nitrous gas compressor include a construction with no dead spaces that might collect and encourage the buildup of ammonium nitrate salt and online washing provisions to remove ammonium nitrate deposits within the compressor. All parts in contact with the gas are made of acid-resisting chromium nickel steels to counteract the corrosive effects of the nitrous gases.

Very important to the design of the NO_x compressor is accounting for the chemical reactions that occur in the compressor. N_2O_4 dissociates to NO_2 and the oxidation of NO converts to NO_2. Heat is released in the oxidation of NO and considerable amount of heat is needed for the dissociation of the N_2O_4. These heat quantities and temperature changes are determined for each individual stage of the compressor.

The toxicity of the nitrous gas requires special sealing provisions and designs to prevent leakage from the compressor.

Figure 6.6 Typical nitric acid PRT including: Expander, centrifugal compressors, gear, and motor.

6.3.2 Auxiliary Drivers

An auxiliary driver is required in each train for starting the train and for providing any power deficiency between the compressors and gas expander. The auxiliary driver selection and sizing is critical to successful start-up and operation of the compressor train. Because the expander does not produce any significant power until the process is light off and running, the auxiliary driver must be used to start up both the train and the process.

During the unit start-up, the auxiliary driver must produce a large amount of power to get the compressors up to the minimum operating speed required to produce sufficient air flow for process light off. In addition, the expander can actually be consuming power during the initial phases of the start-up. Process start-up generally requires the compressors to be operating at least 70% of the normal operating speed, which typically requires 55% to 75% of the compressor(s) rated power. When hot gases begin to be produced, the expander begins to input power to the compressor train and the train is then "boot-strapped" up to the full operating speed and flow conditions.

During normal operation, the expander typically produces 75% to 90% of the compressor full-power requirements. Under normal operation, the power required from the auxiliary drivers is small in comparison to the start-up requirements. Steam turbines and motors are the most common auxiliary drivers. Steam turbines are commonly used auxiliary drivers because the nitric acid process produces a large amount of heat energy that can be conveniently converted to steam. Electric motors are often used in smaller nitric acid plants or where it is advantageous to export surplus steam outside of the plant.

6.4 Expander Selection and Design Considerations

6.4.1 Expander Selection

The type of expander applied to an individual acid plant is highly dependent on the process conditions and the resulting inlet pressure and temperature produced by the process. Depending on the inlet pressure and temperature, the expander can provide a majority of the compressor power requirements. Nitric acid plant expanders are either a single- or multistage axial turbine. Most nitric acid expanders are typically constructed with three to six stages. The number of stages is selected based on the process conditions and in order to achieve a high expander overall efficiency.

Operating Pressures

The expander inlet pressure is determined by the compressor outlet pressure minus the pressure losses in the process. The process losses can range between 4.4 and 29 psig (0.31 and 2.0 kg/cm^{2}g). The typical expander inlet pressure for a nitric acid expander application is between 75 and 250 psig (5.3 and 17.6 kg/cm^{2}g). Unless there is the use of an exhaust heat recovery unit, the expander exhaust pressure is determined only by the downstream duct and stack losses. The exhaust pressure is low and generally ranges between 0.75 and 2.0 psig (0.05 and 0.14 kg/cm^{2}g). This results in expander overall pressure ratios (inlet/discharge) between 6 and 17.

Operating Temperatures

The expander inlet temperature varies significantly in the different plant designs. If the tail gases are heated in counter flow with the hot gases from the compressor discharge, the expander inlet temperature is typically less than 500°F. At these temperatures, the expander is considered a "cool" expander and the construction is similar to that used in industrial steam turbines. Rotor or casing cooling is not required.

In other acid plants, the tail gas is heated by the hot combustion gases discharging from the burner outlet. In these cases, the expander inlet temperature can be as high as 950°F (510°C). High-temperature design upgrades are required for successful operation of these "warm" expanders.

In more modern nitric acid plants, the tail gas is heated with supplemental combustion after the gas passes through the absorption column. This type of operation greatly reduces the NO_x emissions from the plant. The supplemental firing increases the expander inlet temperature into the 1,200 to 1,350°F (649 to 732°C) range. A "hot" expander design and construction is required for these higher temperature applications that used some of the design practices of internally cooled gas turbines. The increased available energy to the expander also increases the expander power output to a point where almost 100% of the compressor power requirements are met by the expander.

Operating Speed

The expander rotational speeds would be considered to be high in the acid plant applications owing to the high-speed requirements of the air or NO_x compressors. Higher speed compressors tend to be smaller in size and have higher flow and pressure rise capabilities.

Smaller expander frame sizes (<6,000 hp) tend to operate at rotational speeds in the 10,000 to 20,000 rpm range. Middle-size expanders (10,000 to 30,000 hp) tend to operate at rotational speeds in the 5,000 to 10,000 rpm range, whereas the highest power expanders >30,000 hp operate in the 3,000 to 4,000 rpm range.

6.4.2 Expander Performance

Nitric acid expanders are designed for a wide variety of operating conditions. To accommodate the many process design requirements, many expander designs have been developed and include single- and multiple-stage designs. The blading can be either of an impulse or reaction design. Multistage shrouded reaction designs are commonly used to maximize the efficiency of the expander.

Flow-Path Design

Because energy costs are a significant part of the overall production costs of nitric acid, a high-efficiency expander is very advantageous to the plant profitability. An expander with a low efficiency may produce only 65% of the power recovery requirements of the compressors, whereas a high-efficiency expander may produce more than 90% of the compressor power requirement.

High efficiency is generally achieved with conservatively designed, lightly loaded constant work reaction stages. The stage energy drops

Figure 6.8 Typical five-stage expander flow path.

in a typically multistage expander is typically very conservative with individual stage energy drops in the 30 to 40 Btu/lbm (32 to 42 to k-joules) range. The resulting velocity ratio (U/C_0) is approximately 0.61, which is at the peak of the efficiency curve for typically used blades.

The expander flow-path design often takes advantage of the high efficiency of a 50% reaction design. Symmetrical reaction blading with equal pressure drop and turning angle in rotor and stator is common. With a 50% reaction design, the stage velocity diagrams are symmetrical with the stator discharge velocity essentially equal to the rotor relative discharge velocity. In addition, the rotor relative inlet axial velocity is essentially equal to the absolute axial velocity exiting the blade. Typical stage velocities are as follows:

Stator absolute discharge	1,050 ft/s (320 m/s)
Blade mean speed	800 ft/s (244 m/s)
Rotor inlet relative	535 ft/s (163 m/s)
Axial (stator and rotor)	535 ft/s (163 m/s)
Rotor relative discharge	1,025 ft/s (312 m/s)

The airfoils typically are of the tapered and twisted design. The twisting from root to tip is required due to the radial peripheral speed changes, the static pressure variation due to the axial velocity change, and the resulting change in gas exit angle. The blade shape and design are used to develop equal work along the length of the blade. The airfoil tapering is used to reduce the operating stresses in the airfoil:

Stage reaction is defined as: $SR = \dfrac{\Delta H_s}{(\Delta H_s + \Delta H_r)}$

SR = stage reaction %
ΔH_s = Enthalpy drop of the stator—Btu/lbm
ΔH_r = Enthalpy drop of the rotor—Btu/lbm

The stage work is often uniform or split evenly across the available stages. A slight biasing of the stage loading toward the discharge is also common as the longer later stage blades have lower profile and leakage losses and therefore produce a higher efficiency.

Performance Estimating

The nitric acid expander performance can be estimated in a similar manner as that described in Chapter 2 for the FCC expander. The basic performance equations are identical even though the stage

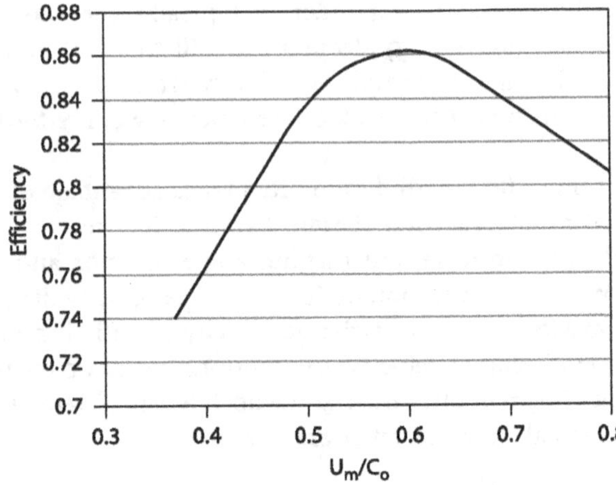

Figure 6.7 Typical stage efficiency for a nitric acid expander design

efficiency is different because of the different blade designs used in the nitric acid expander (Figure 6.7).

Blade Mean Line Velocity Calculation

$$Um = \frac{DN}{229}$$

Um—Blade mean line velocity (ft/s)
D—Mean diameter of the blade (inches [in])
N—Rotational speed (rpm)

Theoretical Nozzle Discharge Velocity Calculation

$$Co = 223.76\sqrt{\Delta H_{is}}$$

Co—Nozzle theoretical sonic velocity (ft/s)
ΔH_{is}—Isentropic enthalpy drop (Btu/lbm)

Isentropic Enthalpy Drop

$$\Delta H_{is} = T_1\left(\frac{k}{k-1}\right)\left(\frac{R}{778}\right)\left[1-\left((P_2/P_1)^{\frac{k-1}{k}}\right)\right]$$

R—Gas constant = 1,545/MW
MW—Molecular weight
k—Ratio of specific heats

T_1—Inlet temperature (degrees R)
P_1—Inlet pressure (psia)
P_2—Discharge pressure (psia)

Isentropic Gas Horsepower

$$HP_{gas} = \frac{m\Delta H_{is}778}{550}$$

HP_{gas}—Isentropic gas power (hp)
m—Mass flow (lbm/s)

Expander Efficiency

The stage efficiency is related to the energy drop across the stage, which is quantified by the velocity ratio U_m/C_0. Each stage design has a different performance characteristic similar to that shown in Figure 6.7. It can clearly be seen that as the stage loading changes, the peak efficiency changes. The stage efficiency is also affected positively (in terms of improved efficiency) by use of flow-path seals and stator and rotor shrouding. The low-stage loadings of the nitric acid expander make for a higher overall efficiency than many other expander designs:

$$HP_{stage} = HP_{gas}\ \eta_{stage}$$

η_{stage}—Stage efficiency

Expander Power

The expander power was then calculated by summing up the individual stage power and subtracting the bearing losses:

$$HP_{shaft} = \left(\sum HP_{stage}\right) - Brg_{loss}$$

Typical Expander Design Point Performance

A simplified example of the typical 500 ton/day nitric acid expander design point can be seen in the following performance estimate:

Inlet Conditions

$P_1 = 110$ psia (7.7 kg/cm^2) $\quad T_1 = 1{,}200$°F (649°C)

Exhaust Conditions

$P_2 = 15.2$ psia (1.10 kg/cm^2) $\quad T_2 = 650$°F (343°C)
$PR = 7.24$

Tail Gas

Mainly nitrogen (96%) $MW = 28.6$ $K_{ave.} = 1.31$

Mass flow = 235,000 lbm/hr (107,000 kg/hr)

Overall Isentropic Enthalpy Drop

$$\Delta H_{is} = 1660\left(\frac{1.31}{1.31-1}\right)\left(\frac{1545/28.6}{778}\right)\left[1-\left((15.2/110)^{\frac{1.31-1}{1.31}}\right)\right]$$

$$= 182.2 \text{ Btu/lbm } (192 \text{ Kj/kg})$$

Assuming a uniform energy drop per stage in a five-stage expander:

Stage isentropic enthalpy drop = 36.4 (38.4 Kj/kg)

$$C_o = 223.76\sqrt{36.4} = 1{,}347 \text{ ft/s}$$

$$U/C_0 = 800/1347 = 0.59 \rightarrow \text{stage efficiency} = 0.86$$

Expander Power

Expander isentropic gas power = 16,820 hp (12,543 kW)

Expander gas power = 16,820 (.86) = 14,465 hp (10,787 kW)

Bearing loss = 70 hp (52 kW)

Shaft power = 14,465 – 70 = 14,395 hp (10,734 kW)

Adiabatic efficiency = (14,395/16,820)(100) = 85.5%

6.4.3 Expander Performance Testing

Factory performance testing of a nitric acid power recovery train (PRT) is difficult for most manufacturers because of a lack of an adequate combustor to power the expander. In addition, the volume and gas mixture is also difficult to replicate for the expander and NO_x compressor. Component and train mechanical testing is routinely performed. Air compressor performance testing is also commonly performed.

Most nitric acid PRT operators do field performance testing to both verify the original performance of the turbomachinery and to monitor changes in process and machinery operation. Performance testing of nitric acid PRT is difficult due to the large number of operating parameters required to calculate the individual component performance. In addition, actual component power is not measured in the PRT. A PRT power balance test is usually used to confirm the component and overall performance of the PRT. A power balance test requires the calculation of the performance of each of the individual components in the train. Theoretically, the power generated by the expander and auxiliary driver should equal the power consumed by the compressors. Instrumentation inaccuracies tend to be the largest source of performance-testing error. A power balance (power generated equals power consumed) that closes within 0.5% would be considered excellent. Closure within 2.0% would be considered acceptable.

The data shown in Figure 6.9 represents the information that is required to perform a PRT power balance analysis:

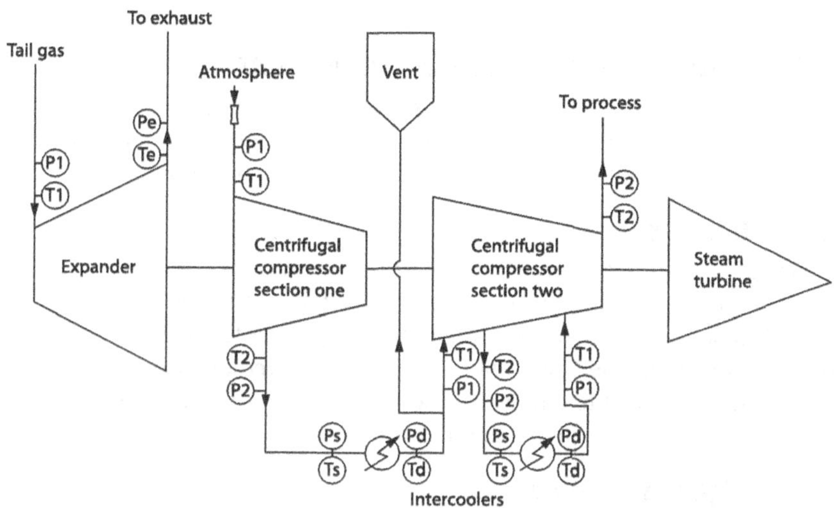

FIGURE 6.9 Performance-monitoring instrumentation.

Location	Required Parameters
Expander inlet	Flow, pressure, temperature, speed, and gas analysis
Expander exhaust first compressor section	Pressure and temperature
Air compressor inlet	Flow, ambient pressure, inlet pressure, temperature, and relative humidity
Air compressor discharge	Pressure and temperature
Cooler water inlet	Flow, pressure, and temperature
Cooler water outlet second compressor section	Pressure and temperature

Location	Required Parameters
Compressor inlet	Pressure and temperature
Compressor discharge	Pressure and temperature
Cooler water inlet	Flow, pressure, and temperature
Cooler water outlet	Pressure and temperature
Compressor inlet	Pressure and temperature
Compressor discharge	Pressure and temperature
Steam turbine inlet	Flow, pressure, temperature, and speed
Steam turbine exhaust	Pressure, temperature, and quality

6.5 Nitric Acid Expander Design Considerations

Nitric acid expanders are special turbines that have a variety of design challenges that include high pressure, high temperature, high and variable speed operation, and potentially corrosive environments.

6.5.1 Inlet Pressure

Nitric acid expanders operate at inlet pressures that vary between 75 and 250 psig (5.3 and 17.5 kg/cm^2). These pressures would be considered to be moderate in comparison to many other turbine designs such as a steam turbine. The expander pressure and temperature drops determine the available energy and are the major considerations in the selection of the type and number of stages.

6.5.2 High Temperature

The expander inlet temperature varies widely depending on the individual plant design and operation. With the inlet temperature varying between 500°F (260°C) and 1,400°F (760°C), very different mechanical designs have been developed by the various manufacturers.

Three different types of expander designs have been developed by the various manufacturers. The different designs have been generically classified as low-temperature, warm, and hot designs.

The low-temperature nomenclature is somewhat confusing since low-temperature expanders are often the radial-type expanders used in cryogenic applications such as those described in Chapter 7. In nitric acid applications, the name or classification refers to the inlet temperature of the expander.

In the high-temperature expander, the inlet temperatures can be high enough to result in major material property reductions, consideration for creep and stress rupture, high-temperature corrosion, in-service aging, and life-limited components. A common design practice is to design the high-temperature components to be capable of continuous operation for 100,000 hr (11.4 years). In addition, process upsets may allow short-term over temperature excursions that dangerously reduce the safety margins of many expander components including the rotor. Transient temperature operation can also significantly reduce the operating safety margins and need to be tightly controlled.

6.5.3 Variable Speed Operation

Most nitric acid expander trains operate under variable-speed conditions that require that close attention be paid to the train and individual body rotor dynamics and component natural frequencies. In addition, the rotating speeds of the nitric acid train would be considered to be fairly high, and most rotors in the PRT operate above the first critical speed of the rotor(s).

6.5.4 Corrosion

Although most nitric acid expanders operate primarily with benign nitrogen gas, conditions can exist that allow the formation of acids that can cause corrosion of the expander components. For this reason, the materials of the expander parts must be corrosion resistant. Further discussion of corrosion can be found in section 6.11 – Common Operational Problems.

Even though all manufacturers offer standard frame sizes of expanders, each application is a custom design and is site specific.

6.6 Expander Construction

6.6.1 Low-Temperature Expanders

Expanders that operate less than 500°F (260°C) are considered low- or cool-temperature expanders. Figure 6.10 shows a cross section of a typical lower-temperature nitric acid expander. The low-temperature expander can be a conventional single- or multistage axial turbine. Steam turbine design principles, construction, and materials are often used in the low-temperature nitric acid expander. The expander is designed to maximize the power recovery from the tail gas heat energy.

Depending on the process operating conditions, the low-temperature expander may be constructed with one to four stages. Both impulse and reaction blade designs are used. The inlet, discharge, and stator casings for the low-temperature expanders are generally cast low-alloy carbon steels. The discharge casings can be cast or fabricated low-alloy steel.

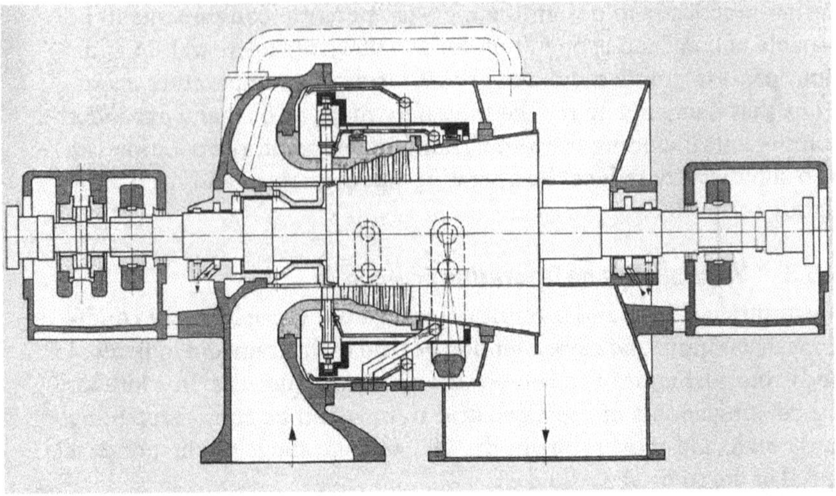

FIGURE 6.10 Typical low-temperature or warm expander construction.

6.6.2 Warm Expanders

Expanders that operate between 500°F and 950°F (260°C and 510°C) are considered to be "warm" expanders. The warm expander is constructed similar to the low-temperature expander, but uses upgraded materials and designs to accommodate the higher operating temperatures. Warm expanders can be a single-stage design, but tend to be multistage axial turbines. High-temperature steam turbine design principles, construction, and materials are used in the warm nitric acid expander designs. The expander is designed to maximize the power recovery from the tail gas heat energy. Both impulse and reaction blade designs are used. The inlet, discharge, and stator casings for the warm expanders are generally cast low-alloy or ferritic 12% chromium steels. The discharge casings can be cast or fabricated low-alloy steel.

6.6.3 High-Temperature Expanders

Nitric acid expanders that operate with inlet temperatures above 950°F (510°C) are considered to be a high-temperature expander. The high-temperature expander is also a conventional multistage axial turbine that includes additional design and construction features that are used to accommodate the higher operating temperatures. Many gas turbine design principles, construction, and materials are used and can be seen in Figure 6.11. Most notable is the use of a high-temperature inlet scroll in the inlet of the expander. Inlet scrolls are used to shield the outer casings from the high-temperature gases. The expander is designed to maximize the power recovery from the tail

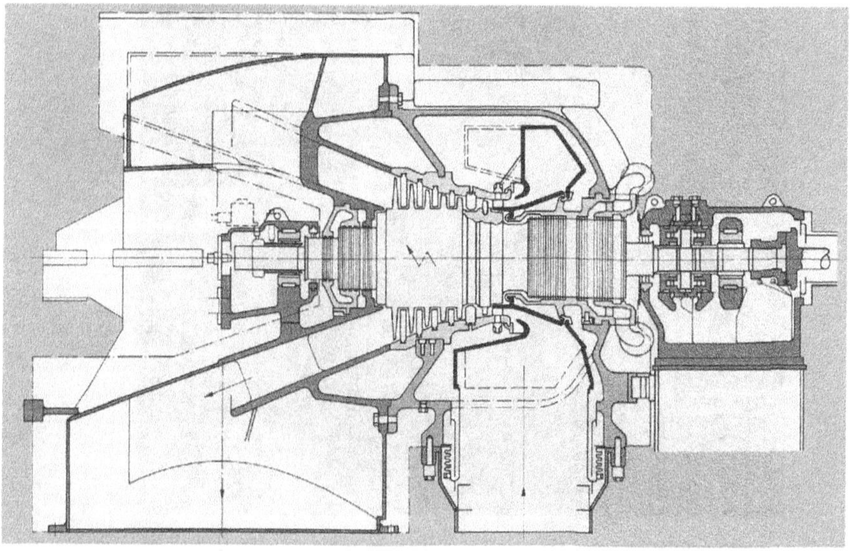

FIGURE 6.11 Typical multistage high-temperature expander.

gas heat energy. Depending on the process operating conditions and high available energy, the high-temperature expanders are usually constructed with two to six stages. The high-temperature expanders require special design features, materials, and construction to accommodate the higher operating temperatures. High-temperature expanders operated with inlet temperatures in the 1,200°F to 1,400°F (649°C to 760°C) range.

6.7 Expander Component Design

6.7.1 Outer Casings

The nitric acid expander is a very compact design that often uses a double-wall casing construction (inner and outer casing). The outer casing is generally considered the pressure-containing casing, whereas the inner casing is a flow guide that accommodates the high-temperature aspects of the tail gas.

The expander outer casing is generally centerline supported to maintain the unit centerline despite multidirectional thermal growths (Figure 6.12). The position and alignment to the driven equipment is controlled by guide keys or pins. The most common support arrangement is to axially fix the outer casing on the discharge end. The transverse expansion is accommodated by transverse guide keys. The axial

FIGURE 6.12 Expander outer casing with thrust-bearing housing.

growth is accommodated by sliding or flex plate (wobble plates) supports on the inlet end of the expander casing. Transverse or side movement is prevented by the use of axial guide keys on both the inlet and discharge ends of the casing. The casing and its nozzles are allowed to grow unrestricted in the vertical direction. The nozzle axial and vertical growth need to be accounted for in the piping system or excessive forces on the casing may be generated.

For assembly purposes, the outer casings are almost always horizontally split. As can be seen in Figure 6.12, the casing may be of a multiple-piece construction with a vertical joint between the high- and low-pressure sections.

The casing nozzle connections (inlet and discharge) are typically located in a radial direction and can be either in an up or down nozzle orientation. Both the expander inlet and discharge nozzles are usually sized for operating velocities between 110 and 150 ft/s (34 and 46 m/s). The inlet flanges are generally 300 lb class, conforming to ASME/ANSI B16.5 piping flanges. The discharge flange is generally a 150 lb class flange, also conforming to ASME/ANSI B16.5 piping flanges. Flexitallic spiral-wound gaskets are commonly used on the inlet flanges, whereas the low-pressure exhaust flanges often use graphite-type gaskets.

Outer Casing Structural Design Considerations

Because the outer casing is considered a pressure vessel, the outer casing is often designed in accordance with the ASME Pressure Vessel

Code. The inlet casing is designed for the full pressure rating of the expander. Outer casing pressure ratings of 250 psig (17.5 kg/cm^{2}g) are common. Outer casings are typically designed for a 20-year life.

Whereas the outer casing is designed to contain the expander-rated pressure, it is not usually subjected to the full inlet temperature of the expander. For ASME code compliance, the outer casing operating temperature must be maintained at 850°F (454°C) or less by use of one or more of the following design features:

- Inlet casing internal insulation
- External insulation of the inlet flow scroll
- No insulation is used on the exterior of the outer casing
- Internal air or tail gas cooling

Casing wall thickness can range from 0.5 to 1.25 in (1.3 to 3.2 cm) and often uses ribs to minimize the casing deflection under pressure. In accordance with the ASME code, the inlet casing undergoes a temperature-compensated hydro test during the manufacture. At a 250 psig (17.6 kg/cm^{2}g) pressure rating, the outer casing receives a temperature-compensated hydrostatic test at 1,125 psig (79 kg/cm^2).

The discharge end of the casing is designed to accommodate the much lower discharge pressures of the expander. Discharge pressures range between 0.5 and 3 psig (0.035 and 0.21 kg/cm^{2}g). The discharge casing design is generally governed by the allowable casing deflections. Excessive deflection can cause flow path and seal rubs as well as bearing misalignments. Discharge casing hydrostatic testing becomes a very debatable issue with many expander manufacturers. If the casing is rated for 150 psi, it should be hydrostatically tested at 450 psig (31.6 kg/cm^{2}g). This may be excessive for a casing that operates at a pressure slightly above atmospheric pressure. Regardless of the criteria used for the design, the hydrostatic test should include temperature compensation.

Outer Casing Materials

The outer casings are manufactured as low-alloy steel castings or as a welded fabrication. The casings are manufactured from high-strength and quality alloy steels or stainless steel. Typical low-alloy steel materials include ASTM A217-WC9 castings or ASTM-387 Gr 22 (2½ Cr, 1 MO) pressure vessel plate. Stainless-steel casings have been made from ASTM A743-CA-15 (12 Cr) castings. The casings are usually supplied in an annealed, normalized, and tempered condition.

6.7.2 Inlet Scrolls

All nitric acid expander designs use an inlet scroll or plenum to turn the high-temperature gas from the radial inlet connection to an axial first-stage nozzle/vanes. The inlet scroll directs and distributes the tail gas around the rotor and into the first-stage nozzles. In cool expanders, the inlet scroll may actually be a part of the internal inner/stator casing. In the higher temperature expanders, the inlet scroll is a separate part generally made up as a fabrication. The shapes are often complex as can be seen in Figure 6.13.

Special seal rings are used at the entrance of the inlet scroll to guide the tail gas into the scroll and prevent gas from bypassing the first-stage nozzles. In addition, the large radial thermal growth of the inlet scroll must be accommodated by the seal rings. These seal rings vary in design from being conventional expansion joints to slip-fit piston rings or top hat designs.

The inlet scroll often supports the first-stage nozzles/stator vanes. Full-admission nozzles are typically used to provide uniform flow and loadings into the rotor blades as well as maintaining a uniform temperature distribution in the scroll and first-stage nozzle.

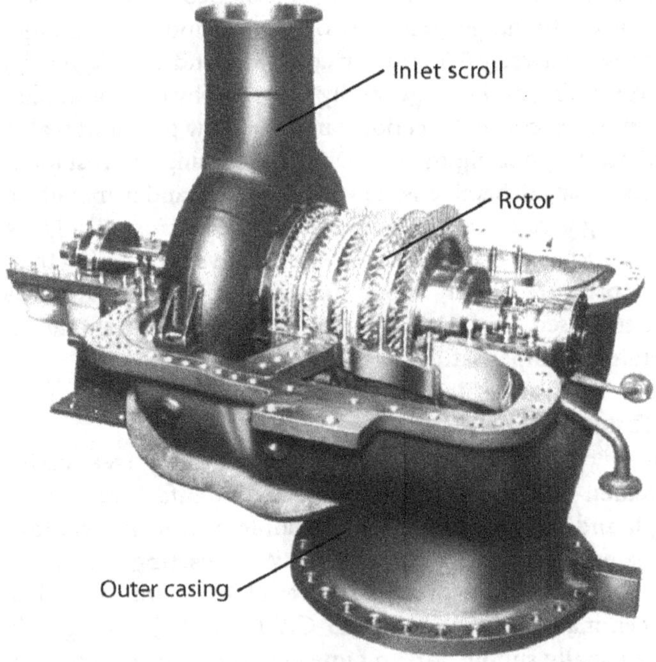

Figure 6.13 Nitric acid expander assembly.

Inlet Scroll Structural Design Considerations

Inlet scrolls are flow guides and are not typically subjected to significant pressure differentials (inside to outside). The low-pressure differential allows the inlet scroll to operate with minimal pressure stresses. The low stresses allow for thin wall construction. The thin wall construction makes for a uniform operating temperature that reduces the thermal stresses as well.

The inlet scrolls operate at the full inlet temperature of the expander. The high operating temperature requires the design to consider a number of structural issues such as reduced material properties, in-service aging, creep, stress rupture, and high thermal displacements. The scroll design must also accommodate a large amount of thermal growth owing to the high operating temperatures and coefficients of expansion of the materials of construction. This thermal growth must be allowed to occur unrestrained and must maintain both the scrolls' axial and radial alignment around the rotor. Centerline support and alignment is maintained by use of keyed support pads and radial pins.

Special seal or expansion joint designs are used to guide the tail gas from the inlet piping into the scroll and to prevent gas leakage around the scroll and first-stage nozzle.

Inlet scrolls may be single piece or horizontally split. Split inlet scrolls tend to experience much more distortion and flange leakage owing to the flanges acting as a thermal mass and heat transfer rib. As you can see in Figure 6.14, the inlet scrolls are complex, multidimensional fabrications. The toroidal-shaped scroll in Figure 6.14 has been formed by a metal spinning process.

Figure 6.14 Typical high-pressure stator casing.

Inlet Scroll Materials

Inlet scrolls are usually fabricated from thin (less than 0.2 in [5 mm]) high-temperature stainless steels or nickel-base alloys. Alloys such as a 300 series stainless steel conforming to ASTM A-240 or nickel-based alloys such as Inconel 600 conforming to ASTM B-168 are commonly used in a solution-annealed condition.

6.7.3 Stator Casing/Inner Casing

The stator/inner casing forms the outer surface of the flow path and positions the stationary blading axially with respect to the rotor element. The stator casings are subjected to both significant pressure drops and temperature differentials. Multiple-stator casings may be used in higher pressure and temperature expanders. Stator casings transmit the stator vane thrust and tangential loadings into the outer casing. With the stage-by-stage pressure and temperature reductions, the stator casing sees significant pressure and thermal stresses. The casing experiences significant radial and axial deflections that must be accommodated in the design to assure that the rotor clearances are maintained. The stator vanes are mounted into the stator casing as horizontally split diaphragms, individual stator vanes, or segmented vane assemblies as shown in Figure 6.14.

The inner casings can be manufactured as castings or welded fabrications. The stator casings are horizontally split to allow rotor installation and may be vertically split as well to provide more uniform expansions. A four-way split caning is shown in Figure 6.15. Specialty high-strength, high-temperature fasteners are used in the casing flanges to maintain adequate sealing pressure and resist the normal pressure and temperature loadings.

Inner Casing Structural Design Considerations

In the high-temperature expanders, the stator/inner casings are often considered to be the temperature casings because they are subject to the full tail gas temperature and need to be designed to accommodate the operating stresses at the higher temperature. The inner casing grows significantly more than the outer casing owing to the higher operating temperatures and coefficient of expansions of the inner casing materials. The differential expansion is usually accommodated by allowing the stator casing to float within the outer casing on radial guide pins or keys. The stator casing is usually maintained in the axial direction in a tongue-and-groove arrangement.

High-temperature stator casings are generally designed to experience less than 0.2% creep in 100,000 hr (11.4 years) of operation. Without a properly designed stator casing, some expanders have

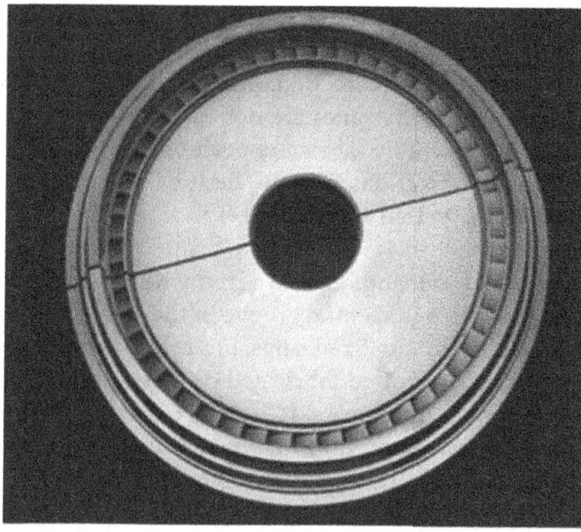

FIGURE 6.15 Fabricated low-temperature stator assembly.

experienced thermal distortion in only one or two 4-year operating campaigns. This distortion generally shows as reduced rotor blade and seal clearances or flange distortions. Casing rework will often restore the condition of the stator casing and allow additional operating campaigns.

Inner Casing Materials

The casings need to be corrosion resistant since high-temperature oxidation can occur at the tail gas normal operating temperatures and acid formations can occur during shutdown or transient operation.

Cool and warm expander stator casings are often manufactured from high-strength and high-quality low-alloy steels such as ASTM A217-WC9 castings or ASTM-387 Gr 22 (2½ Cr, 1 MO) pressure vessel plates. Higher temperature stator casings are manufactured as stainless-steel castings such as ASTM A743-CA-15 (12 Cr) castings. The casings are usually supplied in an annealed, normalized, and tempered condition.

6.7.4 Stator Vanes

The stator vanes in the expander direct the gas flow into the rotor blades. The first-stage vanes are often call nozzles. While some impulse vane and blade designs are used in the first or second stages of some high-pressure nitric acid expanders, a majority of the vane designs are of the reaction design. Fifty-percent reaction designs are common in which one-half of the energy/pressure drop of the stage occurs in the stator and one-half in the rotor blade.

Most nitric acid expanders are constructed with fixed position/geometry stator vanes. The vanes are not adjustable without replacement of the entire assembly. The vane positions are set in the design phase. Several vane positions or vane designs are often part of the expander frame design that allow changes in stage flow, reaction, and loading to be made.

Stator vanes can be manufactured as individual vanes or in assemblies that are slid into a tangential groove in the stator housing or the vanes can be assembled into fixed rings. In the low-temperature applications, stator vanes are often fabricated or cast turbine diaphragms similar to that shown in Figure 6.15. In the high-temperature applications, individual vanes or vanes fabricated in segments as shown in Figure 6.16 are used. These designs allow for the higher thermal expansion of the vanes and housings.

Most stator vane profiles are of a constant cross-sectional design for ease of manufacture. This design is used to produce a free vortex discharge velocity profile. In free vortex flow, the flow velocity is inversely proportional to the radius and the total head in the flow field remains constant:

$$rC_\theta = \text{Constant} \qquad r = \text{Radius}$$
$$C_\theta = \text{Gas tangential velocity}$$

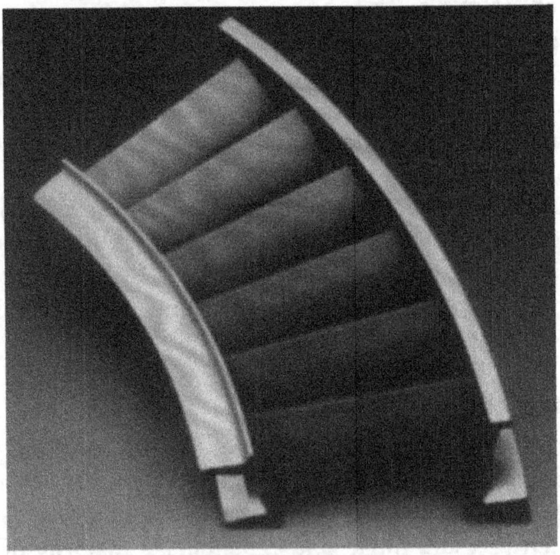

FIGURE 6.16 Typical fabricated stator segment.

Figure 6.17 Double-flow expander with radial and axial stator vanes.

Being typically of the 50% reaction design, the stator vane profiles resemble the mean-line shape of the rotor blades. A number of nitric acid expanders are built as double-flow designs where the flow enters the center of the expander and flows equally in opposite axial directions as can be seen in Figure 6.17. Both radial and axial stator vanes have been used in double-flow designs.

Variable Stators

Some expander manufacturers provide variable geometry stators in the first stage of the expander. The variable stator design allows an optimal matching of the expander flow capacity to that of the process conditions. This can significantly increase the expander efficiency under part load conditions and maximum power recovery over a wide range of process conditions. This is particularly true when used in combination of various guide vane controls of the driven air compressors. Variable geometry stator vanes can typically change the expander capacity by more than ±10% of the rated flow.

Continuously variable stator vanes have been used in a number of low-temperature expander designs. The individual stator vanes are mounted on a shank that protrudes through the stator/inner casing. A cartridge assembly includes guide bearings and purge provisions to prevent the accumulation of debris in the bearings. The individual

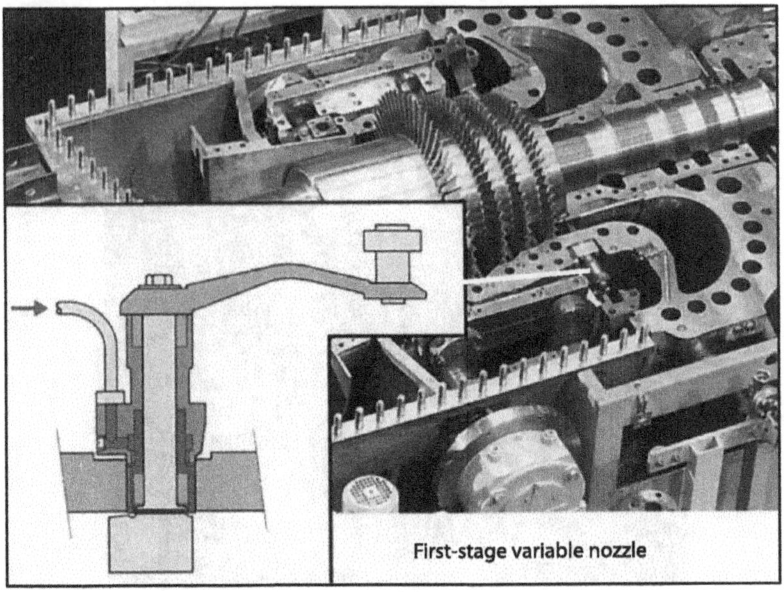

FIGURE 6.18 Variable first-stage stator design.

vanes are held in alignment and moved by a circular drive ring. The drive ring is moved by a series of linkages or levers powered either mechanically, electrically, or hydraulically. Figure 6.18 shows a nitric acid expander with continuously adjustable inlet stator vanes.

Variable geometry stator vanes are not typically used in the high-temperature expanders due to the inherent difficulties associated with the higher operating temperatures. Whereas online continuously adjustable stator vanes do not exist for the high-temperature expanders, some manufacturers offer an adjustable stator design that allows the stator vane setting to be changed during a unit shutdown without replacing the assembly.

Stator Vane Structural Design Considerations

Stator vanes are generally low-stressed components with total stresses generally being less than 5,000 psi (352 kg/cm^2). The highest stressed vanes are the freestanding vanes. Stator vane stresses are low since the vanes are subjected to moderate axial and tangential pressure-bending loads and a small aerodynamic load.

All stator vanes should have their natural frequencies calculated and compared to the rotor blade passing frequency to prevent fatigue failures. Freestanding stator vanes are much more likely to have natural frequencies coincident with the rotor blade excitations.

Shrouded stator vanes typically have very high natural frequencies that are well above the operating speeds of the expander.

Shrouded stator vanes are inherently quite stiff and low stressed. In addition, the shrouding helps improve the stage efficiency typically 2% to 3% and greatly increases the vane natural frequencies.

Stator Vane Materials

Stator vanes can be made from a variety of manufacturing methods including milled bar stock, extruded stock, and castings. The vanes are manufactured from 300 series stainless steels conforming to ASTM A-240 Types 304, 321, and 347. The vanes may be mounted in austenitic cast steel or fabricated in 300 series stainless-steel rings.

6.7.5 Rotor Blading

In addition to the wide variety of rotor constructions, the types of blades and their retention in the rotor vary widely (Figure 6.19). The rotor blade airfoils can be of an impulse or reaction design. The impulse blade design is usually of a constant section and a high degree of turning. Impulse blades are almost always shrouded. The reaction blade usually is of the taper twisted design as shown in Figure 6.20. Reaction blades can be either freestanding or shrouded. Shrouding improves the blading efficiency, reduces interstage sealing, provides

FIGURE 6.19 Low-temperature expander rotor.

FIGURE 6.20 High-temperature stacked rotor with curvic couplings and central tie bolt.

increased dampening for blade vibrations, and can be used for blade-tuning purposes. The effects of shrouding and blade tip seals are significant with the individual stage efficiency being increased by 3% to 5% in a shrouded design.

The type of airfoil used is highly dependent on the energy drop in the individual stages. Both impulse and reaction blade airfoils are used in the nitric acid expander. Figure 6.21 shows the two different airfoil designs. Impulse designs are often referred to as "buckets" and turn the tail gas in the tangential direction with little or no pressure drop across the rotor blading. The reaction blade design is a conventional tapered and twisted design that typically takes half of the energy drop across the stator vanes and half of the energy drop across the rotor blades. With a conventional energy drop per stage, the reaction blading will have the best-stage efficiency. With a high-energy drop per stage, an impulse design will have the best stage efficiency. The impulse designs are usually limited to the highest pressure ratio expanders and used in only the first and second stages of the nitric acid expander.

The rotor blades are mounted and retained in the rotor disc in a wide variety of designs including single- and multiple-tang "T" roots. Lower-stressed rotor blades are generally installed in the rotor in circumferential grooves with single- or multiple-tang "T" roots, internal and external dovetail, and internal and external "Fir Tree" roots. The higher-stressed rotor blades typically use multi-tang axial "Fir Tree" type attachments.

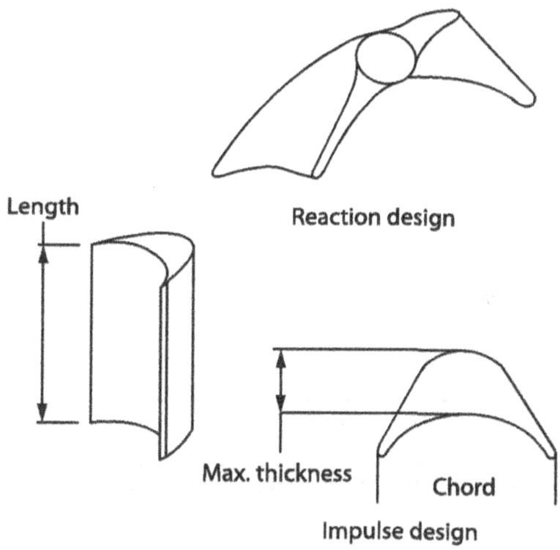

FIGURE 6.21 Rotor blade airfoil designs.

Rotor Blade Structural Design Considerations

The rotor blades are conservatively designed with low-nominal operating stress levels. Nominal/average cross-sectional radial stresses are generally designed to be less than 30,000 psi (2,100 kg/cm^2). Generous attachment fillet radii allow the blade attachment concentrated stress to be less than 90,000 psi (6,300 kg/cm^2). Blade-bending stresses are typically less than 5,000 psi (352 kg/cm^2).

The rotor blade airfoil operating temperature is generally within 10°F (5°C) of the stage inlet gas temperature. The blade attachments are generally within 25°F (12°C) of the stage inlet gas temperature.

In the high-temperature expanders, the elevated temperatures dictate the need for special corrosion-resistant materials. These materials must be creep resistant, have very high endurance limits, and have high elevated-temperature tensile strengths. A convenient method for examining the adequacy of a given material at a given temperature under known stress levels is by the use of a modified Goodman diagram such as that shown in Figure 6.22. The steady-state stresses are evaluated against the tensile capabilities of the materials. At metal temperatures of less than 1,000°F (538°C), the stress is usually evaluated against the ultimate tensile strength. Above 1,000°F, the lesser of the ultimate strength or the 100,000-hr 1% creep or stress-rupture strength is used for the evaluation. Safety factors of 1.25 or higher are often used. The vibratory or alternating stresses are evaluated against the material endurance strength at the operating temperature. The combined stresses are evaluated against a line drawn between the endurance limit and the lesser of the ultimate or

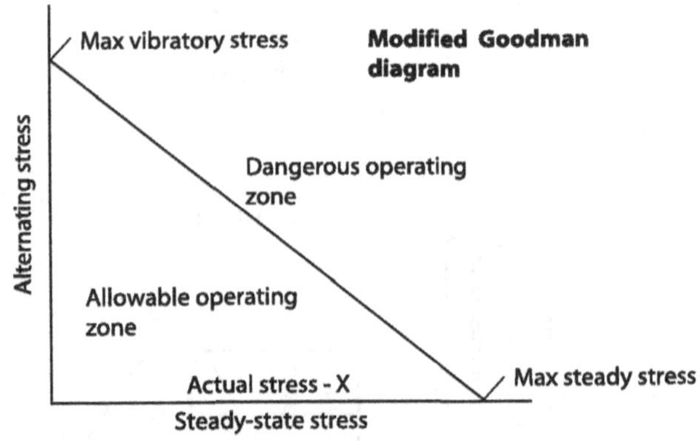

Figure 6.22 Typical construction of a modified Goodman diagram.

stress-rupture strength (see Figure 6.22). Minimum safety factors of 3.0 are often used for the vibratory stress margin.

Rotor Blade Materials

Expander rotor blades have been manufactured from a variety of materials depending on the tail gas temperature. All of the blade materials are resistant to corrosion and provide both high strength and ductility. As the operating temperature of the blade exceeds 1,000°F (538°C), high-temperature strength, creep rupture, and oxidation resistance need to be considered. The following materials are commonly used for rotor blading:

- For blade operating temperatures less than 800°F (427°C), 400 series (12% chromium) stainless-steel materials such as AISI 403 and 410 SS are used.
- For blade operating temperatures between 800°F and 1,000°F (427°C and 538°C), heat-resistant stainless steels such as AISI 422 are used.
- At blade operating temperatures between 1,000°F and 1,200°F (538°C and 649°C), iron base superalloys such as A-286 per AMS 5735 (26 Ni, 15% Cr) or nickel-base superalloys such as Nimonic 80 per ASTM B637 (20% Cr) are used.

6.7.6 Rotor Discs

The expander rotor discs are some of the most critical components in the expander. The discs must accommodate the significant centrifugal and bending stresses of the rotor blades, its own centrifugal loads, tie-bolt loadings, and the power transmission of other stages as well as its own. Where rotor blades may be replaced from time to time due to deterioration, the rotor disc is put back into service time and time again.

A variety of rotor disc mountings are used to make up the rotor assembly. The discs can be mounted as shrunk on a shaft, stacked together with tie bolts, or manufactured as an integral part of the rotor shaft.

Many high-temperature expanders use rotor disc cooling to improve the structural margins and to minimize the corrosion potential. Rotor discs may incorporate seal lips on the faces that serve as radial seals for the interstage stator vanes. The seals can be used to reduce interstage leakage and control rotor cooling.

Rotor Disc Structural Design Considerations

The rotor disc designs in general are conservatively designed with maximum radial stresses in the 25 to 30 KSI (1,760 to 2,100 kg/cm^2) range and an average tangential stress in the 20,000 to 25,000 KSI (1,400 to 1,760 kg/cm^2) range. The maximum tangential stress in the regions of the tie bolt(s) is typically 60,000 KSI (4,220 kg/cm^2).

The rotor disc operating temperature is highly dependent on if the disc is cooled nor not. First-stage rim temperatures are generally within 50°F (28°C) of the stage inlet gas temperature. First-stage rim temperatures in excess of 1,100°F (593°C) range are common in high-temperature expanders. Similarly, the last-stage rim temperature is within 50°F (28°C) of the expander exhaust temperature. Last-row rotor disc rim temperatures often operate in the 500 to 600°F (260 to 316°C) range.

Large axial and radial thermal gradients can exist in the rotor discs with the bore regions often being 300°F (149°C) cooler than the rim region in an uncooled expander disc. Cooled rotors can have disc bore temperatures as much as 500°F (260°C) cooler than the rim. These radial temperature gradients can cause significant radial thermal stresses that equal or exceed the traditional rotor centrifugal and pressure-loading stresses.

Similarly to the rotor blade stress evaluations, the use of a Goodman diagram to evaluate the rotor disc stresses is a common practice. The steady-state stresses are evaluated against the tensile capabilities of the materials. At metal temperatures less than 1,000°F (538°C), the stress is evaluated against the ultimate tensile strength. Above 1,000°F, the lesser of the ultimate strength or the 100,000-hr 1% creep or stress-rupture strength is used for the evaluation. Safety factors of 1.25 or higher are often used for the steady-state stresses. The vibratory or alternating stresses are evaluated against the material endurance strength at the operating temperature.

Rotor Disc Materials

Expander rotor discs have been manufactured from a variety of materials depending on the tail gas and operating metal temperatures. All of the disc materials are resistant to corrosion and provide both high strength and ductility. The materials in the stacked or built-up designs are chosen specifically for the temperature levels that the individual stage operates. Disc materials are generally selected for the operating/maximum temperature in the rim of the disc.

The highest stressed region of both the blade and the disc is the attachment region. Along with the higher stresses found in the attachments, this same region operates at elevated metal temperatures. This combination of stress and temperature requires special materials that have high-strength mechanical properties and endurance limits.

As the operating temperature of the disc exceeds 1,000°F (538°C), high-temperature strength, creep rupture, and oxidation resistance need to be considered. The following materials are commonly used for the rotor discs:

- Disc rims operating at temperatures less than 1,000°F (427°C to 538°C), use heat-resistant stainless steels such as AISI 422 per ASTM A565.
- Disc rims operating at temperatures above 1,000°F (538°C), use iron-base superalloys such as A-286 per AMS 5735 (26 Ni, 15% Cr) or nickel-base superalloys such as Nimonic 80 per ASTM B637 (20% Cr).

6.7.7 Rotors and Shafts

As previously noted, nitric acid expander rotors are manufactured in a variety of ways depending on the individual application, operating conditions, and manufacturer. Built-up, integral, and stacked designs are commonly used.

Built Up

Built-up rotors designed similar to the rotor in Figure 6.19 are commonly used for smaller-power and lower-temperature expander installations. The built-up designs include separate individual rotor disc(s) that are generally shrunk onto a through shaft. The discs are generally keyed or pinned to the shaft for torque transmittal. The shafts are usually shouldered to provide axial locators for the individual discs. Proper disc axial locators are critical to assure proper rotor fit into the casings. Shrunk-on rotors often operate at lower operating speeds and require slower heat-up and cool-down controls, as well as tighter overspeed restrictions since conditions can be created that will allow the rotor discs to lose their required fits to the shaft.

FIGURE 6.23 Integral double-flow rotor assembly.

Built-up rotors are generally less expensive to manufacture since the smaller individual components are easier to manufacture and inspect, and can be processed concurrently.

Integral Rotors

Integral or single-piece rotors are commonly used for low-temperature and warm expander applications. Although more expensive to manufacture, the integral design is generally lower stressed, more thermally stable, does not have the potential for individual disc/stage movement, and has improved rotor dynamics characteristics. Figure 6.23 shows a double-flow integral rotor being installed into the expander casing.

In the warm expander applications, the rotors should be manufactured with heat stability tests to assure that the rotor is dimensionally stable under operation. A heat stability test involves installing the rotor in a lathe with a "heater box" and heating the rotor to the normal operating temperature while it is turning. Stabilized forgings will maintain its geometric centerline throughout the heat stability test. A commonly used heat stability procedure is outlined in ASTM A472.

The higher the operating temperature, the greater is the potential that the rotor will be thermally unstable.

Built-Up Rotors

Built-up rotors are commonly used in high-temperature expander applications. The rotor is subject to large radial and axial temperature gradients that significantly increase the rotor stresses. Distributing the rotor temperature gradients over a number of individual discs greatly reduces the individual component operating stress levels.

The rotor assembly shown in Figure 6.20 is made up of the high-pressure stub shaft with an integral thrust collar and seal region, a spacer sleeve, five individual disc assemblies, a low-pressure stub shaft with the discharge seal and integral coupling hub, and a central tie bolt. One common method of accommodating the differential expansion between the rotor discs is the use of radial expansion pins or Curvic couplings. The Curvic coupling is a radial spline that maintains alignment between the discs, allows for differential thermal expansion, and transmits the torque through the rotor. A single central or multiple tie bolt provides the means of holding these separate components together. Figure 6.24 shows a multi tie bolt Curvic-coupling design.

Figure 6.24 Stacked rotor design using multiple tie bolts and Curvic couplings.

The stacked rotor design allows the rotor disc material to be changed in each individual stage. This allows for the use of specialty high-temperature materials in the inlet stages of the expander and more conventional turbine in materials in the lower-temperature discharge stages. Owing to the special materials and manufacturing techniques, the high-temperature stacked rotors tend to be the most expensive to manufacture. The stacked rotor tends to be more sensitive to assembly practices and in-service creep or relaxation that cause rotor bowing and vibration problems.

Rotor Structural Design Considerations

Unlike the rotor discs and bodies, nitric acid expander shafts and stub shaft ends are conservatively designed and generally have unlimited operating life. The shaft ends are often increased in sized to improve the overall rotor dynamics characteristics versus being sized to just transmit the generated torque. The rotor shaft ends generally have the highest stresses and need to be evaluated for the geometric size changes and resulting stress concentrations. Many shaft ends are designed in accordance with the ASME code.

The basic diameter of the shaft can determined by the following formula:

$$d=\sqrt[3]{\frac{16\times\sqrt[2]{\left(K_m\times M+\frac{1}{8}\times a\times F\times d\times\left(1+R^2\right)\right)^2+(K_t\times T)^2}}{\pi\times S_d\times\left(1-K^4\right)}}$$

d—Diameter (inches)
M—Bending moment (in-lbs)
F—Axial tensile or compressive load (lbs)
a—Ratio of maximum to average intensity of stress resulting from axial load
T—Torsional moment (ft-lb)
S_c—Allowable design stress (psi)
S_s—Allowable shear stress (psi)
K_m—Numerical combined shock and endurance factor
K—Ratio of inside to outside diameter for a hollow shaft
K'—Form stress concentration factor
S_d—Allowable design stress (psi)

$$T=\frac{63{,}000\times Hp}{rpm}$$

Horsepower (hp)
Rotation speed—rotations per minute (rpm)

By restricting this formula to the design of the shaft outboard of the bearings, we have minimal axial and bending loads and the above formula reduces to:

$$d = \sqrt[3]{\frac{16 \times T_{max}}{\pi \times Ss}}$$

The stress at the step in the shaft is determined by:

$$d = \sqrt[3]{\frac{16 \times K' \times T_{max}}{\pi \times Ss}}$$

For known shaft-end geometries, the stresses can be calculated accordingly:

Torsional Shear Stress (straight shaft)

$$St = \frac{16 \times T_{max}}{\pi \times d^3}$$

Torsional Shear Stress (shaft step-up)

$$St = \frac{16 \times K'T_{max}}{\pi \times d^3}$$

K′—Form Stress Concentration Factor from sources such as Peterson's stress concentration factors. Typically between 2.0 and 3.0.

The material properties of the shaft are highly dependent on the material and heat treatment of the shaft. Allowable shear stresses (Ss) (in either torsion or bending) are typically 15 KSI for carbon steel shafts and 25 KSI for low-alloy shafts.

Bending Stress

$$Sb = \frac{32 \times K' \times M}{\pi \times d^3}$$

$$Se = \sqrt[2]{Sb^2 + 3 \times St^2}$$

It is common to assume that bending loads are fully cyclical and fatigue cracking can occur at stress levels significantly less than the materials' yield strength. The bending loads should be evaluated against the endurance limit of the shaft material. Accounting for surface stress concentrations, the corrected endurance limit will be approximately one-quarter of the ultimate strength.

Combined Stress

A modified Goodman diagram is a good tool and is often used for evaluating the effect of steady and alternating stress on a component such as the shaft.

Keyways The strength of a shaft whose cross section contains a keyway needs to be equal to the strength of a shaft whose diameter is equal to the distance from the bottom the keyway to the periphery of the shaft opposite the keyway. If two keyways are present, the strength of the shaft equals the strength of a shaft whose diameter is equal to the distance from the bottom of one keyway to the bottom of the other keyway.

The above general equations and rules of thumb are intended to provide a reasonable estimate of the stresses in a shaft end or to check the required shaft sizing. Shaft ends with large overhung weights, or with stresses exceeding the above guidelines, should be further analyzed with more sophisticated analytical tools such as finite element analysis.

Rotor Materials

The shaft ends of the various expander designs operate at essentially the same metal temperatures since the shafts are bathed in lubricating oil and are fairly removed from the hot gases.

Several different materials have been used for shafting in the nitric acid expander that include carbon-steel shafting per ASTM A321, low-alloy steels per ASTM A193 or A470, and 12% chromium conforming to X20 Cr Mo V 121.

Some high-temperature expander stub shafts can experience significant axial temperature gradients that significantly increase the operating stresses. To reduce the stub shaft operating stresses, the high-temperature section of the stub shaft is manufactured as two components, the stub shaft and a spacer ring or piece. The spacer ring material is usually manufactured from 422 stainless steel per ASTM 565 or A-286 materials per AMS 5735 (26 Ni, 15% Cr).

Tie Bolts

One of the most critical components in the rotor assembly is the rotor tie bolts. Both single centerline through bolts and multiple tie-bolt designs have been used. Stud bolts and through stud designs are most common.

Tie-Bolt Structural Design Considerations

The tie bolt is designed to withstand the stresses owing to the bolt preloading and from the differential thermal expansion between the tie bolt and rotor discs. The preload needs to be sufficient to allow the transmission of the stage torque through the rotor disc interfaces.

Tie bolts should not be used as shear pins to transmit rotor torque. Stage-to-stage or stage-to-shaft torque transmission is accomplished in a variety of methods including disc-to-shaft shrink fits, face-to-face friction, torque transmission pins, and face splines. The most common method of torque transmission is via the face friction. The tie bolts must exert sufficient load on the disc faces to allow full power transmission.

In addition to the torque transmission required, the tie-bolt working stress should be sufficient to withstand a 40G (40 times the rotor weight) unbalance load and to prevent the separation of the discs.

Tie-bolt operating temperatures can be several hundred degrees cooler than the rotor discs because the tie bolts are shielded from the tail gas and often cooled. This temperature differential can cause significant thermal stresses in the tie bolt(s). The thermal differential causes the tie bolt loading to increase under normal operation. These effects can be accentuated during a unit start-up.

Because the rotor tie-bolt lengths can be fairly long and much of the tie bolt is unsupported, tie-bolt natural frequencies need to be investigated. Rotor tie-bolt natural frequencies should be calculated to assure they do not interfere with the running speed of the rotor. Reduced-clearance locating sleeves or lands on the tie bolts are used to limit tie-bolt deflection and increase the bolt natural frequencies. Nonshaft centerline tie bolts are subjected to significant centrifugal loadings that increase bolt defections and bending stresses into the tie bolts.

Tie-Bolt Materials

Rotor tie bolts are made out of a variety of high-strength alloys. The material selection varies from 12% Cr martensitic stainless steels per ASTM A565, specialty 17-4 PH high-alloy steels such as ASTM A564 to nickel-base alloys such as Inconel 718 per AMS 5596.

6.7.8 Rotor Cooling

Many high-temperature nitric acid expanders use air or cool tail gas to lower and maintain the operating temperatures of the expander rotor and casing. The lower operating temperatures increase the structural margins and life of the components. A reduction of only 25°F (12°C) can double the life of a high-temperature component. Rotor cooling is very important since the loss of cooling can result in tie-bolt loosening, as well as rotor blade and disc failures.

Figure 6.25 shows a typical cooling flow through a high-temperature expander. The cooling media may be admitted into the outer casing and

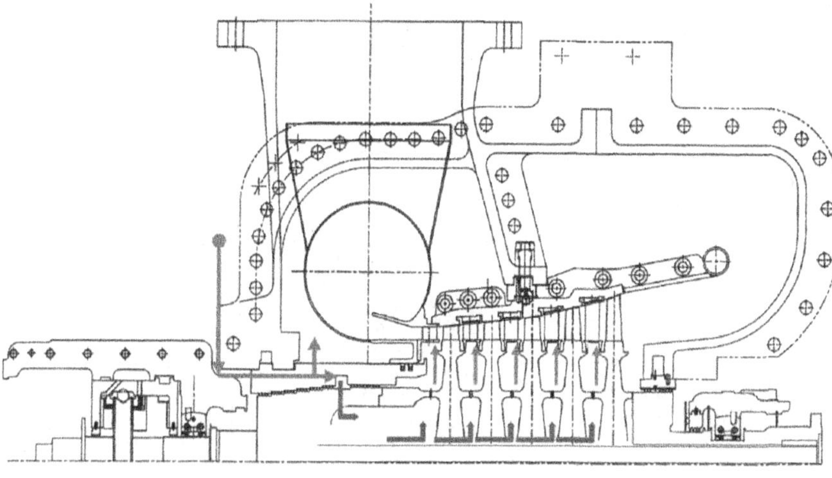

FIGURE 6.25 Cooling flow in a high-temperature nitric acid expander.

is directed into the cavity between the outer casing and the inlet scroll as well as toward the rotor. The cooling media enters the rotor via radial holes in the stub shaft and is directed up the faces of the individual discs and along the rotor tie bolt. The cooling media eventually mixes with the tail gas and carries out the discharge of the expander. The auxiliary equipment used to control the cooling flows is critical in assuring the proper amount of cooling. The cooling flow can be controlled via a differential pressure (supply versus internal supply cavity) or flow controller.

6.7.9 Shaft Seals

Nitric acid expanders have typically used a variety of noncontact seal designs. A majority of the seals are of the labyrinth design. The labyrinths need be thin and tall to dissipate any heat generated in a rotor rub. Owing to the toxic nature of the tail gas, the casing gas/end wall seals are critical and often buffered and vented to a safe location. Both integral and caulked in stainless-steel J strips are commonly used for the gas seals. Figure 6.26 shows the various expander rotor seals.

- *Bearing seals.* Conventional, standard nonsparking labyrinth or floating ring seals are used to control the flow of oil within the bearing housings.
- *Bearing housings.* Expander bearing housings generally use standard nonsparking labyrinth seals to seal the end wall seals of the

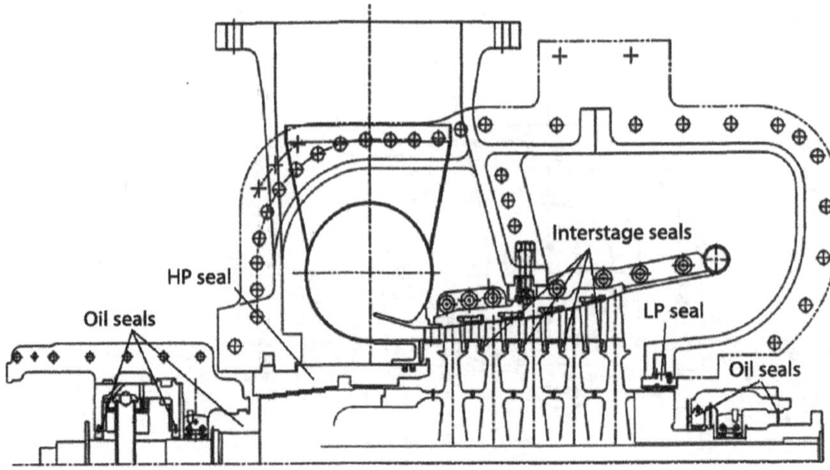

FIGURE 6.26 Various expander seals.

bearing housings. Bearing housing seals are often buffered with filtered and dried air to prevent both the leakage of contaminants into the housing and oil leakage out of the housing. Oil leakage poses a major fire hazard due to the high temperatures of the outer casings.

- *High pressure (HP) seal.* The inlet or high-pressure side of the expander casing prevents tail gas leakage by use of a high-pressure seal box. HP seal leakage represents a major safety hazard since the tail gas can be toxic and the high temperature represents a major burn hazard. In addition, high-temperature leakage of tail gas onto the bearing housing can cause oil coking problems in the bearing housings and represent a major fire hazard.

The HP seal box is generally made up of a number of stepped or high low labyrinth seals as can be seen in Figure 6.27. Tight seal clearances are required to prevent seal box leakage. Radial clearances of only 0.010 to 0.020 in (0.25 to 0.50 mm) are common for conventional labyrinth seals running against solid seal materials. Clearances as tight as 0.003 to 0.006 in (0.076 to 0.15 mm) are common for abradable seal materials. The most common abradable seal material is a stainless steel or Inconel honeycomb material that is configured in a fine honeycomb cell arrangement.

The main tail gas seal is almost always buffered to further reduce seal box leakage and to direct the cooling/seal media into the expander rotor and casing cavities. The seals are most commonly buffered with compressor discharge air or cooled tail gas.

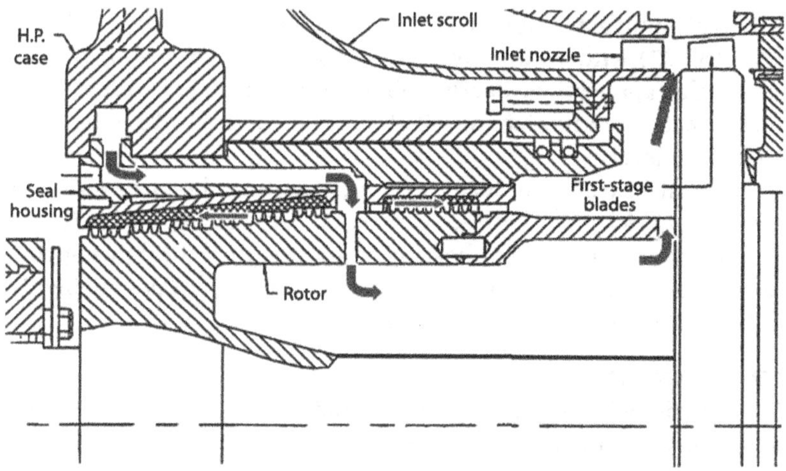

Figure 6.27 HP seal box with typical buffer seal flows.

- *Interstage seals.* Single- or multitooth labyrinth seals are either mounted to or are integral with the inner shroud ring of the stator vanes. The inside diameters of the seal rings are precision machined to provide a radial clearance of approximately 0.030 in (0.00076 mm).
- *LP seal.* The outer casing end wall seal is generally a standard labyrinth seal. The LP seal is a less critical application since it is required to seal lower pressure and temperature tail gas.

6.7.10 Bearings

Nitric acid expanders use standard design pressure lubricated fluid film journal and thrust bearings. The bearing design and selection is critical since these expanders operate at high speeds and often above the first or second critical speeds.

Journal Bearings

Nitric acid expanders have used a variety of journal-bearing designs over time. Both sleeve and tilting pad journal bearings are commonly used. Many early expanders were designed with tapered land or lemon bore sleeve-type bearings. These bearings provided high-load capabilities along with a high degree of fluid damping. Unfortunately, many of these early designs experienced operational vibration problems with sub-synchronous oil whirl.

More modern expander designs are using self-aligning tilting pad journal bearings. All expander bearings are force-fed lubricated with typical turbine mineral oils. An oil viscosity of 150 SSU @ 100°F (ISO VG 32) with an inlet oil temperature of 120°F (50°C) is typical for this application. Bearing oil flows are regulated with supply-side orifices. The lubrication oil should be cooled and filtered to a 10 micron level.

Journal-bearing unit loads are conservative and typically around 100 psi (7.0 kg/cm^2). Surface speeds are typically 150 to 175 ft/s (46 to 53 m/s). The journal-bearing clearances are critical in assuring reliable expander operation. The expander rotor dynamics are highly influenced by the bearing clearances and fits.

Thrust Bearings

Most nitric acid expanders have a unidirectional thrust from the high-pressure inlet toward the discharge. For this reason, the thrust bearings are typically mounted on the inlet end of the expander. The exception is the double-flow expander designs that operate with greatly reduced thrust loadings.

Multipad self-leveling active thrust bearings are the standard design for most nitric acid expanders. The bearings are selected to operate at less than 80% of the bearing manufacturer's calculated load capacity. Bearing loadings range between 300 and 400 psi (21 and 28 kg/cm^2). Mean surface speeds between 200 and 300 ft/sec (61 and 91 m/s) are common.

For some highly loaded or rerated expanders, thrust-bearing upgrades have been required. The use of copper chrome shoes, offset shoe pivots, and directed lubricated bearings are used in these applications.

Inactive thrust bearings are commonly a reduced-size tapered land since most expanders experience little reverse thrust loadings. These bearings are often referred to as bumper bearings and only see loading during unit start-up or shutdown.

6.8 Auxiliary Equipment

6.8.1 Trip or Trip-Throttle Valve

While most nitric acid expander installations use an expander inlet trip valve, many also use a trip/throttle valve. The trip valves are high-speed valves whose primary function is to rapidly shut down the expander under emergency conditions such as PRT overspeed, a coupling failure, or even excessive radial vibration or axial positions. The valves are designed to travel from full open to fully closed in less than one second. Throttle valves are used at times to allow more precise control of the process operating pressure. Expanders operate in the process system as a fixed orifice and control the process operating pressure. Changes in process temperature or flow can cause the expander to be too large or too small and cause the process to operate at a less than optimum pressure. The inlet throttle valve can be used to control the process pressure when the expander flow capacity is larger than the process need.

Expander inlet valves are typically butterfly valves that are sized to have minimal pressure drop (less than 0.3 psi [0.021 kg/cm^2]) across the valve in a fully opened position.

Governor Control Valves

Many of the low-temperature and warm expanders incorporate inlet control or governor valves as part of the expander design. These values are similar to manually operated multivalve steam turbines. The governor valves are typical plug valves that allow gas admission to various arcs of the first-stage inlet nozzle. The use of governor values significantly improves the part load performance and control of the expander.

Bypass Valve

While less common than the use of expander inlet control valves, some installations also have an expander bypass system with a bypass valve. The bypass system allows tail gas to bypass the expander. This can be useful if the expander flow capacity is too small for the process needs. Trying to pass more gas through the expander will increase the expander/process pressure to levels that may not be optimal for the process. By opening the bypass valve, the excess gas can be bypassed around the expander and allow the process to be operated at the desired pressure.

Two operational problems have been experienced with bypass valves. In the fully closed position, bypass valve leakage has been as much as 2% to 5% of the valve full-flow capability. This leakage translates directly to an equivalent loss in expander power. The most common problem has been valve sticking. The sticking is the result of the bypass valve operating for long periods of time in a fixed position at the full tail gas temperature.

Overspeed Trip Mechanism

All nitric acid expander installations are equipped with an overspeed protection system. At the detection of an overspeed, the steam turbine or motor and expander are tripped and the air compressors blow off to the atmosphere, removing the energy from the system. Speed detection is generally performed on both the steam turbine and expander.

Early expanders were constructed with mechanical overspeed devices that tripped a hydraulic circuit to initiate a PRT shutdown.

Although many mechanical overspeed designs have been used, they all work on a similar principle. A bolt goes through the shaft and is retained by a spring that is designed to exert a force equal to the bolt centrifugal force at 10% above the expander operating speed. If the rotor speed exceeds the trip speed, the bolt slides outward and extends outside the shaft to a point where it strikes a mechanical lever that sends a hydraulic signal to the trip valve and shuts down the PRT. Most designs also incorporate an emergency trip button or lever that will allow an operator to locally initiate a unit shutdown.

More modern expanders and upgraded existing expanders use an electronic trip device that has been proven to be more reliable and accurate. Magnetic speed pickups electronically measure the expander speed from a multitooth wheel on the expander shaft. Upon sensing a unit overspeed, the electronic control system initiates a PRT shutdown sequence. Multiple-speed pickups are used to provide system redundancy and prevent instrumentation failure trips. The multiple pick-up outputs are fed into a voting logic that is part of the overspeed control system.

Supervisory Equipment

Most nitric acid expanders are equipped with a full complement of supervisory equipment that includes the following:

- Oil supply temperature
- Oil supply pressure
- Journal-bearing temperature (2)
- Thrust-bearing temperature (3)
- Radial vibration (2)
- Axial position (2 or 3)
- Expander rotational speed
- Expander inlet temperature
- Expander inlet pressure
- Expander discharge temperature
- Low HP seal differential pressure

Turning Gear

Many nitric acid power recovery trains incorporate a turning gear on the train. Mounting is commonly done to the gear box if applicable. The turning gear is used to slow-roll the train at 10 to 50 rpm when the unit is shut down. Turning the train prevents more uniform cooling of the rotating equipment and helps prevent rotor thermal bowing (sags).

6.9 Operational Considerations

6.9.1 Start-Up Considerations

Auxiliary Driver Sizing and Type Part-Speed Operation

In many installations, the auxiliary drivers do not have sufficient power to bring the acid train to full operating speed. Off-design speed operation must be carefully controlled to prevent blade or vane failures in the compressor(s), expander, or auxiliary steam turbine. Blade and vane excitations often occur at part-speed operation and the fatigue failures can occur in even short periods (hours) of operation. The fatigue damage is cumulative so even very short-term operation at detrimental operating speeds can be damaging after a number of unit start-ups.

Burner Light Off The expander can be subject to significant temperature transients during the start-up of the compressor train and process. Initially, the expander is subject to the compressor discharge air. The heat-up rate is generally fairly slow and the maximum temperature achieved is typically only 400°F. When sufficient air and ammonia flow and pressure are achieved, the ammonia burner is lit and the expander inlet gas temperature rises quickly. Most expander manufacturers limit the temperature rise to 400 to 500°F per hour. Unfortunately, during this transient heat-up period, the process has the potential to discharge significant levels of NO_x to the atmosphere. These environmental discharges must be minimized, which can only be accomplished by quickly bringing the process up to its normal operating conditions. Instead of a two-hour warm-up as specified by the expander manufacturer, the expander is often heated up at rates as much as two to four times faster than recommended. These fast heat-up rates have created major mechanical problems in many nitric acid expanders. Some of the mechanical problems include outer and inner casing distortion and distress, inlet scroll distortion and distress, blade or seal rubs, and rotor bowing, among others.

A number of operational controls have been found to be effective in controlling the expander heat-up rate, including maximizing air preheater temperatures prior to burner light off, the use of quench steam to control the tail gas heat-up rate, and bypassing the expander during start-up (this obviously requires a larger auxiliary driver).

Normal Start-Up Sequence The following start-up sequence is commonly used in the start of the PRT:

1. Commission the power recovery train auxiliary systems (lube oil, expander seal, steam turbine seal—if applicable, electrical, etc.). Place the train on turning gear if applicable.

2. Bring the PRT up to its idle/warm-up speed (generally 500 to 1,000 rpm). Perform a walk-around inspection and review all supervisory instrumentation.
3. Increase the PRT speed to the train's lower limit considering critical speeds and resonant frequencies throughout the train. Confirm safe operating speeds on a speed-avoidance diagram.
4. Hold at speed to allow all train bodies to warm up and come to a steady-state operation condition. Perform a walk-around inspection and review all supervisory instrumentation.
5. Accelerate quickly to the minimum governor (lower speed limit) speed of the steam turbine (if applicable) or to motor synchronous speed. Avoid stopping or slow acceleration through train critical speeds and resonant conditions.
6. Hold at minimum speed to allow equipment and process to stabilize. Perform a walk-around inspection and review all supervisory instrumentation.
7. Ignite the ammonia burner and allow expander to heat up to its normal operating pressure and temperature. Perform a walk-around inspection and review all supervisory instrumentation.

Normal Operation Once the process and machinery operation has stabilized, it tends to operate at constant and steady conditions for long periods of time.

One unique operating issue with the nitric acid plant operation is that the catalyst gauze needs to be changed periodically. Approximately every six weeks the plant is shut down for a gauze changeout. The gauze changes are generally quickly performed and take less than 24 hr.

Unit Trip A typical trip sequence is as follows:

1. Steam turbine trip and throttle valve closes in less than one second. Electrical breaker opens on a motor driver.
2. Expander inlet valve is closed in one to three seconds.
3. Expander bypass valve opens and tail gas is diverted to the stack.
4. Production is hauled by closing the ammonia inlet valve at the burners.
5. Compressor surge valves (low- and high-pressure compressors) opens to prevent compressor surge.

6.10 Maintenance

Most nitric acid expanders operate for many years with minimal online maintenance. Major overhauls and turnarounds are generally set by the plant maintenance schedule. Turnaround times of four to eight years between each major unit overhaul are not unusual. The actual time between unit overhauls is very site dependent and is often dictated by the heat transfer equipment or compressor fouling. Rarely does the expander set the plant turnaround time.

6.10.1 Expander Components

Rotors Low-temperature or warm expander rotors can be expected to have component lives of 20 years or more. A majority of these rotors require component replacement prior to 20 years due to part deterioration from wear, corrosion, or mechanical deterioration.

High-temperature expander rotors have life-limited components such as the rotor disc, blades, and spacer sleeves. These components typically have design lives of 100,000 hr (11 years). Detailed rotor inspections need to be conducted after each operating campaign.

Flow-Path Components Although the rotor blades do have a limited life as noted, the stationary components are generally not life limited. These components tend to be high-maintenance components since their high operating temperatures can result in creep deformation. The inlet scroll, high-, and low-pressure stator casing often require major rework during each major turnaround that includes weld repair, heat treatment, and remachining. A realignment of the internal components is generally required as well, which involves the replacement of alignment pins and keys.

The stator vanes generally have long lives and require little rework unless damaged or worn. Stage seal teeth may need replacement if rotor contact has occurred.

Casings The expander outer casings are not life limited but do require periodic rework. The most notable form of distress is casing bore and flange distortion. This is particularly true for the higher inlet temperature expanders. Remachining of the casings is time consuming and needs to be accounted for in any major overhaul schedule.

Wearing Parts Generally all wear parts (bearings and seals) are replaced at each major turnaround of the expander. It is critical that the high- and low-pressure seal boxes be spared since rework requires significant overhaul time.

Nitric acid expanders are manufactured with numerous tight and critical clearances as well as many thermal expansion design features. The maintenance or refurbishment of these units requires significant machinery repair experience, precision machining and assembly techniques, and/or equipment that is not typically found in the field or chemical plant facilities. These capabilities are best found in a turbomachinery repair shop and make it common practice to remove the entire expander as an assembly during a unit turnaround. The casings are generally rigid enough to allow the unit to be removed in a single lift. Since the expander overhaul requires significantly more time than that of the other blower train machinery, some nitric acid expander users have found it to be economically viable to spare the complete expander. Maintaining a spare expander assembly is driven by process shutdown time. The spare unit can typically be swapped out in five days or less, while a unit rebuild generally requires four to eight weeks.

Spare Parts Most users of nitric acid expanders maintain a typical array of spare parts that include those listed below:

1. Class I spare parts—Wearing parts commonly replaced at each unit turnaround.
    - Journal-bearing insert or tilting pads
    - Thrust-bearing pads
    - Thrust-bearing shim packs or locator rings
    - Bearing oil seals
    - Casing low-pressure and high-pressure end seals
    - Interstage seal rings
    - Supervisory instrument probes
        - Vibration probes and proximeters
        - Axial position probes
        - Speed pickups
    - Flange gaskets and sealants
2. Class II spare parts—Assemblies that might reduce turnaround times and the replacement of flow-path components that might be damaged.
    - Journal-bearing shells and retainers
    - Thrust-bearing retainers
    - Internal-bearing housings such as adjusting rings or combined bearing carriers

- HP seal box
- LP seal box
- Complete set of rotor blades
- Complete set of rotor blade locking hardware
- Complete set of stationary nozzle or vanes
- Spare critical casing fasteners
- Loose rotor components
    - Sleeve(s)
    - Thrust collar
    - Balance piston(s)

3. Class III spare parts—Parts that provide the maximum degree of reliability and backup protection due to major machinery problems.
    - Complete rotor
    - Stator casing(s)

Outer casings and inlet scrolls casings are not normally spared.

6.11 Nitric Acid Expander Common Operational Problems

Nitric acid expanders are generally very reliable turbines that operate for many years uninterrupted. Although the overall reliability of the expander would be considered to be equal to or better than most other turbines, there are a number of operational problems unique to the nitric acid expander. These problems tend to occur more in the higher temperature (hot) expander applications.

Case Study 6-1 Rotor Vibration

The most common operational problem with nitric acid expanders is rotor rubbing against the stationary components as can be seen in Figure 6.28. Rotor rubbing causes high radial vibration. These expanders operate with tight clearances in the gas seal regions to prevent the leakage of the hot tail gases. Tight blade tip and interstage seal clearances are often used to boost the efficiency of the expander.

FIGURE 6.28 Expander rotor blade tip rubs.

Piping Strain

As with many turbomachinery designs, piping strains can cause excessive loads to be applied to the expander casing. These strains can cause casing distortions and rotor to casing misalignments that result in rotor rubbing. The hot inlet piping is the primary source of these loadings owing to the higher thermal growths that need to be accommodated. Inlet piping strain is the number-one operational problem of nitric acid expanders. The problem is less of a concern on the expander exhaust, because the exhaust ducting is generally isolated from the expander by use of a fabric expansion joint.

Casing Distortion

Another major cause of rotor rubbing is thermal distortion of the internal flow-path components and seal boxes. The internal casings are significantly hotter than the outer casing and must grow freely. A majority of the internal casings maintain their alignment and thermal growth capabilities by being mounted on radial keys and/or pins. These guide components must allow for expansion of the casings. Distortion of either the casing or guide keys/pins will cause flow-path casing distortion and often results in a rotor rub. The following

Figure 6.29 shows inlet scroll damage due to bound-up expansion pins. At every major turnaround, the expansion keys/pins must be inspected for freedom of expansion. This requires the pin holes to be perfectly straight, round, and in alignment. Similarly the pins must be straight, round, and have adequate clearances (both radial and axial). The selection of the pin material is critical to prevent wear, corrosion, and high-temperature deformation. Alloy steels have been problematic owing to excessive corrosion, whereas stainless steel often does not have adequate strength and tend to gall. Inconel and Stellite expansion pins and keys have provided the best operational experience.

Rotor Distortion

Another common cause of expander rotor vibration is the result of distortion of the rotor assembly. Distortion can result from rotor rubs that cause local hot spots and or mechanical damage. Rotor distortion can also occur from relaxation of the rotor tie bolt(s). Excessive operating temperatures, rapid temperature changes, and/or the loss of rotor cooling can cause the tie bolt(s) to be overstresses and experience loss of tie-bolt load and allow rotor component movements. Large rotor bowing and unbalance results in high rotor vibrations.

Case Study 6-2 Scroll Thermal Binding

The hot internal casings of the nitric acid expander experience higher thermal growth and deflections due to the high operating temperatures. Any constriction of the thermal growth can result in very high thermal stress that cause major casing distortions and failures. The inlet scroll shown in Figure 6.29 is an example of thermal growth constriction. The inlet scroll is mounted to the outer casing of the expander via three radial pins and two horizontal guide keys. In the noted example, the inlet scroll was not allowed to grow freely and major scroll distress was experienced. The photograph shows one of the radial pin guide bosses being broken out of the scroll and horizontal guide key support being pushed inward into the scroll by more than 0.5 in (12 mm).

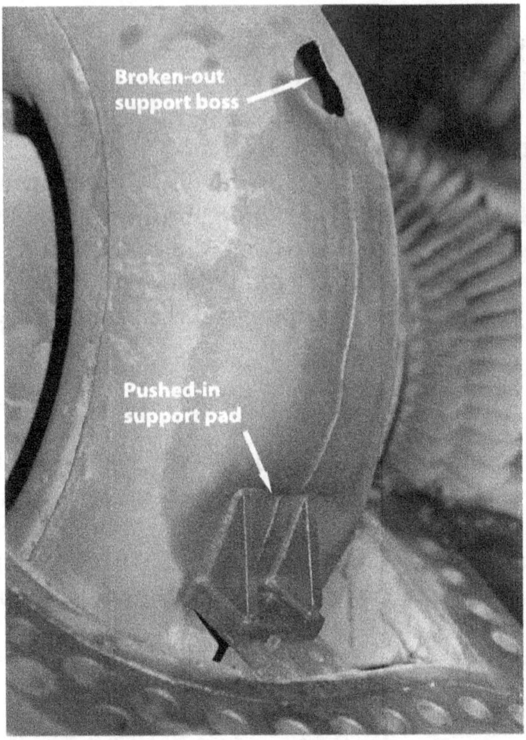

Figure 6.29 Damaged inlet scroll due to thermal binding.

Case Study 6-3 Material Degradation

In-service aging of many of the expander components can occur in the hot expanders. Various types of material aging occur depending on the chemistry, heat treatment, and operating temperature of the various expander components. Several typical aging phenomena affecting expander materials include the following:

Temper embrittlement which can occur in low-alloy steels operating in or passing through the material's transformation range at 700 to 1,100°F (371°C to 593°C).

Creep embrittlement can occur in alloy steels operating in the 800 to 1,200°F (426°C to 619°C) temperature range. While all of the in-service aging involves time and temperature, the creep embrittlement is associated with the operating stress levels.

FIGURE 6.30 Severe in-service failure of expander inlet scroll.

Sigma phase embrittlement can occur in 18-8 stainless steels subjected to operating temperatures between 1,200°F and 1,470°F (650°C and 800°C), with the maximum rate of formation occurring at 1,380°F (750°C).

Embrittlement causes major reductions in ductility and impact properties of the materials and makes them very susceptible to in-service thermal cracking. Once a crack is formed, the cracks tend to run very quickly in the embrittled material. The aged/embrittled materials are difficult to repair due to the high local stresses created during the repair, which tend to drive the existing crack(s). Some success has been achieved after a full solution annealing of the affected part. The inlet scroll shown in Figure 6.30 shows a major failure that is typical of material degradation due to aging.

Case Study 6-4 Excessive Seal Leakage

As can be seen in Figure 6.31, Nitric Acid Expanders use multiple internal seals to control tail gas leakage. Seal leakage can be detrimental to the expander operation. Gas leakage in the interstage seal region can lead to significant power losses (−2% to 5%) inside the expander and increased operating temperatures of the downstream stages.

Excessive high-pressure seal box leakage can cause significant operational problems and represent safety hazards to plant personnel. The leaking tail gas can exceed 600°F (316°C) and is both an irritant and potentially toxic to personnel.

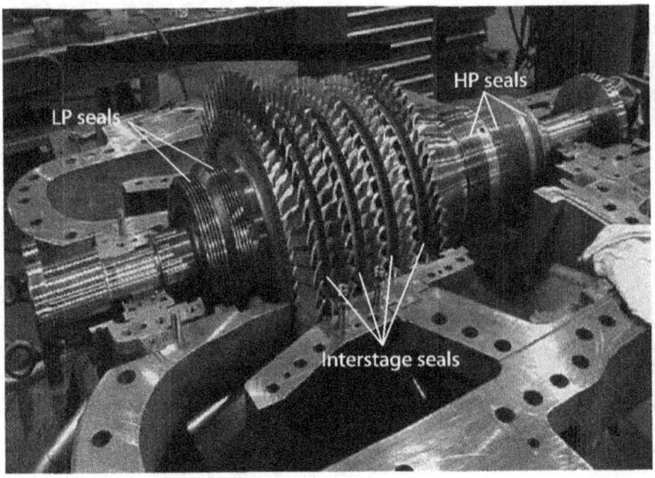

FIGURE 6.31 Expander rotor seals.

Direct impingement of leaking tail gas, from either the HP or LP seal boxes, onto the bearing housing can result in oil coking inside the bearing housing and result in bearing and seal failures. The leaking tail gas is hot enough to be an ignition source for the lubrication oil.

Seal leakages can be the result of worn seal teeth, worn abradable seal material, improperly aligned seal boxes, improperly supported seal boxes, and inadequate buffer seal purge media.

All seals need to be carefully designed to accommodate a rub condition. A large amount of heat can be generated from a seal rub. If this heat buildup is not adequately dissipated, rotor damage can occur. This often results in major rotor bowing and vibration. Rotor vibration often occurs despite the use of rub-tolerant seal materials. The problem is accentuated in the hot expanders because many of the rotor and seal materials have a tendency to gall and build up in a rub condition.

Case Study 6-5 Thrust-Bearing Failure

Understanding the expander thrust loadings is critical to the selection of proper thrust bearings. The expander operates with relatively small axial clearances and a bearing failure is likely to result in extensive rotor damage as can be seen in Figure 6.32. This expander rotor thrusted downstream and rubbed four stages of rotor blades and discs as well as damaged the HP and LP stub shafts. Very few parts of the rotor were salvageable.

FIGURE 6.32 Expander thrust-bearing failure: Note extensive blade and disc rubs.

The cause of the failure was an undersized active thrust bearing. The expander had been rerated several times over the life of the installation (over 20 years). While the bearing was adequately sized for the early years of operation, the bearing was pushed beyond its safe operating capacity with the multiple rerates. Prior to the bearing failure, the steady-state operating temperature exceeded 230°F (110°C). Without warning, the temperature spiked to over 280°F (138°C) and the rotor thrusted toward the active thrust bearing. The installation did not have an automatic unit trip system for bearing over temperatures.

The solution to the bearing overloading was to upgrade the active thrust bearing with copper chrome shoe material, offset pivots, directed lubrication, and an improved oil-drain configuration.

Case Study 6-6 Erosion

Even though the tail gas in most nitric acid expanders is clean and dry, some installations experience flow-path erosion due to entrained particulate. The particulate is often the result of dislodged piping and process vessel oxidation. As can be seen in Figure 6.33, the erosion is most severe in the inlet stages of the expander owing to the higher

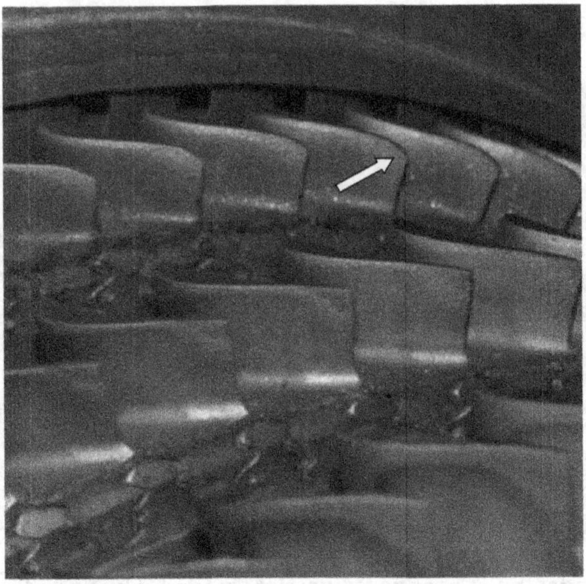

FIGURE 6.33 Rotor blade erosive wear.

gas velocities and the tendency for the particulate to get centrificated outward in the flow path. The leading edge and suction-side erosion is typical of large (>10 micron) particle erosion.

Little can be done in the expander to prevent flow-path erosive wear. The solution to blade erosion is to eliminate the source of the particulate.

Case Study 6-7 Corrosion

Most nitric acid expanders operate with tail gas that is composed of more than 95% nitrogen and a small amount of carbon dioxide, argon, and water. The gas is dry with less than 3.5% water and little in-service corrosion occurs to the expander flow path and rotor.

Whereas the normal operation has a low corrosion potential, the expander does operate during start-up and transient conditions with hot air that contains significantly higher oxygen and water levels (21% oxygen and 1% to 5% water). The oxidation potential is significant at temperatures above 800°F (427°C) for many of the iron base materials. Figure 6.34 shows heavy oxidation of stator casing fasteners subject to high temperature and the oxidizing environment. Stainless steel or

FIGURE 6.34 Heavily oxidized stator casing fasteners.

nickel-based fastener materials must be used to prevent excessive oxidation and in-service failures.

Whereas in-service corrosion is not common, shutdown corrosion can be a problem in many expander installations. Occasionally, nitrous oxides can be carried over and in the presence of water form very aggressive acids. The acids can cause corrosion-pitting damage as can be seen in Figure 6.35. Keeping the expander dry during shutdowns and transient conditions is important to prevent aqueous corrosion problems.

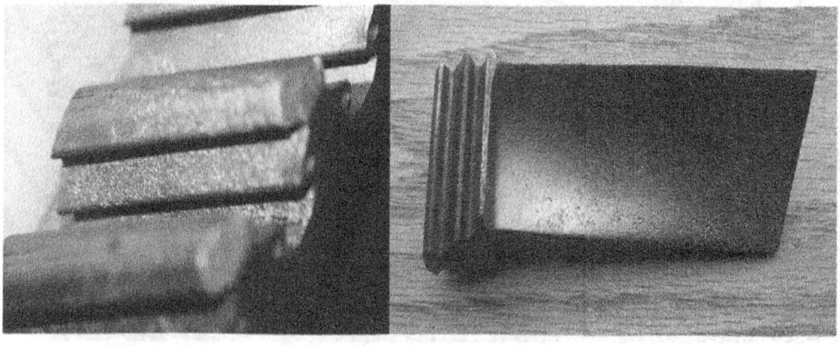

FIGURE 6.35 Rotor disc and blade corrosion pitting.

Companders in Nitric Acid Plants

In the early development of the nitric acid power recovery train, a number of machinery manufacturers developed single-casing machines that incorporated both the process compressor and gas expander. A typical compander can be seen in Figure 6.36. One manufacturer included a start-up steam turbine as part of the single-casing design. This type of power recovery train was mainly applied to nitric acid plants of less than 300 TPD. Although these compressor/expander combinations have a lot of appeal from an initial and installation cost perspective, they generally had operational problems that included rotor dynamics issues and casing thermal distortion problems.

The compander design reemerged in the 1990s as multishaft compressor and expander technology developed. These machines resulted from the development of even more compact and economical compressor and expander designs. The integration of the expander stage(s) into the gear unit configuration of the multishaft compressor leads to the rebirth of the compander. The gear-driven multistage compressor and expander design have grown in size to service intermediate single-pressure plants as large as 500 tons per day.

The compander design has been applied to other applications including naval ship boiler superchargers. In this application, a single-body steam turbine-compressor-expander is used to provide pressurized air to a shipboard steam boiler.

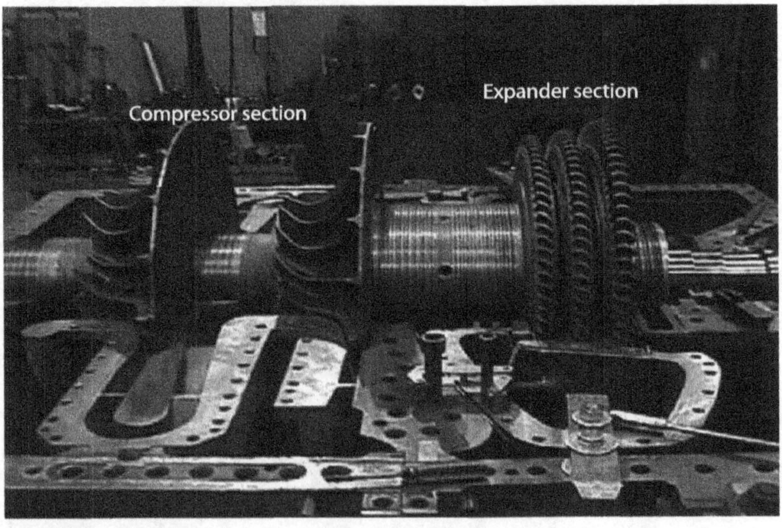

Figure 6.36 Compander with expander and compressor sections in a single casing.

CHAPTER 7

Turboexpanders/ Cryogenic Turboexpanders

7.1 Introduction

There is a family of single-stage, generally high-speed turboexpanders used to recover power in a wide range of processes. The design of these machines is radial inflow expanders connected to either a compressor or generator to recover power. For the family of power recovery expanders, applications as turboexpanders/cryogenic turboexpanders are the most numerous. They are often referred to as cryogenic turboexpanders because a major application is in refrigeration systems.

These turboexpanders are radial inflow machines in which gas enters at the outer diameter of the impeller through inlet guide vanes that are most often variable and exiting at the eye of the impeller. Power is generated by the expansion of the gas and the reduction of the outlet temperature of the gas stream. In addition to the recovery of power, these turboexpanders are used in a variety of oil and gas processes to augment very low-temperature refrigeration streams. This effect can also be used to separate liquid components from saturated streams, thereby recovering some heavier product.

Turboexpanders are found in ethylene plants, air separation facilities, gas processing plants, refineries, and gas liquefaction facilities.

Applications of these machines have been expanding in number and size. Two recent applications were in the industrial hub of Ras Laffan with a 15,100-horsepower (hp) expander compressor combination, and in Qatar with a 13,900-hp combination. The demand for increasing sizes is expected to grow as the application of

FIGURE 7.1 A large turboexpander mounted on a factory test stand. (courtesy of GE Oil & Gas.)

modern design technology and tools permits the design of even larger machines with sufficient reliability. Figure 7.1 shows one of the large turboexpanders mounted on a factory test stand.

7.2 Historical Background

The concept of using a centrifugal machine as an expander for refrigeration purposes goes back to Germany with Siemens in the mid-1850s. The development proceeded in fits and starts until the 1930s when Guido Zerkowitz, a Linde engineer, completed a design. In the late 1930s, the Russian Pyotr Kapitsa designed and built a centrifugal expander whose basic design has evolved as the basis for today's machines. Kapitsa was a low-temperature physicist who designed the machine for his physics work. In 1978, he received a Nobel Prize in Physics.

The demand for these machines grew dramatically in the 1970s in natural gas processing plants to extract ethane, propane, and butane from oil-condensate and gas streams. They were also employed in ethylene in the 1970s and beyond to replace the Joule-Thompson (J-T) valves. As the applications grew in numbers, new and improved engineering designs and manufacturing tools enhanced the machines, making them more capable and reliable.

In the 1980s and 1990s, industrial demands, particularly in the oil and gas industry, led to advanced designs that were better able to provide operability and maintainability in these machines. The ready availability of computers to apply FEA and CFD to the designs permitted the designers to develop better, more predictable designs.

A seminal advance occurred with Dr. Judson Swearingen, who patented a design in 1991 that was built by his company, Rotoflow. That company spurred the application and the growth of the usage of these machines in hydrocarbon applications. This was paired by the work of Sharif Mafi with his company, Mafi-Trench. A principal focus was natural gas and ethylene plants.

In more recent years, mergers and buyouts have changed the names of the previous manufacturers. Current manufacturers include GE Oil & Gas, Atlas Copco, Cryostar, Ebara, and others.

Earlier designs experienced issues with impeller failures due to resonances and some rotor dynamics issues. In addition, there were a number of impeller failures as the result of problems in fabrications of the complex shapes. Advancements in technology, including the application of finite element analysis, computer-aided fluid design, advanced measurement applications, and magnetic-bearing applications, have improved the designs. In addition, advances in fabrication technologies, inspection, and testing have been shown to be the most valuable. The market is growing both in numbers and capacity.

7.3 Machinery Design and Configuration

7.3.1 Turboexpander Design and Construction

The overall design and construction of a cryogenic turboexpander is rather straightforward. Gas enters at the outer diameter of the machine. It is turned and directed by the inlet guide vanes into the outer diameter of the complex impeller. In most cases, the guide vanes are adjustable and are used to optimize the efficiency of the expansion for variations in operating conditions. A requirement in the American Petroleum Institute (API) 617, Centrifugal Compressors for Axial and Centrifugal Compressors, is that turboexpander compressors specify that all inlet guide vanes be variable. (API 617, Seventh Edition, 2002).

Gas travels through the impeller, which is typically a three-dimensional, twisted, tapered configuration that exhausts at its inner diameter. The gas is exhausted axially at a lower pressure and temperature through a vaneless diffuser in most cases. Depending on the service, the gas may then go through a separator to the exhaust gas delivery system. This may include exhausting into a compressor for use in a refrigeration system or delivery into a fuel gas system.

Figure 7.2 shows a cutaway sketch of a turboexpander as viewed from the impeller discharge nozzle. It shows the impeller with the inlet guide vanes. Visible are the variable-guide vane control arms. Figure 7.3 shows the inlet guide vane/impeller configuration from both the front and back sides.

Turboexpanders are designed as reaction machines. That is, a portion of the pressure reduction is accomplished across the inlet guide vanes (stationary element) and a portion of the pressure reduction is accomplished across the impeller vanes. The design maximizes the expansion efficiency by matching the nozzle exit velocity and the impeller relative speed. The configuration of the radial inflow expander permits a near isentropic expansion of the gas by transferring the energy so efficiently. Normal designs have reaction ratios that range from 50% to 65%.

FIGURE 7.2 Cutaway sketch of a turboexpander looking at the impeller and inlet guide vanes from the discharge side.

FIGURE 7.3 Front and back views of a turboexpander impeller and adjustable inlet guide vanes.

Gas entering the turboexpander is often saturated and at a relatively high pressure. The nozzle exit velocity and the impeller relative speed are matched carefully. This technique is used to achieve very high efficiencies in the expansion. To maintain the proper velocity ratio, the inlet flow rate needs to be controllable and adjustable to accommodate changes in operating point including changes in gas condition. Figure 7.4 shows a typical plot of isentropic efficiency versus velocity ratio. Note that in the desired operating range, the efficiency in is excess of 90% approaching 95%:

U_2 = Blade velocity at the impeller outer diameter in ft/s
C_0 = Stage isentropic enthalpy change expressed in fps
dhs stage = Stage isentropic enthalpy change, energy per unit mass in Btu/lbm

$$C_0 = 223.8\sqrt{dhs\ stage}$$

C_0 is sometimes called the isentropic spouting velocity and is the calculated isentropic enthalpy drop across the stage expressed in velocity (fps).

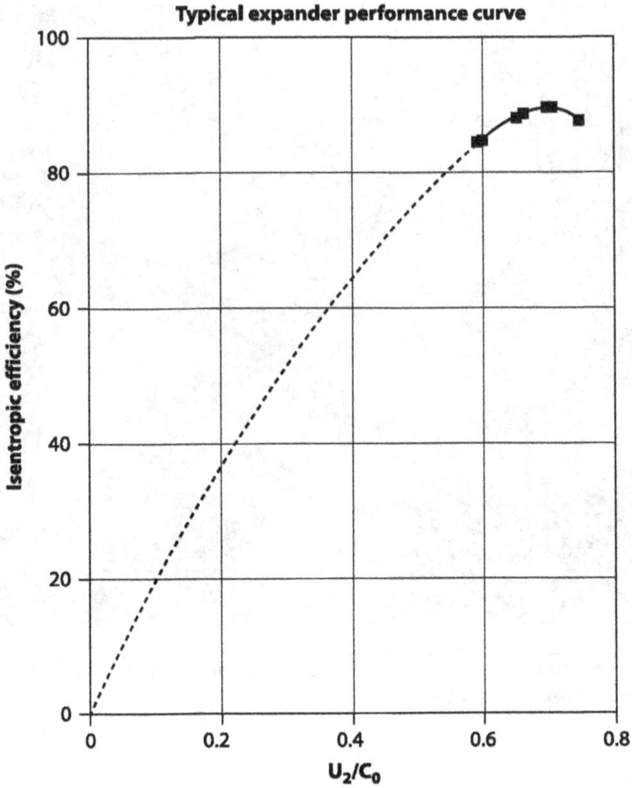

FIGURE 7.4 Efficiency curve for a typical turboexpander Jumonville (2011).

7.3.2 Impeller

The impellers used in these turboexpanders are three-dimensional twisted, tapered, aerodynamic designs. A lot of design effort has been expended over the years to:

- Optimize the aerodynamics for performance and efficiency.
- Permit the impeller to carry the stress levels transmitted to it.
- Calculate structural resonances so that these may be avoided in normal operation.

Earlier applications of turboexpanders had problems with failures of the impellers due to overload and/or resonant excitation of the vanes.

In addition, most of the impellers are single piece, which means they were milled out of a single piece of material. Early on there were problems with failures at the attachments when the impeller vanes were welded to the wheel.

FIGURE 7.5 A typical turboexpander impeller.

Figure 7.5 shows a typical impeller. It is of a single piece, is unshrouded, and has a typical twisted, tapered shape. This is the most common design applied to these machines. However, vendors may have a variety of designs available for specific applications, including shrouded and scalloped designs.

Figure 7.6 shows a damaged impeller (Cencula et al. (2006); Singh et al. (2007)). In this case, the failure was caused by resonant excitation exacerbated by cracks in the protective coating on the blade. This type of damage on the tips of the exhaust end of the impeller was one of the major failure manifestations experienced with the turboexpanders. Careful design work on the part of the manufacturers has limited this failure in the most recent applications. More detail on the analytical tools available to evaluate the impeller designs is presented in Chapter 9.

Current technology advances in the analytical tools available, fabrication and machining enhancements, and testing capabilities have significantly reduced this type of problem. They may still be encountered but are far less likely to be encountered.

7.3.3 Shaft

In the case of the turboexpander driving a compressor in a refrigeration-type service, the shaft is very often common to the turboexpander and compressor. Figure 7.7 shows a drawing of this class of machine. The flow that exhausts from the turboexpander may pass through a separator or other treating steps and is then input to the compressor wheel. In many/most of these installations, the compressor

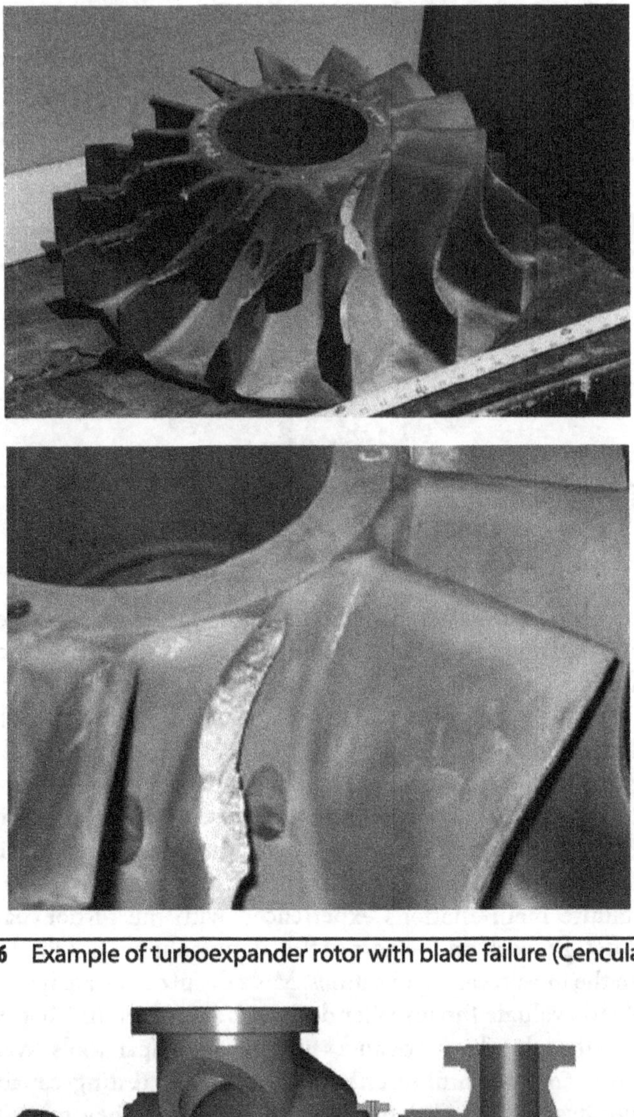

Figure 7.6 Example of turboexpander rotor with blade failure (Cencula et al.).

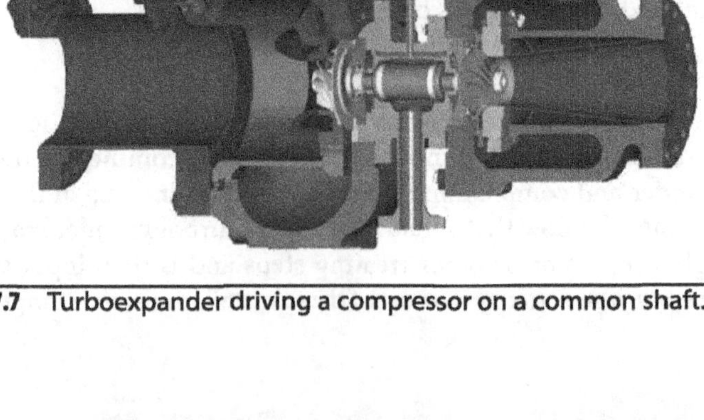

Figure 7.7 Turboexpander driving a compressor on a common shaft.

is a single-stage design with an aluminum impeller manufactured in a similar fashion to the turboexpander impeller. In most applications of a turboexpander, a single-wheel compressor is able to produce sufficient head for the service.

In the case where the turboexpander is driving a generator, there will be a gearbox interposed between the turboexpander and the generator.

7.3.4 Bearings

Turboexpanders were initially designed with hydrodynamic bearings. Many of the earliest designs used multilobe bearings. Those bearings have good damping characteristics but modest stability margins. They also can cause manufacturing headaches, particularly in smaller sizes.

Current designs will often use tilting pad bearings. There was an initial early effort to design turboexpanders as "stiff shaft" machines. That means that the critical speeds are above operating speeds. However, the high speeds required to achieve the capacity and efficiency desired in the machines precluded that design option. Rotor dynamics issues were experienced. A comprehensive discussion of one of the earlier issues is shown in Davidson et al. (1992). These have since been addressed and managed and have been experienced relatively infrequently with current generation machines.

Many if not most current design turboexpanders use magnetic bearings. A major driver for this application is the very damaging effect of oil leaking into the exhaust gas stream on turboexpanders used in refrigeration systems. This oil becomes cooled, thick, and sticky and plates out on the downstream exchangers. Heat transfer can be very severely degraded in this fashion. It has also been stated that it is not possible to get all of the oil out of the cold box in several refrigeration services. Because the turboexpanders are used so extensively in refrigeration services, this was a significant driver to move to magnetic bearings. The application list for users of turboexpanders with magnetic bearings is quite extensive.

Figure 7.8 is a cutaway sketch of a typical turboexpander compressor configured with magnetic bearings. The two radial bearings are located inboard of the impellers with an overhung expander and overhung compressor. The thrust disc is located mid-span on the shaft with a magnetic thrust bearing on either side of the disc. Magnetic bearings do not require oil lubrication so there is no need for an oil supply. This eliminates the issue of leakage into the downstream exchangers. This design has become very nearly a de facto standard by this time.

Although there has been considerable experience with the application of these machines with magnetic bearings, the application also

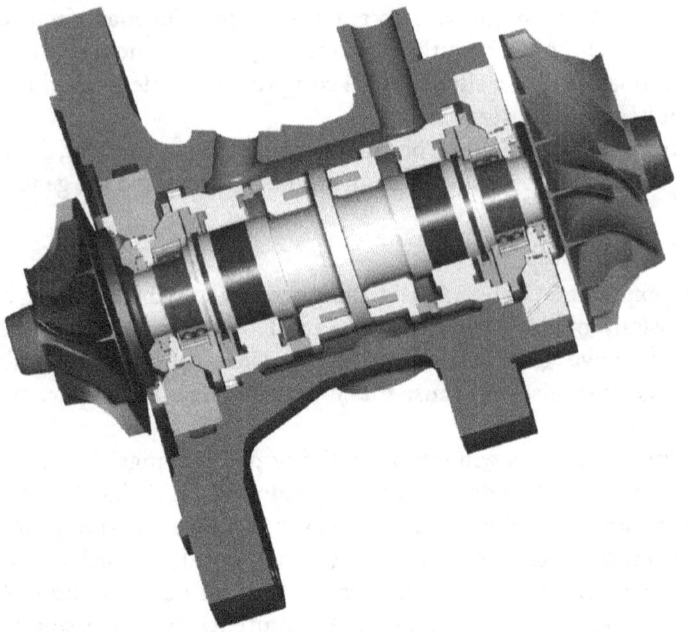

FIGURE 7.8 Cutaway sketch of a turboexpander with magnetic bearings. (courtesy of Atlas Copco.)

brings some of its own issues. One of those is that magnetic bearings are "softer" than hydrodynamic bearings, that is, the stiffness achievable is less than is possible with hydrodynamic bearings. This means that the critical speeds will be reduced from that achievable with hydrodynamic bearings and must be accommodated. A second issue is that the magnetic bearings have less load-carrying capability. This makes the thrust balancing of the turboexpander more critical.

Another issue with magnetic bearings is that they cannot operate without a feedback control loop governing their performance. Figure 7.9 shows the magnetic bearings inside a control loop. Even though the figure looks simple, inside the "blocks" are complex control devices that must be tuned. The position is measured, and the feedback signals manage the strength of the magnets to position the shaft. In addition, the stiffness and damping characteristics of the bearing are controlled independently by the electronics in the control loop. On some magnetic-bearing controllers, there may be embedded adaptive control to adjust the characteristics as the machine runs and degrades, in effect counteracting the impact of unbalance changes on the rotor.

Among the additional design features that must be employed with magnetic bearings is the need for backup power. And backup

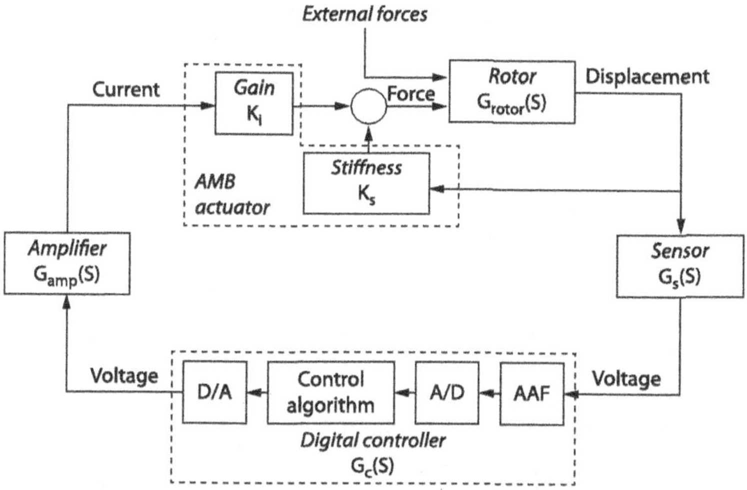

Figure 7.9 Controls configuration for the magnetic bearings.

bearings must be included on the shaft in case the magnetic bearing levitation is lost. There has been considerable development of these backup bearings. Initially, they were loose-fitting antifriction bearings. However, there were severe limits on the number of "drops" that could be experienced without having to tear down and replace the "backup" bearings. Special designs have been developed and employed by the magnetic bearing manufacturers. Magnetic bearings have been recognized by API and are addressed extensively in the current revisions of API 617 and API 684 currently nearing completion.

As an example of the type of problems that may be encountered, Cencula et al. published a case study discussing his and his colleagues' problems with the magnetic bearing controller. In this case, the problem was the result of initial inadequate settings of the transfer functions on the magnetic bearing controller that led to vibration excursions of the machine.

The experience with magnetic bearings has been largely positive with magnetic bearings in turboexpanders. In fact, it is most likely that it is this class of turbomachinery that has the most active magnetic bearings installed in operating units. There are a number of users who wouldn't consider another configuration.

7.3.5 Shaft Seals

The basic design of a turboexpander is as a hermetically sealed system. However, the product gas may have particles in it as well as being at significantly reduced temperatures. As a result, there is a need to provide

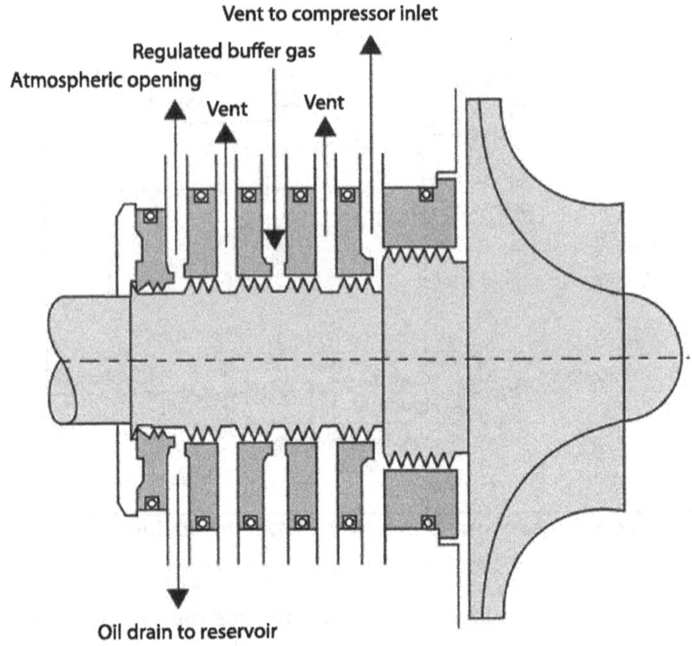

FIGURE 7.10 Straight labyrinth seal. (courtesy of Atlas Copco.)

sealing and a seal gas system for the machine to protect the bearings from the gas, both with hydrodynamic bearings and magnetic bearings. Hydrodynamic bearings may experience contamination of the bearing fluid as well as viscosity changes due to the cooling effect. Magnetic bearings have many components that may not be consistent with the contact.

The majority of turboexpanders use a basic labyrinth seal as shown in Figure 7.10. This shows a stepped labyrinth with seal gas entering in the middle of the labyrinth and multiple vents. A variation of the labyrinth seal is sometimes used with a tapered stepped labyrinth. This may be used since it has some capability to deal with axial movement of the shaft.

In addition, the other common compressor seals may be specified and applied. Figure 7.11 shows the dry gas seal. In some cases, a floating carbon bushing seal or oil seal may be used.

7.3.6 Thrust Balancing

Turboexpanders attempt to preserve the maximum capability of their thrust bearings through use of an automatic thrust equalization system. Figure 7.12 is a schematic of a typical system used on almost all machines. Signals from the bearings give the equalizer the

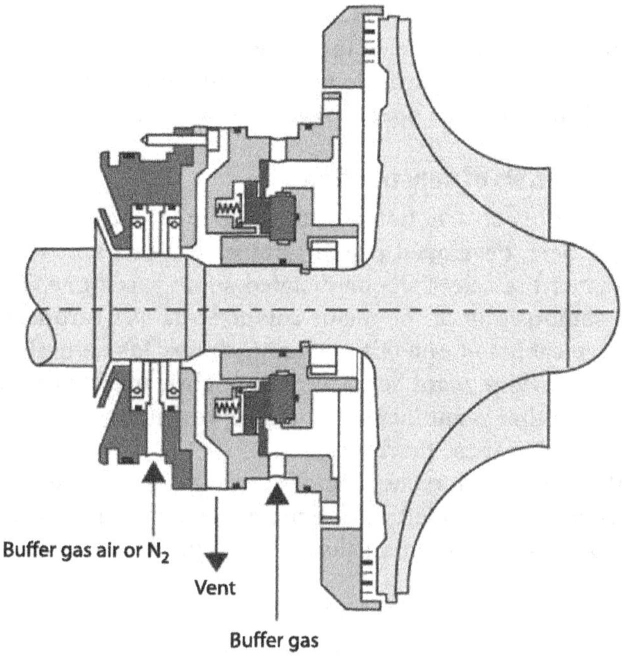

FIGURE 7.11 Single dry gas seal. (courtesy of Atlas Copco.)

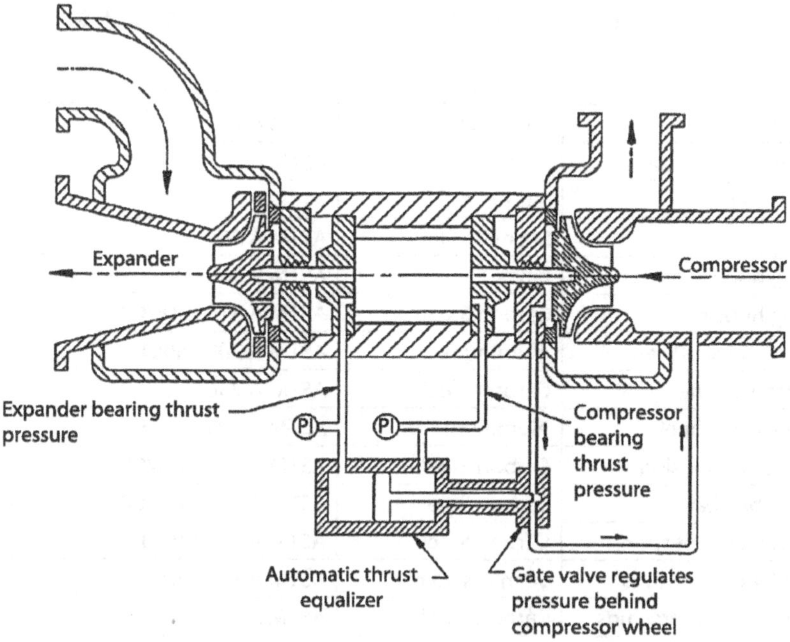

FIGURE 7.12 Automatic thrust balancing (Jumonville, 2012).

indication of the shaft position. This causes a valve to move, changing the pressure behind one of the impellers to change to balance the thrust. In this way, the thrust load on the bearing is maintained at a low level throughout the operating range of the machine.

7.3.7 Materials of Construction

The materials of construction of a turboexpander are not very exotic. They have been developed over a number of years and experience has validated the selections or dictated some variations. Table 7.1 lists the composition of the major components of a turboexpander. It is the typical list of one of the manufacturers for natural gas processing or ethylene plant services. Although it may not be exactly the same as other manufacturers, it is representative of the type of materials used in these machines.

In the case of more nonstandard operating requirements, the materials may be upgraded. For example, the standard wheel material shown is a 7075-T6 aluminum alloy which is used up to around 225°F.

Description	Nominal Material	Typical ASTM Specification
Expander housing	Stainless Steel	ASTM A 351 GR CF8
Inlet guide vane, linkages, and housing	Stainless Steel	ASTM A 276 416 and 410
Nose cone	Aluminum	ASTM A 352 GR LC3
Expander wheel	Aluminum	ASTM A 209 - 7075
Expander wheel seal	Brass	ASTM B 36
Heat barrier	MICARTA	
Shaft	Stainless Steel	ASTM A 564 - 630
Bearing body	Brass	ASTM B 36
Bearing housing	Carbon Steel	ASTM A 216 GR WCB
Compressor shaft seal	Aluminum	ASTM A 209 - 6061
Compressor wheel seal	Carbon Steel	ASTM A 236
Compressor wheel	Aluminum	ASTM A 209 - 7075
Compressor housing	Carbon Steel	ASTM A 216 GR WCB
Diffuser bracket	Carbon Steel	ASTM A 216 GR WCB
Compressor inlet spool	Carbon Steel	ASTM A 516 GR 70
Expander casing studs	Stainless Steel	ASTM A 320 GR B8
Compressor casing studs	Carbon Steel	ASTM A 193 GR L5

TABLE 7.1 Typical Materials of Construction for a Turboexpander

At higher temperatures, the casing is changed from stainless steel to carbon steel unless the gas is corrosive. The impeller may change to titanium alloy 6 AL 4V, which is sometimes required because of rotor dynamics issues. If the operating speed is low and rotor dynamics issues aren't present, 17-4PH stainless may be used. At higher powers, the expander wheel may be upgraded to titanium or 17-4PH. Break points on these applications will vary between the vendors based on their experience and design calculation results.

7.3.8 Driven Equipment

When a turboexpander is used in a refrigeration system, it will drive a compressor. In many if not most of the designs, this compressor is on the same shaft as the expander. Configured in this way, the unit is often called a compander. Figure 7.13 shows a compander configuration.

Figure 7.14 shows a figure of a turboexpander driving a generator. Note the gearbox included in the generator drive configuration as the turboexpander operating speed is normally higher than that of most generators.

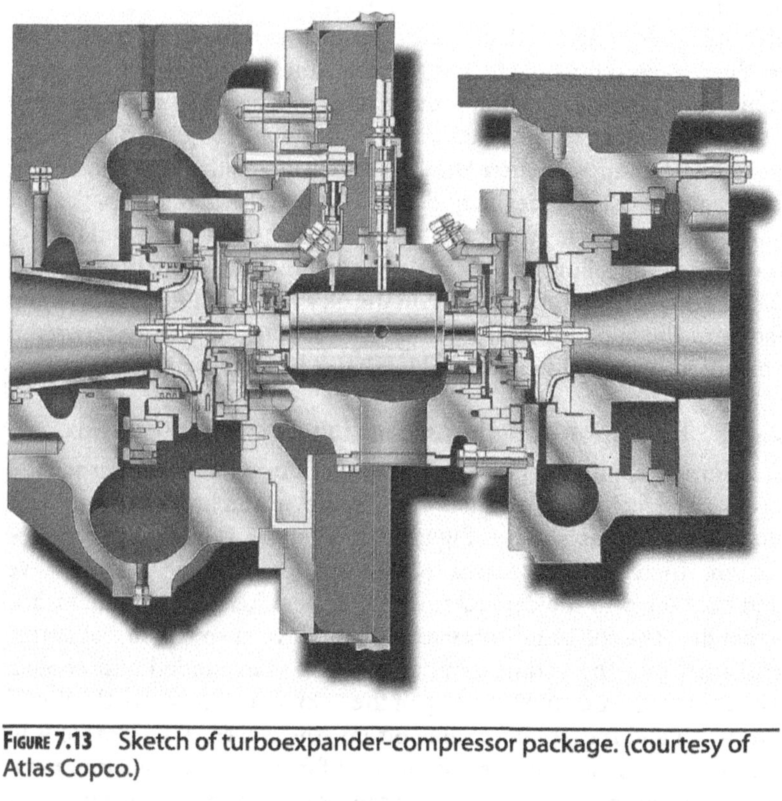

Figure 7.13 Sketch of turboexpander-compressor package. (courtesy of Atlas Copco.)

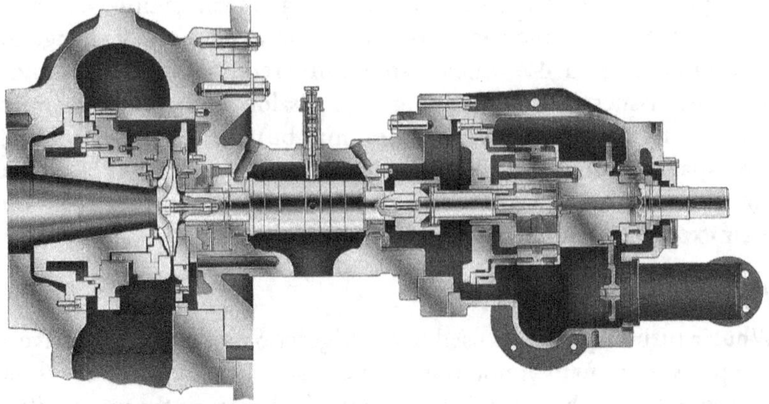

FIGURE 7.14 Sketch of a turboexpander-generator package. (courtesy of Atlas Copco.)

7.4 Process Descriptions and Applications

The turboexpander has been used in many processes where gas is available at higher pressures than needed for further processing. In addition to the simple reduction in gas pressure, the temperature reductions achieved in such a reduction provide the opportunity to generate a significant refrigeration effect and/or the dropout of liquid components from the stream. Anywhere that a J-T valve is required in a process, a cryogenic turboexpander may be considered. The specific application affects the manner in which the turboexpander is used and will impact the design. The following discussion shows how and where the turboexpander is typically employed within the oil and gas industry. A deciding factor in the application is whether the value of the energy recovered will offset the cost of the installed hardware and its sustained operation.

7.4.1 Extraction of Hydrocarbon Liquids from Natural Gas

A common application of a turboexpander in oil and gas–associated production facilities is extractions of hydrocarbon liquids from a produced natural gas stream. Figure 7.15 shows a typical process flow design. There are two passes for the gas, one through a mole sieve and exchanger to the demethanizer and the other through the turboexpander. The cold, high-pressure stream flows through an exchanger and then into the turboexpander where it is expanded and cooled. The cooled stream enters the top of the demethanizer where the liquid drops out and is carried out the bottom. The dry gas then proceeds to the compressor where it is compressed for delivery at pipeline pressure. The compressor is driven by the expander-recovered power. In

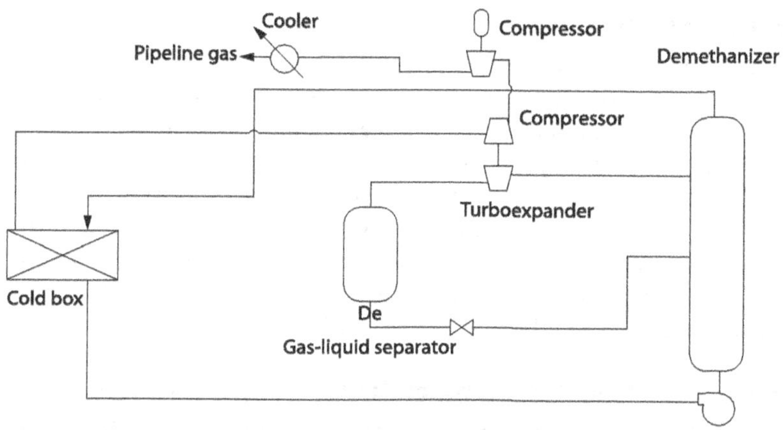

Figure 7.15 Process flow diagram of a turboexpander used in a natural gas extraction plant.

this design, a common shaft is often employed with the compressor directly coupled to the expander.

7.4.2 LNG Liquefaction Plant

Cryogenic turboexpanders are used on the back end of liquefied natural gas (LNG) liquefaction plants to recover the last bits of LNG from the fuel gas stream as well as to provide heat recovery. Figure 7.16 shows an APCI LNG process, one of the more prevalent process designs. The stream from the top of the main cryogenic heat exchanger goes to a turboexpander. This is likely coupled to a generator to recover power. The outlet of this turboexpander goes to a heat recovery unit. What is at the top goes to fuel gas and what is at the bottom is liquid LNG, which goes to storage.

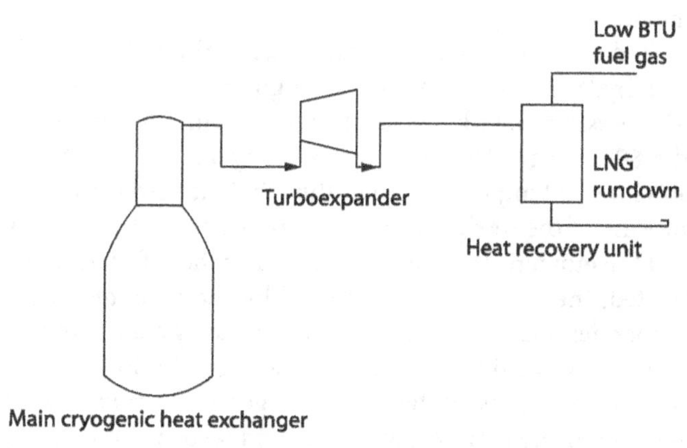

Figure 7.16 Liquefaction process flow diagram incorporating turboexpander (as per APCI type process).

7.4.3 Power Recovery

A general application of a cryogenic turboexpander is simply as a power-generation device when a gas stream must have its pressure dropped for further processing. This is seen in a number of services in many manufacturing facilities, in particular, in refineries or chemical plants. In this case, a turboexpander will be installed coupled to a generator, typically through a step-down gearbox. The generator can then be tied into the grid to recover the power generated. In this case, there may be a need for additional separation downstream of the turboexpander if the inlet stream is saturated and liquid dropout occurs.

7.4.4 Ethylene Plant Refrigeration and Power Recovery

The application of a turboexpander to replace the J-T valve in an ethylene plant has been a long-lived application. Every ethylene plant has one or more streams off the cold box that need to be expanded. This stream is cooled in that expansion and is then reinjected to achieve colder refrigeration levels. In the past, this was accomplished with the J-T valve. In that case, the recovered power was thrown away and as streams became larger, the valves became expensive and were troublesome in large pressure drop services. As a result, application of cryogenic turboexpanders became more and more popular.

These applications are among the most numerous applications of cryogenic turboexpanders. As the use grew, design improvements were demanded. These applications became one of the largest users of magnetic bearings in turbomachinery. They were desired to provide an "oil-free" design. Oil had been a particular issue as any leakage into the refrigeration system would plate out on heat exchange surfaces, causing a reduction in the heat-transfer capability. This directly translated into financial losses. Almost all current applications are "oil-free" designs using magnetic bearings.

The feed stream to the turboexpander is the tail gas, which is a mixture of hydrogen and methane at a high pressure and low temperatures. The pressure is reduced isentropically with efficiencies that can be in the 80% range. This produces a cryogenic stream that may be 40 to 50°C lower temperature than the ethylene refrigeration stream. This lowering of the refrigeration temperature permits the recovery of the last of the ethylene from the tail gas. After the refrigeration service is completed, the tail gas is compressed by the compressor powered off the expander and is delivered to the fuel gas system of the facility.

There are two basic ethylene plant designs: liquid feed or ethane feed. These have process differences that generally require a different configuration of the turboexpander. Liquid feed (primarily naphtha) only requires a single turboexpander stage to perform the necessary expansion. Figure 7.17 shows the layout of this installation.

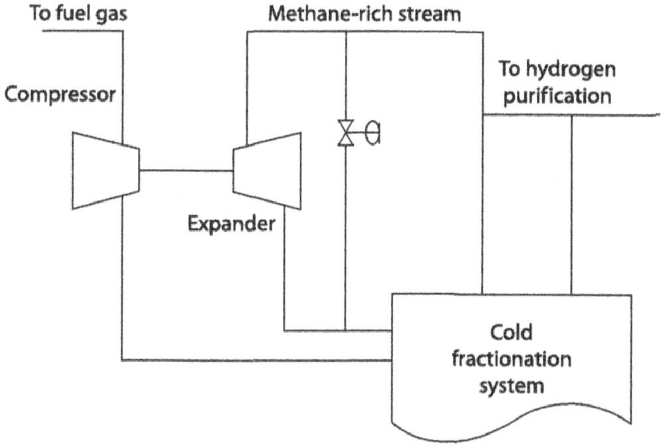

Figure 7.17 Turboexpander in an ethylene plant with liquid feedstock.

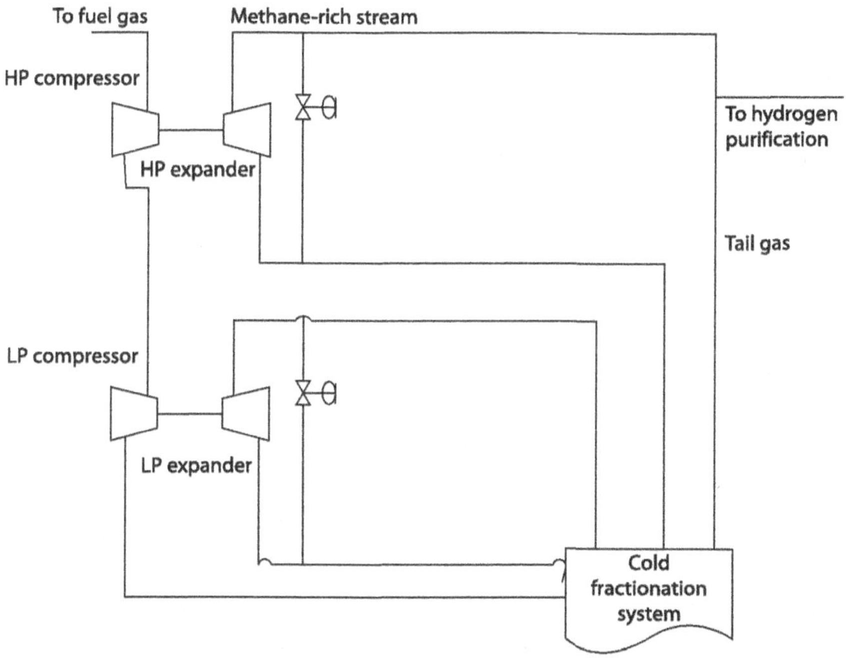

Figure 7.18 Turboexpanders in an ethylene plant with ethane feedstock.

An ethane cracker produces a high ratio of hydrogen to methane. To produce the desired refrigeration along with the recovery of ethylene from the tail gas can require two levels of turboexpander as shown in Figure 7.18.

CHAPTER 8

Rotor Dynamics

8.1 Introduction

It is relatively easy to see if a machine design has flaws in dynamic performance on the test stand or in field operation. Measured vibrations will be high and/or there will be obvious evidence of bearing damage, internal rubs, and the like. However, it may be difficult to see the potential dynamic faults early enough in the design phase to redesign and remanufacture. If the faults are identified on the test stand, there may be insufficient time available to correct the problem, and a compromise is often reached to apply the best flawed design. True and complete mechanical reliability of rotating equipment depends heavily on decisions made by both the purchaser and the vendor early in the design prior to equipment manufacture.

To address this issue, rotor dynamics analysis tools have been developed and are continuing to evolve. The purpose of these tools is to provide predictive capability for the motion of a spring-mass-damper system that represents the machine. They are widely used and are one of the design tools that have helped improve the reliability of rotating machinery over the last 40 to 50 years.

In the late 19th century, there was a widespread belief that machines could not be operated above the first critical speed, and designs focused on machines which operated below that critical. Once it was proven that operation above the first critical was possible, it was still difficult to execute the analyses because of the complexity of the analysis and the limits of computational power.

In the mid-20th century, critical breakthroughs were made in the analytical tools. Once this was coupled with the explosion of computational power, analysis methodology began to experience significant growth. Today, very comprehensive analyses may be performed on computers, and analysts are continuing to develop more accurate and complex modeling techniques. In addition, enhanced measurement capabilities are providing the opportunity to improve this modeling work.

These analysis tools and procedures have become a fundamental design tool for most turbomachinery. Most manufacturers have quite good capabilities for their machines. With the considerable experience they have from their testing, they are able to verify and validate those results. However, problems can nevertheless still occur and the analyses need to be challenged and reviewed on a consistent basis.

Among the current family of challenges attacking expanders are demands for increasing power and speeds, demands for increased reliability which translates often into longer intervals between maintenance, more accuracy in prediction so that margins can be shaved, challenges to machine footprints (particularly on offshore applications), and the application of dry gas seals and magnetic bearings, particularly in turboexpanders. For this reason, users will need to be able to review the analyses and determine when they should be challenged and audited.

By and large, the expander family is not one for which rotordynamics response is an overriding concern. However, there are a few issues peculiar to expanders that need to be addressed.

The required analyses are defined in the API Machinery Standard Paragraphs and are detailed in the standard that covers expanders—API Standard 617 (2002). Considerable more detail and comprehensive tutorial data is included in API Recommended Practice 684 (2005), which details API Standard Paragraph 6.8, Dynamics (2011). Of interest is that all three of these documents have been recently revised but have not yet been published in total. API 617 and RP 684 are expected to be published in 2014. The standard paragraphs have already been incorporated in both documents. There will be a number of items of interest to expander users, including details on at-speed balancing, minimum reporting requirements for rotor dynamics studies, and an annex on active magnetic bearings (AMB).

The material that will be presented in this chapter is an overview of the subject and will include specific features and analysis protocols that should be used with expanders based on their particular design features. Many more comprehensive references may be accessed by users in API RP 684 as well as a large number of rotor dynamics books from very credible authors. A few of these books have been included in the Bibliography.

8.2 Introduction to Dynamics

There is a particular language associated with rotor dynamics. It builds on the basics of vibration theory. This material is covered in every basic vibration theory text and is covered well in API RP 684. Thus, it is not included in this book.

8.3 Procedural Steps in the Analysis

A standard rotor dynamics analysis will attempt to satisfactorily address the following items:

- Identification of rotor-bearing system-critical speeds (damped natural frequency)
  - First critical speed
  - Second critical speed
- Response of rotor to unbalance (forced response)
- Bearing characteristics and sensitivities
- Rotor-bearing system stability (transient response)
- Train torsional analysis

There are specific requirements and analysis protocols identified in the API requirements. Depending on the class of expander and the specific components, these will be more or less important.

To satisfy the requirements, several steps are employed. These start with modeling of the rotor and bearings. These models are then used in an undamped critical speed analysis, which provides an understanding of the potential dynamic performance of the rotor. An unbalance response provides a forced response of the rotor to synchronous unbalance forces. Then, a stability assessment will be performed. Two levels of this analysis could be performed depending on the results achieved. Finally, if deemed necessary, the torsional analysis will be performed.

8.3.1 Modeling

To perform the analysis, a mathematical model of the rotor must be developed, which is then used with the standard rotor dynamics tools. The steps to be followed are to:

1. Generate a mass-elastic lateral model of the rotating assembly from the rotor drawing.
2. Calculate the static-bearing reactions coming from rotor reaction and gear loading.
3. Calculate the linearized fluid film-bearing coefficients and/or active magnetic bearing characteristics.
4. Determine the dynamic support stiffness.
5. Calculate the linearized floating ring oil seal coefficients (if present).
6. Calculate all other excitation mechanisms, such as aerodynamic effects and labyrinth seal effects.

The rotor is broken up into sections located when there is a change in cross section of a device or attachment and/or a location of interest for predicted values. For example, there is normally a station established at the centerline of each bearing and seal, at each external element, like an impeller or coupling hub, and at the locations of any measurement sensors. These elements are then broken up into lumps for which mass, inertia, and flexibility may be calculated. Finally, the model is assembled as a set of spring and mass elements that may be analyzed by the calculation routines. Again, considerably more detail may be found in the References and Bibliography.

Several unique features are associated with expanders, some for Hot Gas Expanders and some for turboexpanders.

In the case of both the Hot Gas Expanders and the turboexpanders, the rotors have significant overhung impellers. This can introduce two issues into the rotor dynamics: (1) overhung rotors can experience some "strange" whirling motion in the rotors; and (2) it is difficult to explicitly calculate the transverse and polar moments of inertia of the discs and impellers accurately. For the greatest accuracy, a test may be necessary.

Figure 8.1 identifies the areas of a Hot Gas Expander that have unique rotordynamic characteristics.

Hot Gas Expanders

This is an overhung rotor. To determine the static load imposed on the bearings by the rotor, accurate measurements are required. In one of the early-design Hot Gas Expanders, the "fat" shaft center section was missing and the disc-end bearing was actually loaded on the top shell. That caused both strange response characteristics and concern

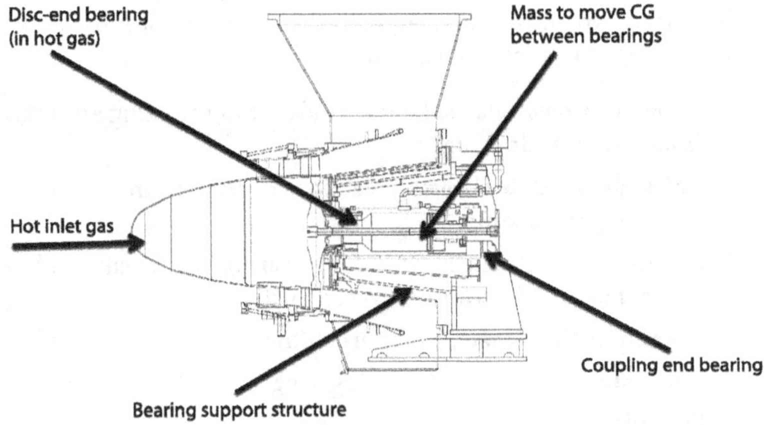

Figure 8.1 Identification of rotordynamics issues on Hot Gas Expander body.

when the shaft end was seen to rise almost to the bearing clearance. To avoid this phenomenon, the center section weight has to be sufficient to maintain an appropriate bearing load in spite of weight that may be accrued on the blades owing to catalyst deposition. Most current Hot Gas Expanders have a design that is quite stable with respect to load bearing with a well-defined bearing static loading.

Another modeling issue associated with these machines is bearing support stiffness, which is lower than the general centrifugal compressor. The bearing carriers and support structure on a Hot Gas Expander will produce a support stiffness that is 25% to 50% lower than the more typical stiffness of a classic compressor. This reduces the overall bearing composite stiffness and can reduce the first critical speed. In many conventional centrifugal compressors and turbines, the support stiffness is so much greater than the bearing stiffness it doesn't impact the composite stiffness and may be ignored. Studies have been performed by the manufacturers with FEA to calculate the baring support stiffness. There should be an assurance step to ensure that this parameter is included in the rotordynamics modeling for the user's expander. Older generation Hot Gas Expanders may be more prone to be affected by this issue than more current ones.

The disc-end bearing of a Hot Gas Expander operates in an environment that experiences a significantly higher temperature than most hydrodynamic bearings. This will cause the bearing to "grow" when operating temperatures are achieved in the expander. This bearing must have the thermal growth included when modeling the bearing. The operating clearance that can be significantly larger than the cold, set clearance, and the stiffness and damping will be lowered.

A general modeling issue with Hot Gas Expanders is that the gas path temperature is of a higher temperature than most compressors or steam turbines. The modulus of elasticity of all metallic components in the gas path may be reduced by the temperatures. The variation needs to be included for accuracy.

Turboexpanders

There is less of an issue maintaining a downward force on the bearings owing to the size and the fact that many if not most of the turboexpanders have a common shaft for the expander and compressor. However, care must be taken in the modeling to ensure that the impeller inertias are calculated accurately in order to generate the correct gyroscopic effects. The polar moments of inertia must be calculated accurately. One issue with this is the only way to check the results is with a specific dynamic test. Reference to this issue and potential results are included in Davidson et al. (1992).

An issue can and has occurred with bearing loads associated with the gears used on generator drives. Inasmuch as the static load produced by a gear is a function of torque transmitted, the bearing load and stiffness goes through a transient state on start-up and shutdown. This can cause rather dramatic changes in the bearing characteristics. The critical speeds will track the load generation to some extent. This must be incorporated in the analysis so that critical speeds don't encounter operating speeds during the start-up or shutdown.

At this time, many turboexpanders use active magnetic bearings. Modeling of these devices requires incorporation of the control system along with the bearings themselves. Magnetic bearings will be considerably softer than hydrodynamic bearings. This will reduce the first few critical speeds of the machine. There are complexities in evaluating active magnetic bearing systems with respect to rotordynamics (Swanson et al., 2008).

Seal modeling is not very critical with these machines.

8.3.2 Undamped Critical Speeds

The first step in a rotordynamic analysis is the calculation of an undamped critical speed map. This is the simplest analysis and the one that has been performed the longest. While it is not as sophisticated as some of the analyses to be discussed later, it has significant value in describing how a particular machine will perform.

In order to perform this analysis, the rotor model is required. Then a series of bearing stiffness are input and the natural frequencies of the rotor calculated. The stiffnesses can be simply the baring stiffness or they may include the support stiffness. Figure 8.2 shows a typical plot. As shown the vertical lines are the natural frequencies plotted as a function of the support stiffness. This particular undamped critical speed map shows the first three critical speeds. Then, the bearing characteristics are overlaid on the plot. Points of intersection indicate about where the critical speeds will occur. If the intersections occur near operating speeds, this is a forewarning of potential difficulties to be evaluated.

In addition to the approximate location of critical speeds, undamped mode shapes can be calculated from this analysis. These will show evidence of potential issues. For example, large modal amplitude at a coupling or disc indicates the need for care in balancing should be employed. If there is limited motion at the bearings, the rotor will likely have low damping as damping force is generated only when there is a velocity present at the damper. And, if there is a node at a displacement probe location, the measurements at that sensor will not be of great value. These give an early warning in the analysis of potential problem areas.

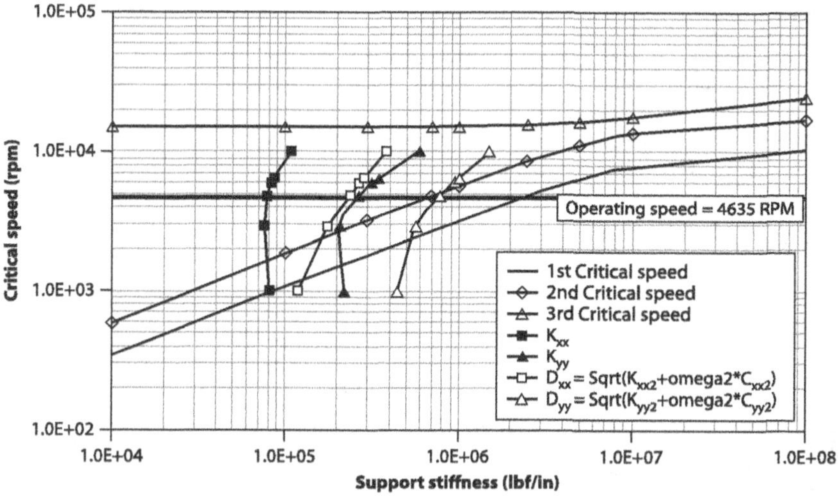

FIGURE 8.2 Typical undamped critical speed map.

8.3.3 Unbalance Response Analysis

The unbalance response analysis should be the most representative of the machine response the user sees in operation, as long as the modeling is accurate. For an unbalanced response analysis, the rotor-bearing system model is assembled and unbalances are placed at various rotor stations among the model. The locations are those that would be most typical of seeing unbalance during normal operation. This will typically include a distribution across the impellers and at the coupling and thrust disc. The results are calculated through a speed range of interest, normally from zero to about twice operating speed. The results are a prediction of motion at selected rotor stations versus speed. The basic computation will generate maximum and minimum axis amplitude for the two-dimensional orbit of the shaft. This is often translated into the vertical and horizontal motions, the maximum motion in any direction, and so forth based on the user's choice. This analysis is used to identify the critical speeds of the rotor and the amplification factor produced at those critical speeds. Amplification is approximately proportional to the damping factor of the rotor. It should be noted that all amplitudes are a function of the unbalance located in the model. Since the modeling is linear, variations in the unbalance amplitude proportional changes in the predicted response.

Figure 8.3 shows a typical unbalance response result. The peak amplitudes correspond to the rotor-bearing critical speeds.

Critical speed problems as identified by unbalance response analysis are not very common for Hot Gas Expanders that operate with

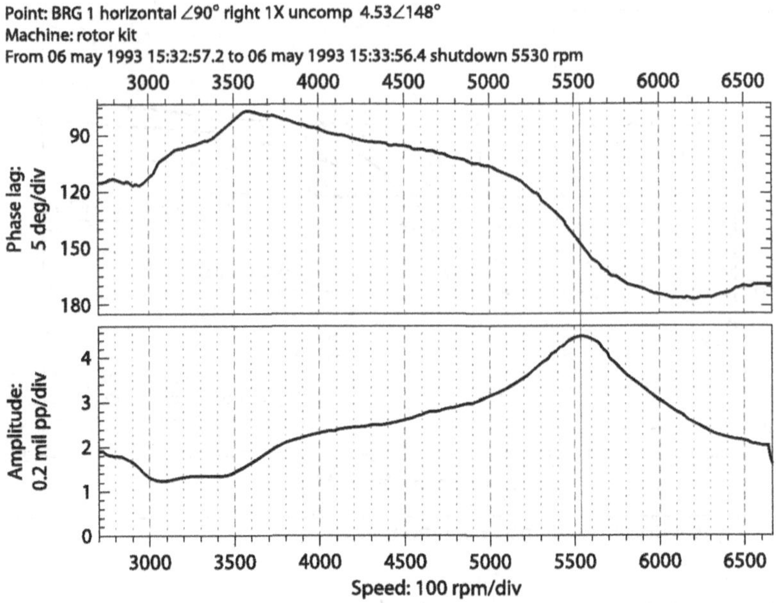

FIGURE 8.3 Typical unbalance response prediction.

synchronous generators. There may be a natural frequency near the operating speed, but it is typically so highly damped as to not be a concern. To predict this frequency, the correct bearing housing stiffness must be used.

Historically, there have been some issues with turboexpander critical speeds, particularly in cases with turboexpanders that operate at elevated speeds, some of which may operate in excess of 50,000 rpm. In the few reported cases, it was a second or third critical speed that was the issue (Davidson et al., 1991). In addition, there have been some issues with generator drives that include gearing. Since the gear reactions are proportional to power generated, the bearing loads will run through a range on start-up and shutdown. It may be possible to cause the critical to traverse that operating speed.

8.3.4 Stability Analysis

The stability analysis is the most complex of the analyses. These analyses have two levels in evaluation, Level I, which uses an approximate Alford force as the excitation. If the Level I criteria are not met, a Level II analysis must be performed. This requires a much more comprehensive assessment involving very detailed modeling of all the seals including shaft seals and tip seals.

The concept of instability has not been reported in a Hot Gas Expander and will not be pursued.

There has been a report of a potential instability in very high-speed turboexpanders (Davidson et al., 1991). However, other reports have not been found of that phenomenon.

It is quite possible that there may be some very high-speed turboexpanders or potentially those with active magnetic bearings that encounter a control system instability driving the rotor response. However, these have not been reported in the literature.

CHAPTER 9

Bladed Disc Vibration and Reliability

Fatigue failures in blade and disc of expander are attributed to vibration-related phenomenon owing to possible resonant excitation. The resulting damage to blades and disc is due to high-cycle fatigue. At resonant excitation the dynamic stress amplitudes increase as the exciting frequency approaches the resonant speed and the response decreases after passing through the resonant speed. The design decision is dependent on the identification and possible avoidance of resonant frequencies of the bladed disc system.

It has been realized that estimation of the resonance characteristics and its effect on reliability of blades is a multidiscipline activity. Information about unsteady aerodynamic forces over a wide variety of flow conditions may be needed within the operating range. Structural vibration characteristics, that is, frequencies and associated mode shapes within the operating range, need to be estimated with appropriate damping in the system (structural as well as aerodynamic). Material properties, for example, fatigue properties and mechanical either estimated or actual, at the temperature in the operating range should be available. Well-defined manufacturing and quality assurance processes should be followed. Each of these issues is essential to assess the reliability and, in turn, this may require a separate technical discipline to support an evaluation.

As a first approximation, a spring-mass-lumped model was used to study the vibrational behavior of a blade. This early approximation led to the concept of resonance, that is, the coincidence of frequencies (forcing frequency and natural frequency of blade) produced the largest amplitude of response. This response may be many folds larger than if the force was independent of time. Next, the behavior of a single cantilevered beam was considered as a continuum. Theories of bending and vibration of a beam were applied. It was realized that, in the case of banded (shrouded) blades (as used in steam turbines), there existed more natural modes and natural frequencies than those estimated from a single beam model. This was due to the existence of coupling among the blades of a packet through the shroud band. The study (Weaver and Prohl, 1956; Prohl, 1956) of this phenomenon identified the existence of many modes and frequencies in a packet of blades not detected by a single-blade analysis. Blades are coupled to each other through the shroud band. Analytical results, as well as test data, show that many modes and frequencies are also influenced by the stiffness of the disc. Packets of blades are coupled to each other through the disc. These additional modes are not calculable through the analysis of a single packet of cantilevered blades and it required the next level of complexity. Singh and coauthors (1982, 1988, 1989, 1994, 2008, 2011) described various aspects of the vibrational behavior of packeted bladed discs and introduced the concept of the Safe diagram. Ewins (1970, 1973) described the behavior of a bladed disc that formed some of the basis for packeted bladed disc analysis. Singh (2013) described the history and various applications of the Safe diagram for tuned and mistuned systems.

This chapter deals with vibration behavior of the bladed disc system of an expander. It is known that fatigue damage may occur owing to vibration of the blades. A large alternating stress develops due to this phenomenon, which may lead to fatigue damage.

9.1 Brief Description of Analytical Methods

The last seven decades have seen phenomenal theoretical development toward the understanding of the behavior of blades and bladed disc systems during vibration. This has been aided by significant development in the availability of powerful computers.

Theory to evaluate the behavior of mechanical structure is estimated by using Newtonian mechanics. Newtonian mechanics considers mass and energy to be separate from each other and conservative.

Some concepts that are useful in the interpretation of the vibration of blades and bladed disc systems are briefly discussed. A mechanical structure such as a blade or bladed disc has infinite degrees of freedom (DOF). The number of natural modes of vibration of a mechanical structure equals the number of DOF. By use of generalized coordinates, each mode of vibration is decoupled for

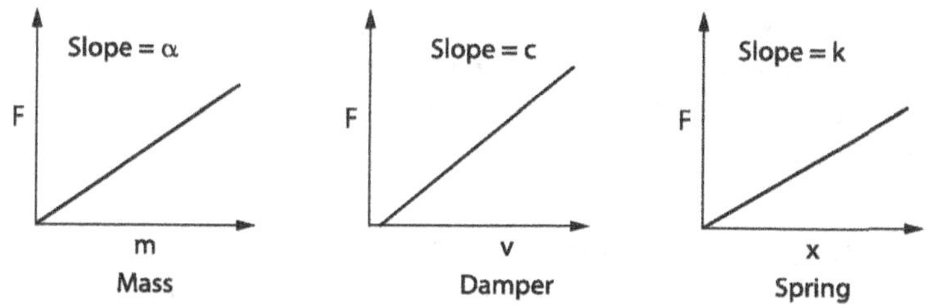

FIGURE 9.1 Behavior of spring, viscous damper, and mass under applied load.

a multidegree of freedom system. Each isolated natural mode of such a structure by itself (in a generalized coordinate system) can be considered as a single spring-mass system. Therefore, the study of the dynamics of a single spring-mass-damper system highlights the behavior of a complex system, such as a bladed disc.

Three different parameters are considered for analysis of a mechanical structure, namely stiffness, mass, and damping. A building block approach helps to understand and to define the behavior of mass and the role stiffness and damping play under an applied force. Three laws as enunciated by Newton are known as Newton's law of motion. When force is applied to a spring (representing stiffness), a mass and a viscous damper, their responses are depicted graphically separately in Figure 9.1. Any mechanical structure can be estimated as a combination of numerous springs (stiffness), dampers, and masses.

Response of mass, viscous damper, and spring (stiffness) under an applied force is given by the following equations:

$F = m\alpha$, α is acceleration
$F = cV$, V is velocity
$F = kX$, X is displacement

The spring-mass-damper system, an elastic beam, collection of beams, and circular plates that help in the understanding of vibrational characteristics of a bladed disc will be discussed. The reason for the discussion is that a bladed disc is composed of these items and the modes demonstrated by the bladed disc are expected to have a combination of behaviors exhibited by these individual components.

9.1.1 Spring-Mass and Damper Systems

Analysis of a discrete system composed of mass, spring, and damper provides relevant information with simple mathematics. It sheds light on what happens to a component that is made of materials with infinite degrees of freedom during vibration. This section includes the development of analytical expressions for natural frequency and response under the action of a sinusoidal force of a single mass, spring, and damper system.

Single-Mass, Spring, and Damper System

Analysis of a spring-mass system (Figure 9.2) aids in the understanding of the vibration characteristics of a mechanical system. The applicationof Newton's third law of motion yields the equation of motion for the above system Eq. (9.1).

$$m\frac{d}{dt}\left(\frac{dx}{dt}\right)+c\frac{dx}{dt}+kx=F(t) \tag{9.1}$$

where m = mass

$\frac{d}{dt}\left(\frac{dx}{dt}\right)$ = the acceleration of the mass

c = viscous damping

$\frac{dx}{dt}$ = velocity of the mass

k = stiffness

x = vertical displacement

$F(t) = F_0 \sin(\omega t)$ is the time-dependent force

The absolute magnitude of the response of the system after assuming a harmonic solution for the displacement, $x(t) = A_0 \sin(\omega t + \phi)$ is expressed as magnifier and is given in Eq. (9.2).

$$\text{Abs}\,(A_0k/F_0) = 1/((1 - r^2)^2 + (2\zeta r)^2)^{1/2} \tag{9.2}$$

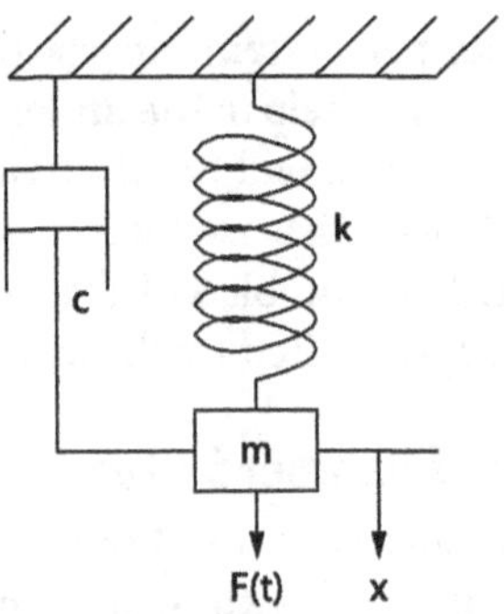

Figure 9.2 Single spring-mass-damper system.

where $\omega_n = (k/m)^{1/2}$, natural frequency of vibration
$r = \omega/\omega_n$
$\zeta = c/(2\omega_n)$, percent of critical damping

A_0 is the amplitude of displacement (response) during vibration whereas F_0/k is the steady state displacement due to steady force, F_0.

Singh et al. (1982, 1988, 1989) discussed the resonance of a circular symmetric structure and the resonance is considered as a state where built-up maximum energy is in the system. In other words, in the state of resonance, the magnitude of the response (stress, displacement) of the system attains its maximum value (Figure 9.3). At the frequency ratio of $r = 1$, the exciting frequency equals the natural frequency and the magnifier is maximized.

The response Eq. (9.2) of the system is a function of the ratio of frequencies. It should be noted that the response (x) of the mass (m) and the direction of the force (F(t)) are in the vertical direction. Magnitude of response can be different if the same force is applied other than vertical direction. Singh and Lucas (2011) provided the analysis for such a case. This important finding is from a very simple analysis about the importance of mode shape (displaced pattern of mass) and the direction (shape of applied force). This point becomes clearer in the later part of the chapter. They applied the force at an angle (α) from a vertical direction. The equation of the response is then given by Eq. (9.3).

$$\text{Abs}\,(A_0k/F_0) = \cos(\alpha)/((1 - r^2)^2 + (2\zeta r)^2)^{1/2} \tag{9.3}$$

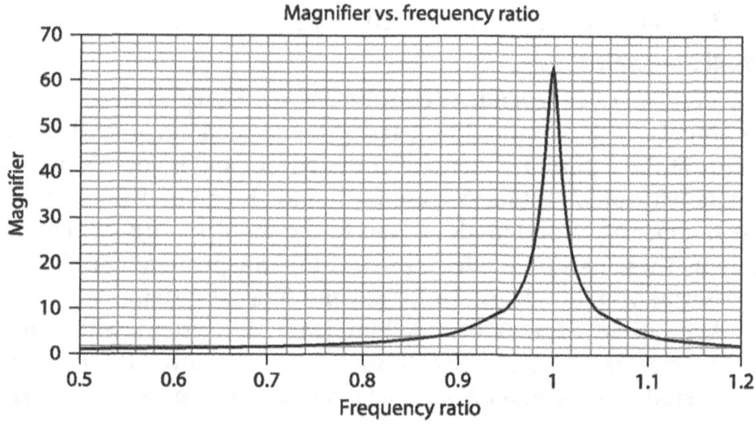

Figure 9.3 Absolute value of magnifier vs. frequency ratio.

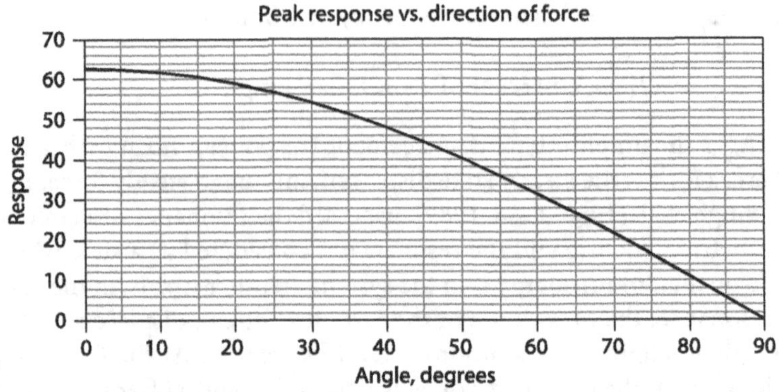

FIGURE 9.4 Peak response vs. angle of the force.

Variation of the peak response Eq. (9.3) of the mass as a function of the direction of the force (α) is plotted in Figure 9.4. The degree of freedom of the system is the displacement in the vertical direction, that is, the mass can move only up and down. Thus, the natural mode shape of this system is in the vertical direction. The system should thereby respond only to the vertical component of the force, F. The response is greatest when the full force acts in the vertical direction (angle, α, equal to zero), that is, the shape of the force matches with the shape of natural mode. The response is zero when the vertical component of the force is zero (angle, α, equal to 90°) that is, shapes do not match. This is an important result that will help us to understand the response of a multidegree freedom system, such as a bladed disc.

9.1.2 Multiple Degrees of Freedom System with Equal Frequencies

The number of independent coordinates necessary to describe the motion of any system should equal degrees of freedom. Free vibration occurs at the natural frequencies of the system and it goes through a simple harmonic motion corresponding to the natural frequencies. The number of natural modes equals to the degrees of freedom of the system. These natural modes represent the independent coordinates to describe vibration behavior for any mechanical system.

A multiple degrees of freedom (MDOF) system with equal frequencies is described by Johnson and Bishop (1956). They considered n equal masses, m attached to a single mass, M, through n identical springs. Spring constant of these springs are equal to k. Mass, M, is attached to the ground through a spring. This system

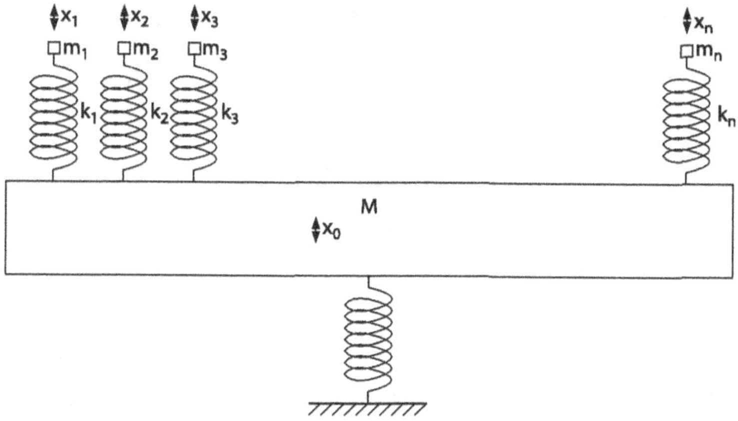

FIGURE 9.5 System with equal mass and stiffness.

resembles a bladed disc system of an expander. Analysis of this simple system helps to understand the behavior of a bladed disc system. Small mass m represents the mass of the blade and the blade's stiffness is k. They used Lagrange's method to analyze this discrete system. They showed that there are a number of equal natural frequencies in this system, but the natural mode shapes are different. The system is expected to have (n+1) natural frequencies, and these are equal to the degrees of freedom.

It is important to start the discussion with each mass and springs different from each other as shown in Figure 9.5. It will become clear later that owing to the manufacturing process, each blade in reality is different even though each is within the specified tolerance. Basic equation of motion is given below for this case following the method given by Johnson and Bishop (1956).

The kinetic energy and potential energy of the system is given by the Eqs. (9.4) and (9.5):

$$2T = M(dx_0/dt)^2 + m_1(dx_1/dt)^2 + m_2(dx_2/dt)^2 + \cdots + m_n(dx_n/dt)^2 \tag{9.4}$$

The potential energy is given by the following:

$$\begin{aligned} 2U &= k_1(x_0 - x_1)^2 + k_2(x_0 - x_2)^2 + \cdots + k_n(x_0 - x_n)^2 \\ &= (k_1 + k_2 + k_3 + \cdots + k_n)x_0{}^2 + k_1x_1{}^2 + k_2x_2{}^2 + \cdots \\ &\quad + k_nx_n{}^2 - 2k_1x_0\,x_1 - 2k_2x_0\,x_2 - \cdots - 2k_nx_0\,x_n \end{aligned} \tag{9.5}$$

Lagrange's method is used to develop a set of differential equations.

The resulting (n+1) equations of motion are listed below:

$$M(d^2x_0/dt^2) + (k_1 + k_2 + k_3 + \cdots + k_n)x_0 - k_1x_1 - k_2x_2 - \cdots - k_nx_n = 0$$

$$m_1(d^2x_1/dt^2) + k_1x_1 - k_1x_0 = 0$$
$$m_2(d^2x_2/dt^2) + k_2x_2 - k_2x_0 = 0$$
$$\cdots$$
$$m_n(d^2x_n/dt^2) + k_nx_n - k_nx_0 = 0$$

Natural frequency equations are obtained by assuming a simple harmonic motion of the system. By setting the determinant equal to zero, the nontrivial solution is obtained. A special case of the above situation is when each spring has the same spring rate (stiffness) and magnitude of each small mass is the same. It will yield the same equation as given by Johnson and Bishop (1956).

Substituting

$$k_1 = k_2 = k_3 = \cdots = k_n = k$$

and

$$m_1 = m_2 = m_3 = \cdots = m_n = m$$

the resulting characteristic Eq. (9.6) (Johnson and Bishop, 1956) is given by the following:

$$\omega^2(k - m\,\omega^2)^{n-1}\,(Mm\omega^2 - (M + nm)k) = 0 \tag{9.6}$$

Solution of the above equation yields (n + 1) natural frequencies, and these are given in Eq. (9.7) below.

$$\begin{aligned} &\omega_0^2 = 0 \\ &\omega_1^2 = \omega_2^2 = \omega_3^2 = \omega_4^2 = \cdots = \omega_{n-1}^2 = k/m \\ &\omega_n^2 = k(M + nm)/Mm \end{aligned} \tag{9.7}$$

There are many identical frequencies whereas the principal mode corresponding to ω_0 is a rigid body motion. They demonstrated that there is a pair of modes having the same shape but out of phase. One is of sine form while the other shows cosine forms. Mode shape is named after complete cycle in the displacement pattern. A pair of third- and fifth-order modes is listed below.

Third harmonic sine and cosine modes are given in Eq. (9.8).

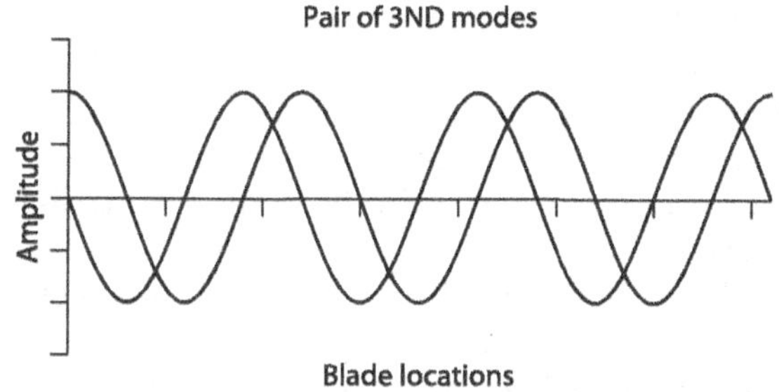

FIGURE 9.6 Pair of 3ND modes (Three sine and cosine waves).

$$x_0 : x_1 : x_2 : x_3 : \cdots x_n :: \sin 3\pi/n : \sin 6\pi/n ; \cdots \sin 3\pi$$
$$x_0 : x_1 : x_2 : x_3 : \cdots x_n :: \cos 3\pi/n : \cos 6\pi/n ; \cdots \cos 3\pi \quad (9.8)$$

Third harmonic sine and cosine modes are plotted in Figure 9.6.

9.2 Vibration Behavior of Bladed Disc System

The behavior of the bladed disc as one system is presented without complex mathematical expressions. There is a coupling between blades through the disc depending on the stiffness of the disc that may change natural frequencies. There is also a coupling between the blades through the shroud at the tip of the blades (if used); this may also influence the magnitude of the natural frequencies of a certain mode. The following two points have been helpful in understanding the vibration characteristics of a bladed disc system:

1. The displaced form of the blades on a disc should be similar to the displaced form exhibited by a single blade in a similar mode of vibration.
2. A disc should exhibit similar behavior with or without blades.

Results of modal analysis conducted using the FEA method of a bladed disc of an expander is given next. Two types of constructions have been analyzed. For this analysis the number of blades on the disc is taken as 90. First analysis is conducted on the bladed disc containing 90 blades with no shroud. A second analysis is performed with the same bladed disc with shroud.

9.2.1 Bladed Disc Containing 90 Blades with No Shroud

Expanders may be designed and manufactured with stage in which blades are not connected through their shrouds. In one design, the shroud is used as a cover for individual blades, but it does not connect blade to blade. The second type of construction does not have any shroud. If the same disc is used to mount these two types of blades, the natural frequencies are expected to be different, but the mode shapes will be similar. When blades are mounted on a disc, their natural frequencies are influenced by the stiffness of the disc and by the shroud's stiffness. Statistically, the total number of modes displaying one type of blade bending mode (e.g., first bending) will be equal to the number of blades on the disc.

Finite element model of the disc for freestanding blades is shown in Figure 9.7. Some mode shapes of the first family of natural modes are given in Figures 9.8 through 9.11. The natural mode as shown

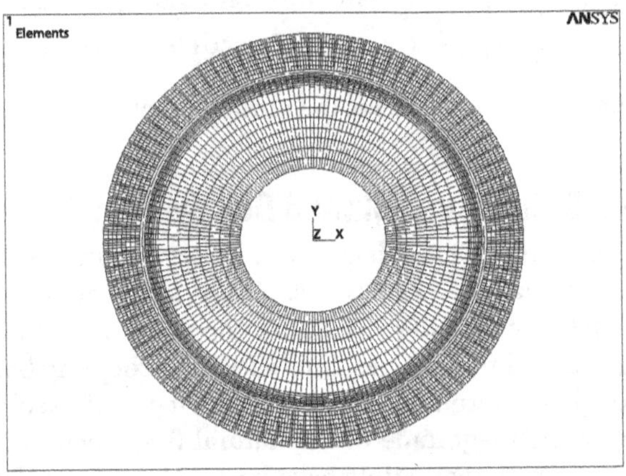

FIGURE 9.7 Finite element model for free standing blades.

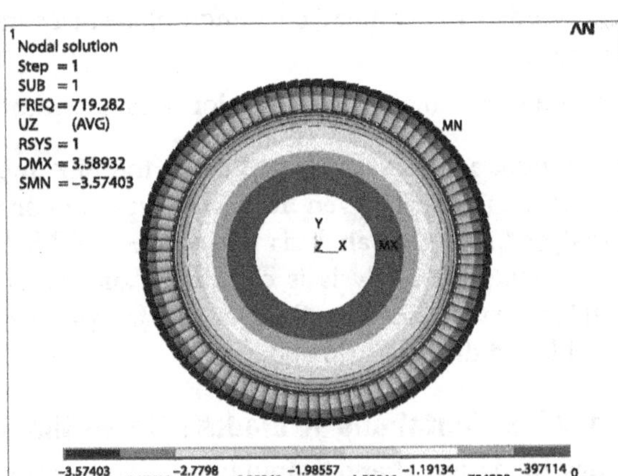

FIGURE 9.8 Plot of 0 ND mode shape for first family.

in Figure 9.8 is 0 ND (nodal diameter), and it is also known as umbrella mode. It should be noted that there is only one mode of this kind. Natural modes as shown in Figures 9.9 to 9.11 for 3 ND, 6 ND, and 14 ND, respectively occur in pairs. Note that they descriptively have the same nodal diameters description but differ from each other by a phase.

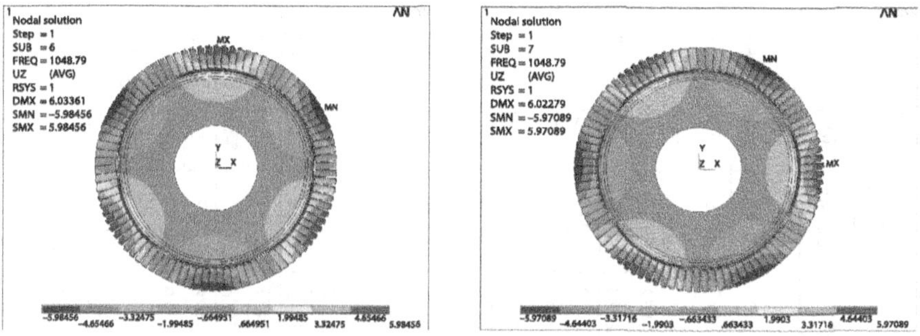

FIGURE 9.9 Plots of the pair of 3 ND mode shapes for first family.

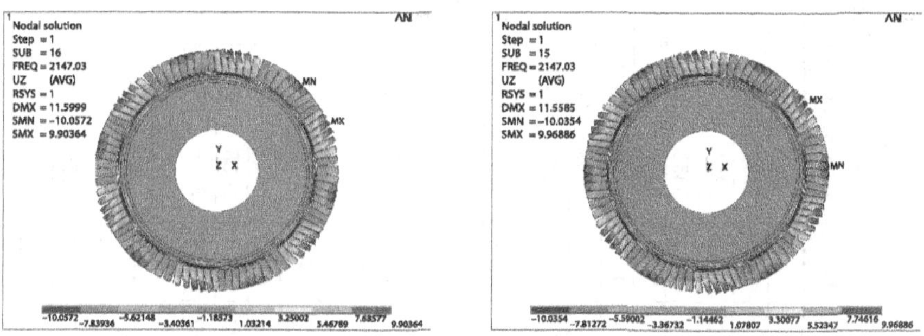

FIGURE 9.10 Plots of the pair of 6 ND mode shapes for first family.

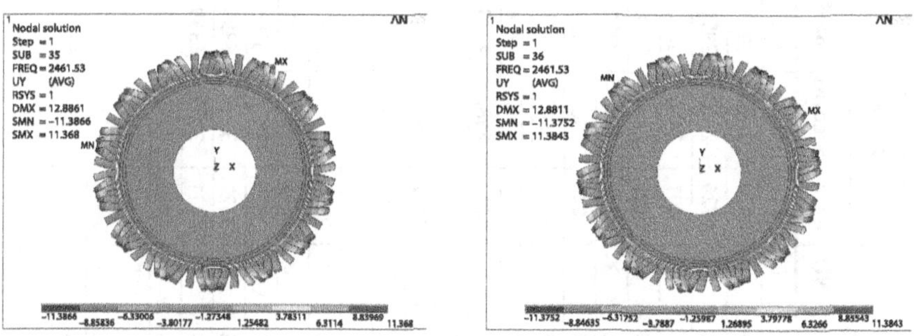

FIGURE 9.11 Plots of the pair of 14 ND mode shapes for first family.

Mode #	ND	Natural Frequencies, Hz — Family			Mode #	ND	Natural Frequencies, Hz — Family			Mode #	ND	Natural Frequencies, Hz — Family			Mode #	ND	Natural Frequencies, Hz — Family		
		First	Second	Third			First	Second	Third			First	Second	Third			First	Second	Third
1	0	719	1909	2762															
2	1	726	2309	2951	24	12	2455	3367	6571	46	23	2471	3743	6790	68	34	2480	3844	6854
3	1	726	2309	2951	25	12	2455	3367	6571	47	23	2471	3743	6790	69	34	2480	3844	6854
4	2	801	2460	3080	26	13	2459	3433	6610	48	24	2472	3757	6798	70	35	2481	3849	6857
5	2	801	2460	3080	27	13	2459	3433	6610	49	24	2472	3757	6798	71	35	2481	3849	6857
6	3	1049	2505	3333	28	14	2462	3489	6643	50	25	2473	3770	6806	72	36	2482	3854	6860
7	3	1049	2505	3333	29	14	2462	3489	6643	51	25	2473	3770	6806	73	36	2482	3854	6860
8	4	1440	2542	3734	30	15	2464	3536	6670	52	26	2474	3782	6814	74	37	2482	3857	6862
9	4	1440	2542	3734	31	15	2464	3536	6670	53	26	2474	3782	6814	75	37	2482	3857	6862
10	5	1843	2587	4328	32	16	2465	3576	6692	54	27	2475	3793	6820	76	38	2483	3860	6864
11	5	1843	2587	4328	33	16	2465	3576	6692	55	27	2475	3793	6820	77	38	2483	3860	6864
12	6	2147	2662	5080	34	17	2466	3610	6712	56	28	2476	3802	6826	78	39	2483	3863	6866
13	6	2147	2662	5080	35	17	2466	3610	6712	57	28	2476	3802	6826	79	39	2483	3863	6866
14	7	2314	2787	5792	36	18	2467	3640	6729	58	29	2476	3811	6832	80	40	2484	3866	6868
15	7	2314	2787	5792	37	18	2467	3640	6729	59	29	2476	3811	6832	81	40	2484	3866	6868
16	8	2388	2936	6186	38	19	2468	3666	6744	60	30	2477	3819	6837	82	41	2484	3867	6869
17	8	2388	2936	6186	39	19	2468	3666	6744	61	30	2477	3819	6837	83	41	2484	3867	6869
18	9	2421	3073	6359	40	20	2469	3688	6758	62	31	2478	3827	6842	84	42	2484	3869	6870
19	9	2421	3073	6359	41	20	2469	3688	6758	63	31	2478	3827	6842	85	42	2484	3869	6870
20	10	2438	3190	6455	42	21	2470	3708	6769	64	32	2479	3833	6846	86	43	2484	3870	6871
21	10	2438	3190	6455	43	21	2470	3708	6769	65	32	2479	3833	6846	87	43	2484	3870	6871
22	11	2448	3287	6521	44	22	2470	3726	6780	66	33	2480	3839	6850	88	44	2484	3870	6871
23	11	2448	3287	6521	45	22	2470	3726	6780	67	33	2480	3839	6850	89	44	2484	3870	6871
															90	45	2484	3871	6871

Table 9.1 List of Frequencies with Corresponding Nodal Diameter for Freestanding Bladed Discs

Natural frequencies with corresponding mode shapes are listed in Table 9.1. It should be noted that there are 90 modes for each family that is equal to number of blades on the disc. Reason being that displacement of each blade is taken as one degree of freedom, therefore, total degree of freedom is 90 (number of blades). Natural modes will equal number of degree of freedom. It should also be noted that maximum designation of nodal diameter is 45 which is half of the number of blades (90). This is one of the reasons that on Safe diagram the maximum number of nodal diameter is 45.

9.2.2 Completely Shrouded Bladed Disc

When the expander stage has blade tips joined by a single-shroud band with the two ends of the shroud joined together, it is called a "completely shrouded" (360° shroud) design. The behavior of a z-lock shroud in the limit may be considered as a completely shrouded bladed disc. The vibration characteristic of this design is similar to that of the freestanding blade design. The difference is that there is a coupling between blades through the shroud band and the shroud provides stiffness to the system, raising the frequency. The frequencies of a completely shrouded design are higher than those exhibited by a freestanding design.

9.3 Evaluation Concepts for Blade Resonant Vibration

Estimation of natural frequencies and natural mode shapes are important because an applied dynamic force may increase the static stress by many folds. This gives rise to alternating stress. Fatigue failures in blade and disc of expander are attributed to vibration-related phenomenon due to possible resonant excitation. The resulting damage to blades and disc is due to high-cycle fatigue. At resonant excitation the dynamic stress amplitudes increase as the exciting frequency approaches the resonant speed and the response decreases after passing through the resonant speed. The design decision is dependent on the identification and possible avoidance of resonant frequencies of the bladed disc system.

9.3.1 Campbell Diagram

The strategy for blade design is to avoid the coincidence of the harmonics of running speed, nozzle-passing frequencies, and any distortion of flow due to any interruption with the natural frequency of the blade (Figure 9.12). The following example of the Campbell diagram (Singh and Lucas, 2011) is instructive to demonstrate that if (a) the shape of the force and mode shape are identical and (b) the

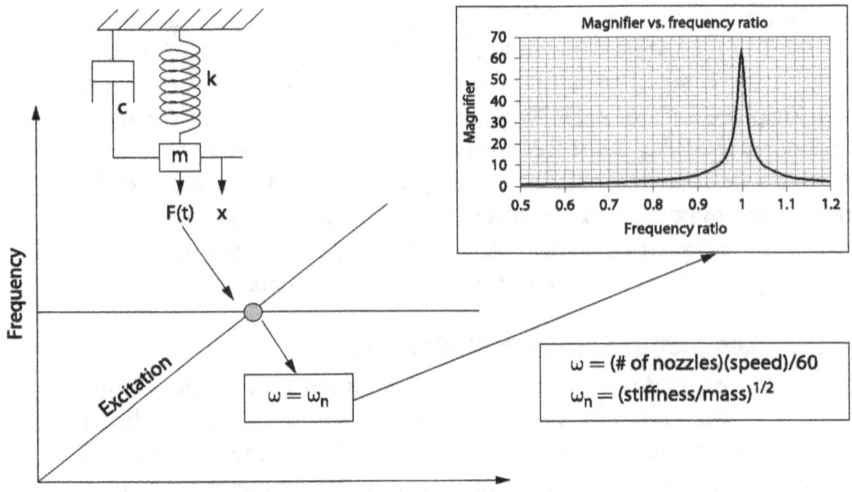

FIGURE 9.12 An explanation of a Campbell diagram.

natural frequency is equal to the frequency of the exciting force, the magnitude of the response will be large. However, the blade will also respond to forces even if the frequencies are not equal, but clearly, the response will be very small. Resonant response is also inversely proportional to damping magnitude. The key point is that when the frequency ratio is equal to unity, the response of the system is the largest. The concept of resonance where the mode shape and the shape of the force also play a defining role will be explained later.

9.3.2 Work Done by an Applied Cyclic Force

General expression of work done by a force during displacement of a mechanical body is expressed as

$$W = \int_0^x P dx$$

In the case when force is sinusoidal acting on the mode of a vibrating body, the expressions for the k_{th} harmonic of the force acting and m nodal diameters mode shape of a bladed disc are given below:

$$\begin{aligned} P_k(\theta, t) &= P_k \sin(\omega_k t + K\theta) \\ X_m(\theta, t) &= -A_m \cos(\omega_m t + m\theta) \end{aligned} \tag{9.9}$$

General expression of work done by the k_{th} harmonic of the force acting on m nodal diameters mode shape of a bladed disc in one complete period, T, can be expressed as follows:

$$W = \int_0^{2\pi}\int_0^{T} P_k(\theta, t)\frac{d}{dt}X_m(\theta, t)\frac{N}{2\pi}\,dt\,d\theta$$

$$= \int_0^{2\pi}\int_0^{T} P_k \sin(\omega_K t + K\theta)\,\omega_m A_m \sin(\omega_m t + m\theta)\frac{N}{2\pi}\,dt\,d\theta$$

$$= N\pi P_k A_m \quad (\text{for } m = K \text{ and } \omega_m = \omega_K)$$

or

$$= 0 \quad (\text{For either } m \neq K \text{ and/or } \omega_m \neq \omega_K) \qquad (9.10)$$

- The first expression of Eq. (9.10) is the positive work ($N\pi P_k A_m$) done by the force. This is true only if the nodal diameter, m, of the mode shape is the same as the shape of the k_{th} harmonic of the force and the natural frequency (ω_m) of vibration is equal to the frequency, ω_K, of the force.
- Second expression of Eq. (9.10) states that zero work results when either the nodal diameter, m, of the mode shape is not the same as the shape of the k_{th} harmonic of the force or the natural frequency, ω_m, of vibration is not equal to the frequency, ω_K, of the force.
- The above argument is the basis of the Safe diagram. It points towards the need to examine the frequency of vibration, the mode shape and the frequency and the shape of the force when considering resonance.

Singh and Lucas (2011) presented a simple way to examine the implications of Eq. (9.10) by considering the behavior of a multiple spring-mass system. They were able to show that for three degrees of freedom system of spring-mass; the only possibility of resonance is when mode shapes and frequencies matches.

9.3.3 Safe Diagram

The Safe diagram was originally introduced for the analysis of a packeted bladed disc as used in steam turbines in late 1970s. Since that time it has found application and use in other bladed turbomachinery. The frequency of the exciting force is a function of the speed of rotation and the number of interruptions in the flow field. It is implied in the Campbell diagram that the shape of the force is identical to the natural mode shape. This assumption might not be realized in each instance for coincidence of frequencies as might be inferred from the Campbell diagram for a bladed disc. The basic justification was provided by Eq. (9.10), which has been discussed above.

The following are two essential conditions as represented by Eq. (9.10) for a true resonance to occur:

1. The frequency of the exciting force must be equal to the natural frequency of vibration.
2. The shape of the applied force must have the same shape as the mode shape associated with that natural frequency.

This is true for a tuned system where each component is identical. Because of manufacturing tolerance or geometrical construction (called mistuned system), there might be response but not true resonance. This topic is discussed later.

For resonance to occur, both conditions must be satisfied at the same time. The time-varying periodic forces experienced by rotating blades can be resolved in harmonics. The resolution of harmonics is accomplished by performing the Fourier decomposition of the periodic force shape. The frequencies of the harmonics are an integer multiple of the speed of rotation. In general, the force experienced by the blades of an expander disc during a complete revolution is the consequence of any circumferential distortion in the flow field.

Expression of frequency of the excitation due to nozzle vanes, struts, or any obstruction in the flow field is given by Eq. (9.11).

$$\omega_K = (K.N)/60 \tag{9.11}$$

where ω_K = frequency of the exciting force, Hz
K = number of distortions in the flow per 360°, e.g., number of nozzles, or number of struts in the flow field, etc.
N = expander speed, RPM
K = shape of the excitation at the k_{th} nodal diameter

Singh and Vargo (1989) described a method of using this information to create a Safe diagram for a packeted bladed disc of steam turbine. The Safe diagram combines natural frequencies, mode shapes, excitation frequencies, and operating speeds on one graph. Throughout the years, it has been used as an excellent guide to determine the potential of exciting a particular mode of vibration. In any vibration cycle, certain parts of a mechanical structure remain stationary. For a circular symmetric system, these points fall on the radii (always an even number) or circles. A pair of radii that appear in the mode shape is considered a diameter. Mode shapes are characterized by specifying the number of nodal diameters (ND) and nodal circles (NC). Nodal diameters are plotted on X-axis of the diagram and the frequency is plotted on the Y-axis of the Safe diagram. A nonresonant design is expected to yield a Safe diagram indicating no coincidence of possible excitation force with a natural mode of the bladed disc.

A Safe diagram is a graphical description of Eq. (9.10). Eq. (9.9) describes the vibrating characteristics of a bladed disc by its natural frequency, mode shape. Owing to the influence of speed, the natural frequency can vary with the speed of expander. Therefore, to represent mode shapes of a bladed disc, three axes are required, namely, frequency, shape, and speed. This is represented on the left side of Figure 9.13. This is essentially a 3D representation of the mode shapes of bladed disc showing a surface. Eq. (9.9) describes the forcing function that may act on the moving blades. This equation also contains the same three variables: frequency, shape, and speed. The right side in Figure 9.13 is the representation of forcing that also shows a three-dimensional surface with the same description of three axes.

The probable points of resonance can be found by superimposing these two surfaces. The intersection points do represent the condition described by Eq. (9.10). As these two intersecting surfaces are three dimensional, the points of intersection and its locus will be a 3D curve. Figure 9.14 shows the three planar views (two vertical and one horizontal) of these points for a 40-bladed disc. The first plot consists of two axes—frequency and speed; these are used to plot a traditional

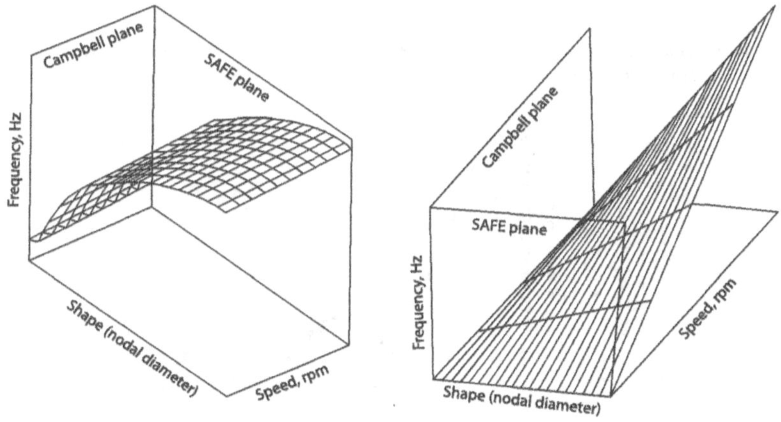

Figure 9.13 3D representations of modes of a bladed disc and of excitation force.

Campbell diagram. The second plot consists of shape and frequency as axes; this plot is named Safe diagram. The third one consists of shape and speed, but it does not have information about frequency, which is essential to ascertain resonance. These are the resonance points that satisfy Eq. (9.10). Frequencies are equal at these points; also the mode shapes are identical to the shape of the force.

As a planer graphical projection is presented for the user to examine the resonance condition, a choice is to decide on which of the projection methods to use. It is clear that for higher order of excitations the resonance points get cluttered on a Campbell plane (frequency and speed plot). The horizontal projection becomes difficult to ascertain the resonance points for a higher order of excitation. On the other hand, the projection on the Safe plane is very symmetric. Once the resonance points for excitation are in order, up to half of the blade count is determined, and then the resonance points for any order of excitation can be drawn. This is a very important result. This became the basis of choosing the Safe diagram to examine resonance conditions of a bladed disc.

For true resonance, the coincidence of frequencies is required and the shape of the exciting force must match the mode shape at the same

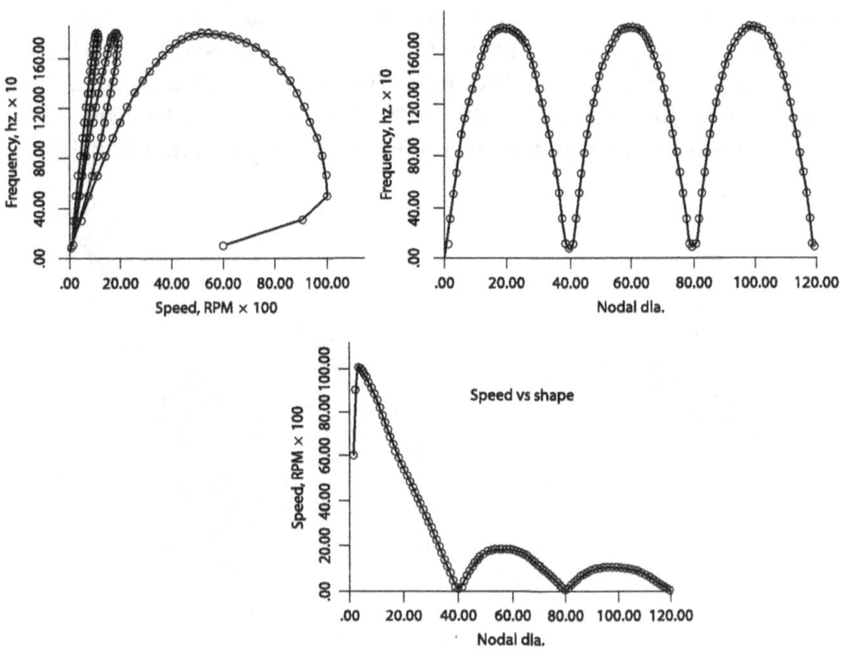

FIGURE 9.14 Three planar projections of resonance points (completely shrouded or free standing blades).

time. As usually drawn, the examination of a Campbell diagram can only infer coincidence of frequency. Thus, the Campbell diagram is overly conservative and can predict the intersection of excitations that is not physically possible. Avoidance of such intersections may restrict the design without any real value.

A Safe diagram can be constructed as follows:

- The mode shape as nodal diameter is plotted on the horizontal axis and the natural frequencies are plotted on the vertical axis.
- The equation of the force (mentioned earlier $\omega = (KN)/60$) is used to draw the frequency of the excitation line for a given speed as a function of the nodal diameter. This is a radial straight line starting from the origin.

Safe Diagram When Harmonics of Excitation Are Larger Than Half of the Number of Blades

The Safe diagram plots modes up to half of the number of blades. There are situations where the harmonic of excitation may be larger than the maximum mode shape plotted on the diagram. For example, when the number of nozzle is 36, then 2xNPF excitation relates to a harmonic of excitation of 72 ND. For a bladed disc containing 90 blades, the maximum mode shape displayed on the diagram is 45 ND. The question is whether this excitation may excite any mode displayed on the diagram.

Two sine waves Sin(14θ) and Sin(76θ) up to 90° are plotted in Figure 9.15. The subtended angle between blades is 4 degrees for a stage that has 90 blades. These two waves intersect at precisely 4°. The implication is as follows: if each blade is instrumented to measure and thereby to determine the shape of the fluctuating force in each revolution, then both waves are equally probable. This means that a 14-nodal diameter mode of the bladed disc may be excited by a force that has a 79-nodal diameter shape, provided that the frequencies are the same. There are many combinations for which this assertion is true, but the combination is unique for each individual bladed disc.

A mode shape and the shape of the excitation should be the same for resonance to occur. There are instances where the shape of excitation force (ND) is greater than the half count of blades. The question then arises, will the system respond to the force containing greater excitation shape? Let us consider a case where the nodal diameter (the shape of the excitation) is larger than the maximum nodal diameter of the mode shape as discussed above. For example, let us assume that the number of nozzles used for the design is 42 and the shape

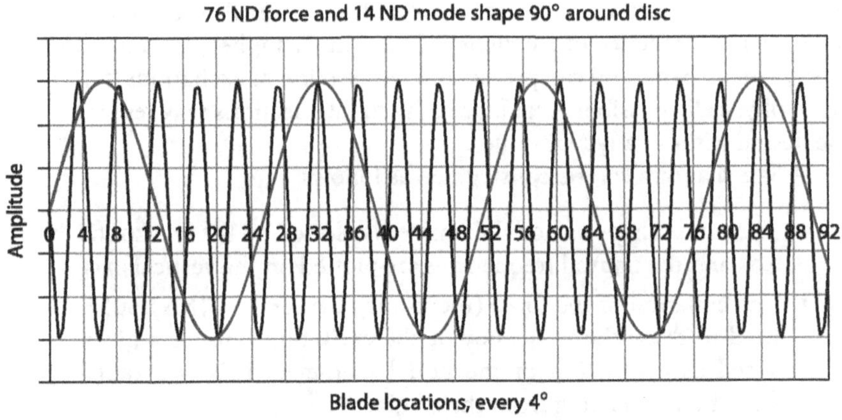

FIGURE 9.15 Sin(14θ) and sin(76θ).

of the 2xNPF excitation is 84. The question is will the force resulting from 2xNPF excite any bladed disc mode that is less than 45 nodal diameters? Even though 45 ND is the maximum shape representation required for a bladed disc vibration of 90 blades, the shape of the exciting force may be larger than 45 ND. A 6 ND mode shape may be excited provided frequencies are equal. This phenomenon (aliasing) is taken care of on the Safe diagram by folding the radial speed lines (see Figure 9.16).

The following relationships (Eqs. (9.12) and (9.13)) provide equations for possible combinations between mode shape and the harmonic of force for resonance for a tuned system.

$$K = \text{abs}(n.L + M) \tag{9.12}$$

$$K = \text{abs}(n.L - M) \tag{9.13}$$

where K = shape of the force (nodal diameter)
n = 0, 1, 2, 3, …, L
L = total number of blades
M = nodal diameter of mode shapes

The Safe diagram (Figure 9.17) based on the results of FEA presented in Table 9.1 is constructed.

Frequencies are plotted on the vertical axis and the corresponding mode designations (e.g., 1 ND, 2 ND, etc.) are plotted on the horizontal axis. The excitation lines are the radial lines representing the number of

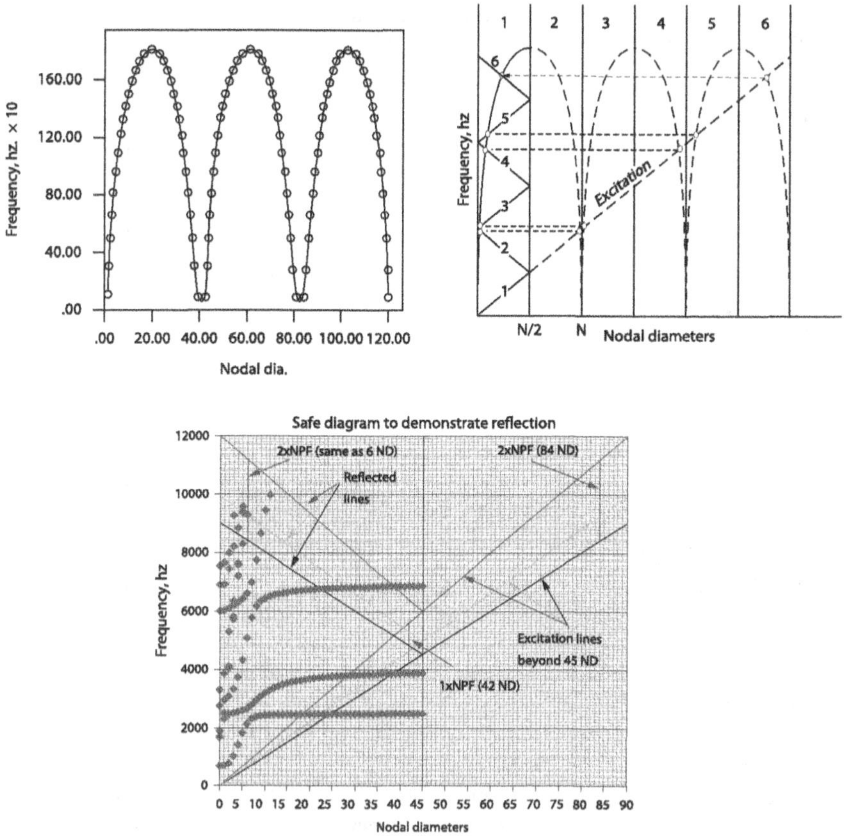

Figure 9.16 Excitation shape larger than blade mode shape with reflected speed lines.

nozzles (42) and expander speeds (minimum, design, and maximum) using the following equation. The excitation force felt by blades due to nozzles are drawn as vertical lines between the speed lines at 42 ND and 6 ND (representing 84 ND) for 1xNPF and 2xNPF, respectively.

A Campbell diagram (Figure 9.18) is plotted using the frequency data of Table 9.1 for a bladed disc with freestanding blades. Many excitation lines due to running speed have also been drawn. It is clear from the diagram that there are many possible resonances within speed range of operation that should be of concern if Campbell diagram is used. However, the Safe diagram (Figure 9.17) of the same data displays no resonance condition within a speed range of up to 1xNPF excitation. The reason for this is it considers a frequency match as well as a match of the shape of modes with shape of the force.

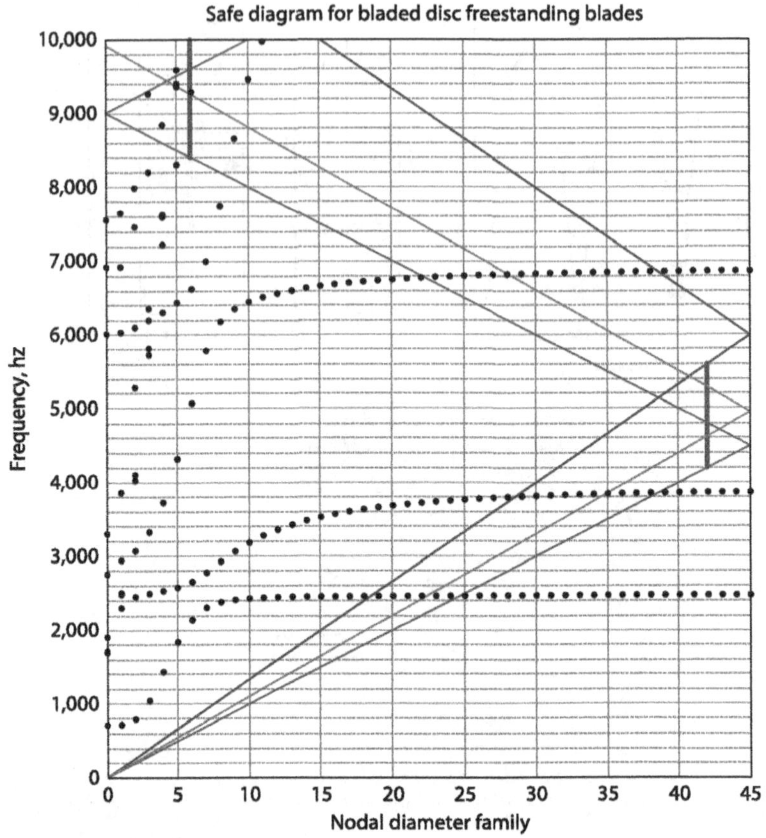

FIGURE 9.17 Safe diagram for bladed disc (free standing blades).

On the Safe diagram the natural frequency of disc, single blade, and a packet of all blades are shown (Figure 9.19). The frequency of the disc will increase as the order of nodal diameter mode increases. Magnitude of natural frequencies for cantilevered blades for each mode shape will not change. The reason for this is that the base of the blade is anchored to the ground (i.e., the base has infinite stiffness). But it will show a displacement pattern where blades may be in phase or out of phase. For blades mounted on a very stiff disc, their characteristics are expected to be similar in the in-plane vibration. This assumption is nearly realized when the vibration of blades on a disc is considered in the tangential direction. The single-blade-mode frequency seems to be the asymptote of the frequencies of a freestanding bladed disc. On the other hand, the frequency of a single packet of all blades seems to be the asymptote of the frequencies

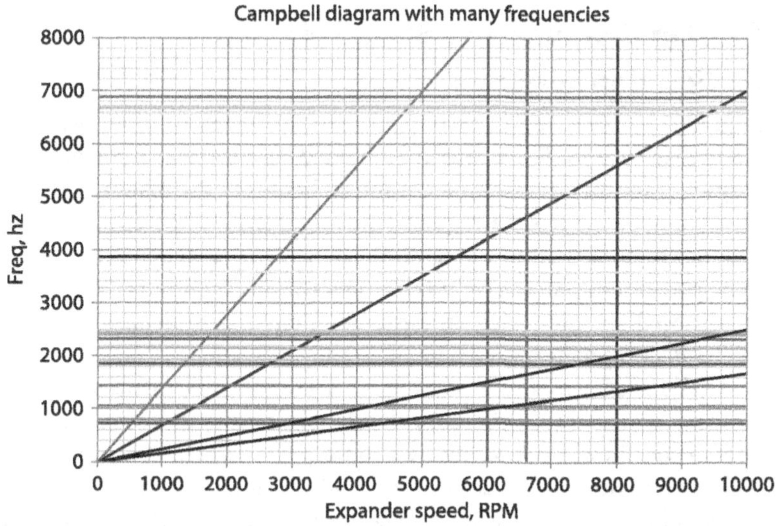

FIGURE 9.18 Campbell diagram for bladed disc (free standing blades).

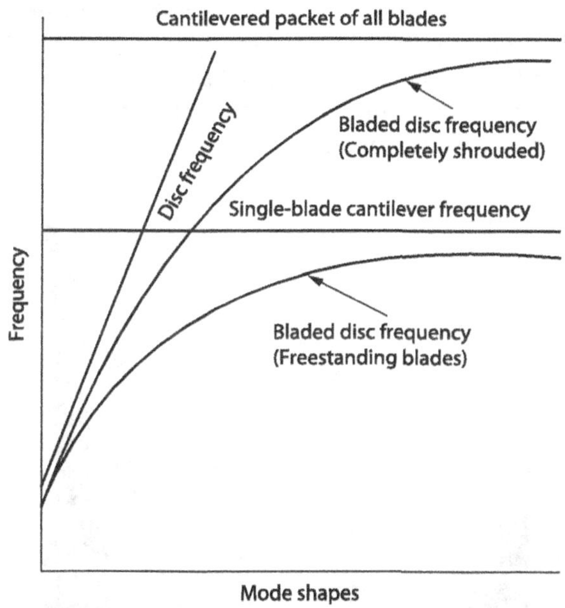

FIGURE 9.19 Representative behavior of disc alone, blade alone, and bladed disc.

of a shrouded bladed disc. In low-order nodal diameter modes, the disc seems to control the behavior, whereas at higher order modes, the blade seems to control the behavior. There is a transition zone where the relative stiffness between the disc and the blade seems to play the controlling role.

9.4 Effect of Mistuning of a Bladed Disc System on Vibration Response

Singh (2013) discussed various aspects of the behavior of a mistuned bladed disc system. For a mistuned bladed disc system, the vibration energy is unevenly distributed among all blades of the disc. The mode shapes may be far from a regular sinusoidal shape. The examination of modes in such a situation is done using Fourier decomposition of the displacement pattern by extracting harmonic contents. The mode is traditionally named with the largest content of the harmonics of the Fourier decomposition in a Safe diagram.

Analysis of a mistuned system is a complicated process, but it may have considerable effect on the dynamic response of blade. The papers listed in the Bibliography should provide greater insight into this phenomenon and analytical methods for interested readers.

Including geometrical variations among blades was the next logical advancement in the analysis of the dynamic response of blades. This type of variation may be owing to manufacturing tolerances, but more importantly it occurs due to erosion, corrosion, and deposition of airfoil and stator vanes in the expander. Figure 9.20 shows deposition of catalyst on blades that, in essence, makes the system mistuned, and the vibration pattern will be quite different from that of tuned case as assumed during design process.

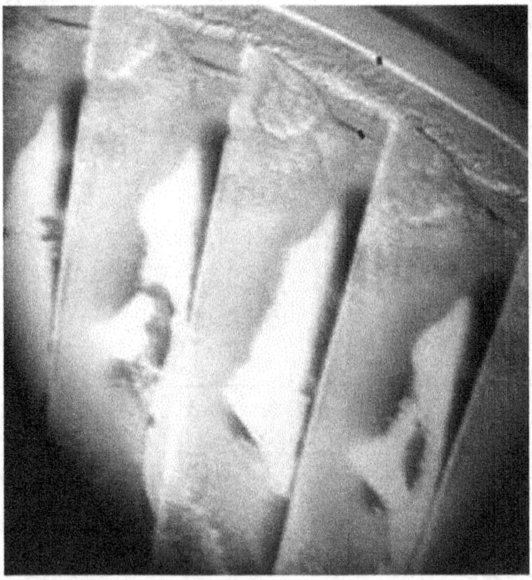

Figure 9.20 Deposition on blades after operation.

It has been shown that geometrical variations influence mode shapes and frequencies and, in turn, the response of a bladed disc system. In a tuned system where each blade is identical, modes in general occur in duplicate. There are two modes that differ by a phase angle, but these modes have identical frequencies. However, when symmetry is disturbed through variation from blade to blade, these modes tend to split in two frequencies. Also, the shape of these modes get distorted from the tuned case, which is a phenomenon called "mistuning." Ewins (1976) has extensive discussions of this phenomenon and the response of the mistuned case was found to be different from the tuned case. This was attributed mainly due to a change in the mode shape and may pose a serious decision point for designers to account for variability among blades. Ewins specifically dealt with completely shrouded or unshrouded bladed disc constructions.

Irregularity in the bladed disc system makes the response of the system asymmetric. These irregularities can be caused by structural and/or aerodynamic means. Structural irregularities arise due to variations in geometrical dimensions from blade to blade; they can also be complicated due to variations in the boundary conditions and due to variations in the fixity of blades to the disc or shrouding. Frequencies may be different from frequencies in the tuned case and the mode shapes are nonsinusoidal. Irregularities in the aerodynamic forces come from variation between nozzle openings. This causes force around the periphery of the disc to be of a nonsinusoidal form. Mode shapes and forces though the nonsinusoidal are periodic in nature. Fourier decomposition of these with respect to the angular position around the disc will yield harmonics in addition to that of the tuned case, which results in a different response of the blade compared to the tuned case.

The analysis of a mistuned bladed disc system becomes complex. Due to manufacturing variations, blades on the same disc will be different from each other even when they are within manufacturing tolerances. Their contact condition in the disc slot may also be different. Therefore, assembly of these blades on a disc will be random unless each blade is individually measured and its location on the disc noted. Sometimes the location is chosen in order to balance a disc containing long blades. However, even known locations pose difficulty in analysis. Each bladed disc produced in such a manner from the same lot of blades will be different and, therefore, their vibration characteristics will also be different. The complexity of the analysis is compounded by the presence of a large number of blades on the disc.

Ewins (1969) reported that for given excitation and damping conditions, the response in one mode (or pair of modes) depends on the "split" of the two natural frequencies. The presence of slight mistuning, small blade imperfections, can cause resonant stress levels of up to 20% above the optimum. This kind of increase depends on the specific arrangement of the blades around the disc. Ewins (1976) reported the results of a study of a slightly mistuned bladed disc. The study considered a group of nominally identical blades. Various dimensional variations in blades nominally result in about ±2% to 3% variation in a 1F frequency. Even though there is a slight change in frequency the result showed that the effect on mode shapes is large. This phenomenon of variation in the mode shape has considerable effect on the response of the blade. The main interest in the design of a blade is its response under load; hence, mode shape should be critically examined. Singh and Ewins (1988) and Singh (1992) presented probabilistic analysis results by considering a random arrangement of blades on the disc utilizing a Monte Carlo simulation to arrive at the results.

The mode shape for a mistuned system gets distorted depending on the type and amount of mistuning, whether it is deliberate or random. A packeted bladed disc, balance holes in the disc, odd numbers of blades on a disc, or missing blade space for closing are examples of deliberate mistuning. Random mistuning may occur due to manufacturing tolerance of the blades. Blades may be within a specified tolerance, but each blade may be different if they are not individually measured and assembled randomly on the disc. In these cases, mode shapes are distorted and certain modes might contain many harmonics. It was found in the case of a packeted bladed disc that even in such a situation, contribution of a particular harmonics will be largest. For plotting a Safe diagram, this has been named after that harmonic with the understanding that there may be other harmonics in the shape.

9.4.1 Splitting of Modes in Two Frequencies

In the case of random mistuning it is expected that frequency of each mode will split in two frequencies and both modes become distorted (i.e., these are not symmetrical) as shown in Figure 9.21. The resulting response curve is the summation of the response of these modes separately.

Figure 9.22 shows such a feature just for two modes for demonstration purposes. It is expected that for random structural mistuning, nearly all modes may split. Examination of this figure shows that response owing to excitation will occur at the mistuned frequencies.

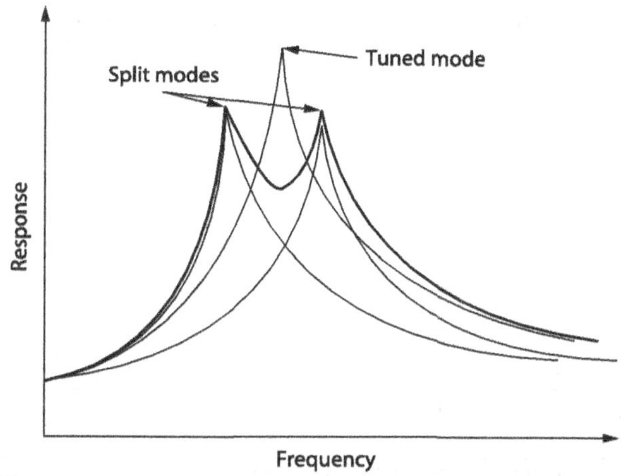

Figure 9.21 Split frequencies due to mistuning. (courtesy: Safe Technical Solutions.)

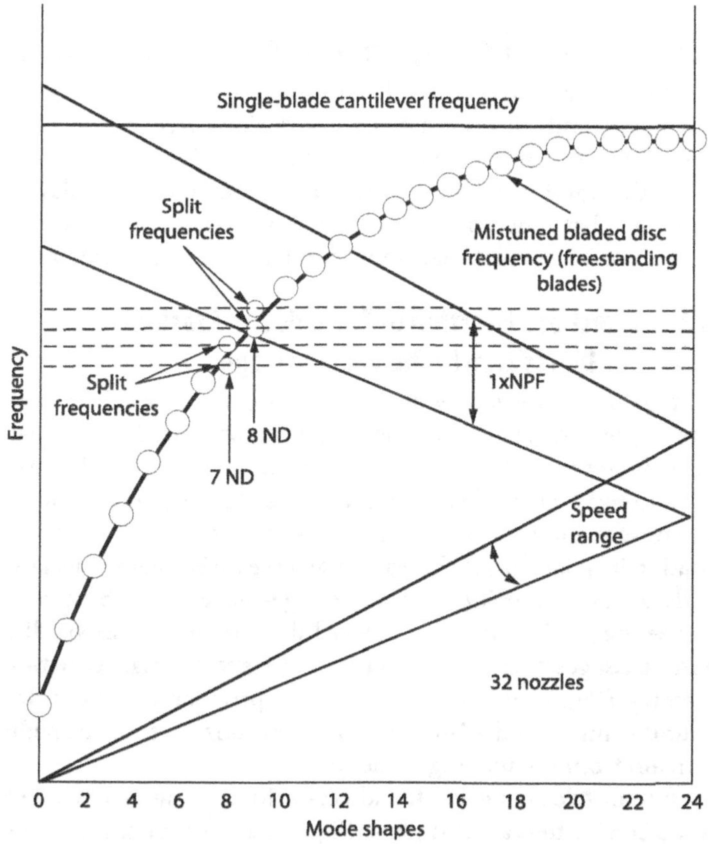

Figure 9.22 Split modes plotted on Safe diagram (shown for just two modes of 7 ND for explanation). (courtesy: Safe Technical Solutions.)

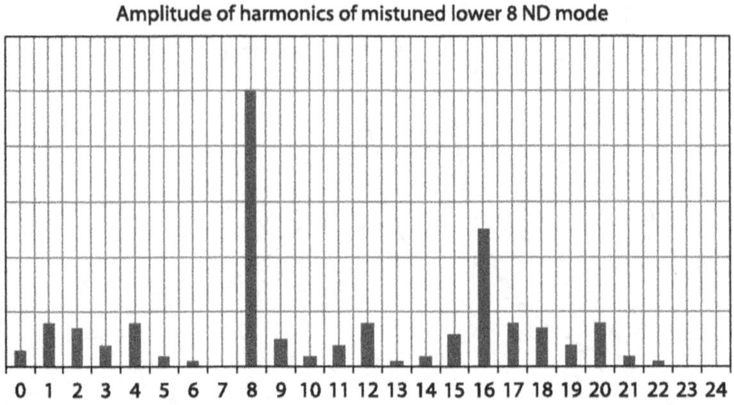

FIGURE 9.23 Harmonic contents for lower 8 ND mode due to mistuning. (courtesy: Safe Technical Solutions.)

9.4.2 Random Mistuning (Structural): Erosion and Corrosion

Structural variation may be due to manufacturing tolerances but, more importantly, it occurs owing to erosion or corrosion of airfoil in expander. When the frequencies split, the other effect of splitting is that the modes become distorted. These modes display many harmonics that otherwise contain only one harmonic for the tuned case. Figure 9.23 depicts harmonic contents of the distorted mode.

9.4.3 Effect of Distortion of Force due to Erosion of Vanes or Deposition in the Vane Passage

This type of variation may be due to manufacturing tolerances, but, more importantly, it occurs due to erosion of vanes in the expander. Unsteady forces due to effects from the rotating blades passing the stationary nozzle partitions are possibly the most important, and consequently the most studied, source of excitation for blade vibration in axial-flow turbines. These forces arise mainly because the nozzle vanes' or partitions' trailing edges have finite thickness, and because the gas is discharged toward the moving blades via discrete nozzle passages that are subject to a variety of small variations in geometry (Figure 9.24). In turn, these produce a variety of flow nonuniformities and disturbances at the nozzle exit, immediately upstream from the moving blade row.

Since nozzle characteristics always exhibit variations as a result of manufacturing tolerances (refer to Figure 9.25), variations in the nozzle exit area, circumferential pitch or spacing, and setback in otherwise

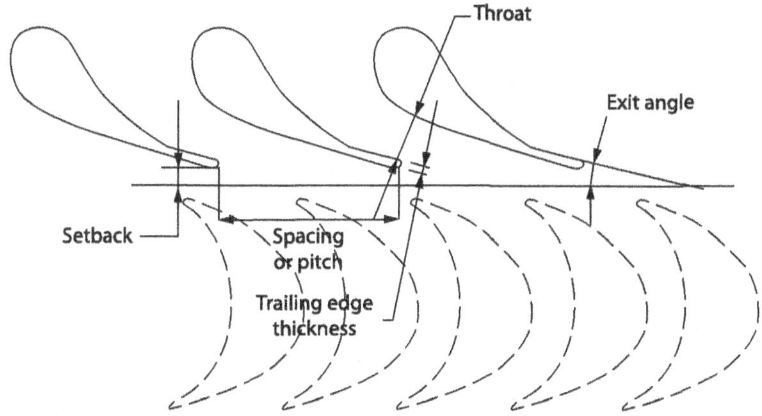

FIGURE 9.24 Nozzle geometry definitions.

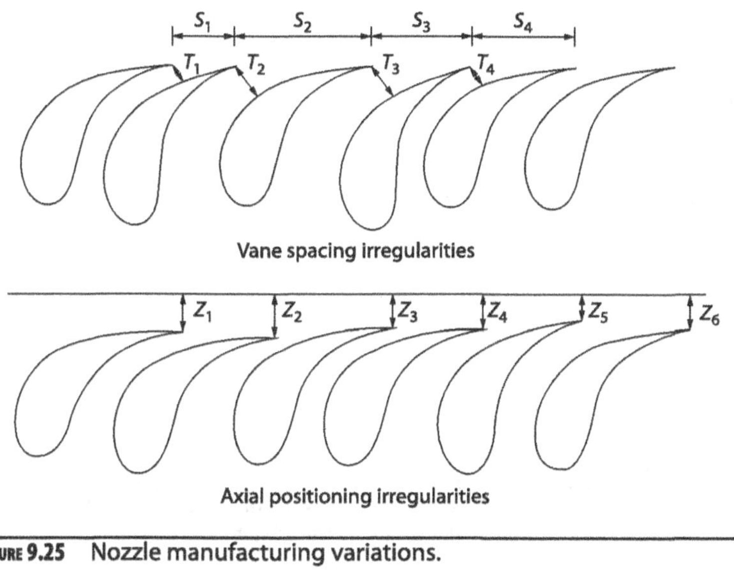

FIGURE 9.25 Nozzle manufacturing variations.

pure signal at 1X nozzle passing frequency, and superimpose other harmonic frequencies on the predominant 1X nozzle passing frequency "signal" or forcing function (Figure 9.26). Although this is primarily concerned with variations from design values due to manufacturing tolerances, the inevitable variations that originate from erosion, wear, and foreign-object damage during turbine operation can also produce potentially destructive unsteady forces on the rotating blades.

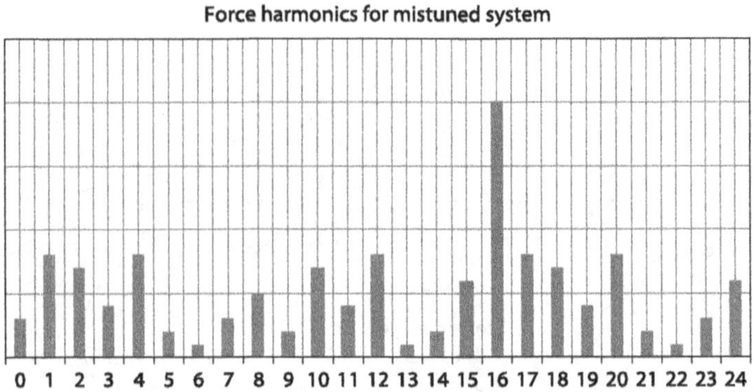

FIGURE 9.26 Harmonic stimuli from vane spacing irregularities. (courtesy: Safe Technical Solutions.)

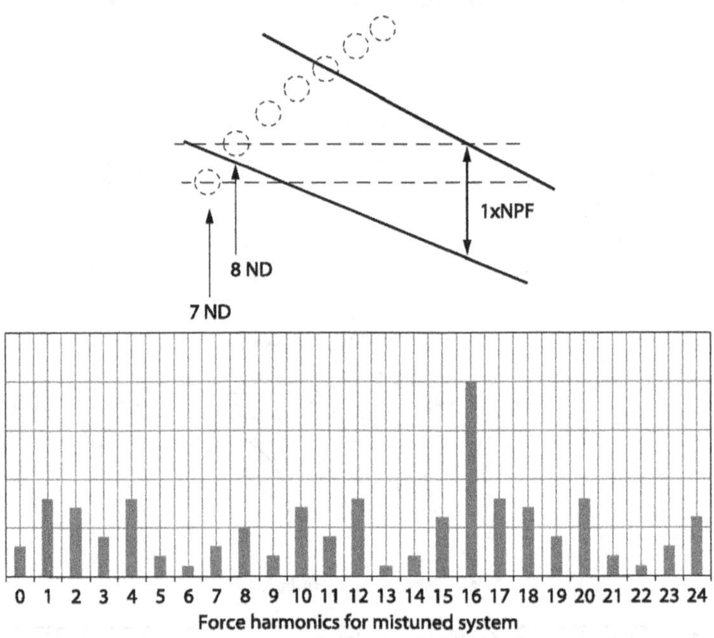

FIGURE 9.27 Safe diagram representation of harmonic contents of 1xNPF excitation for mistuned forcing. (courtesy: Safe Technical Solutions.)

9.4.4 Safe Diagram for Structurally Tuned System with Force Mistuning

Many instances occur in the design where a system is almost completely tuned, but the force excitation is not. A Safe diagram for such a system for the modes of interest is drawn, and it is shown in Figure 9.27. Again it shows that natural frequency matches the frequency of excitation, but due to mistuning of force only the harmonic contents of the force that match with the harmonic of the mode will contribute to the response of the system. An equation developed earlier for work done shows that there will be response, but it will not be due to the 16th harmonic of the force. Many times the forces are mistuned deliberately to take advantage of this phenomenon.

9.5 Special Topics Related to Reliability of the Expander Blade

Some special topics are included in this section that may influence the final reliability of blades. Some of the topics are also found to be important in explaining premature failure of blades during operation.

9.5.1 Effect of Blockage of Inlet Passages

Commonly in a turbine stage, steam is introduced only in a portion of the complete rotation of the blade, known as a "partial admission stage." It was found that the response of a blade due to the steam forces is different from where steam is admitted to the stage completely. This phenomenon may also occur in expander due to deposition of fines that may block inlet vanes passage of gas. This makes the stage similar to a partial admission stage of a steam turbine. Concept of partial admission may be used to analyze the blockage of passage in an expander due to deposition.

The following situations may arise from the blockage of passages:

1. There may be complete blockage in a few passages, similar to the partial admission of the steam turbine stage.
2. There may be a uniform partial blockage in all passages; this will change the magnitude of forces in all blades.
3. The blockage might be different in different passages; this is the case of mistuning of force.

Many years ago a simple method was described by Kroon (1940) and Allen (1940). Pigott (1980) presented a comprehensive treatment, in which the combined effect of shock load and nozzle wake was considered (Figure 9.28). The vibratory response of the blade for these two effects was analyzed. Pigott considered a packeted-bladed-disc assembly and the effect of the disc was not included in the analysis. By separating the effects of two types of load, Pigott demonstrated the dependence of blade response on the natural frequencies, running speed, number of nozzles, number of blades, number of blades in a packet, and damping. As a single-packet-vibration independent of the disc's influence, considering the

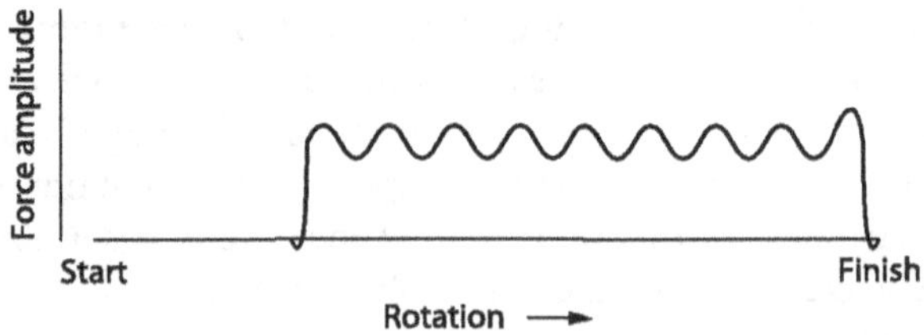

FIGURE 9.28 General forcing for partial admission includes nozzle wake force.

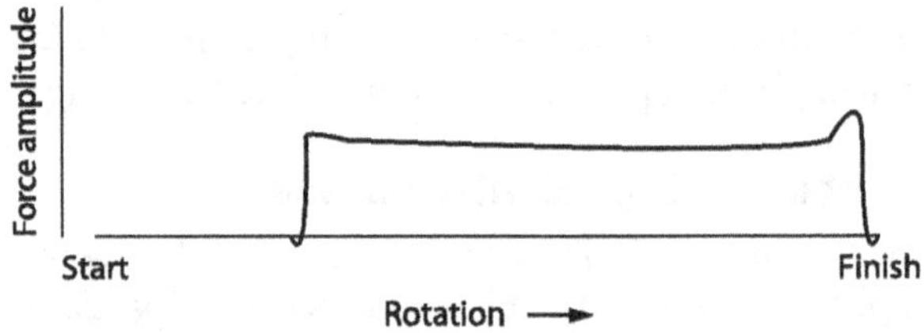

FIGURE 9.29 General forcing for partial admission only.

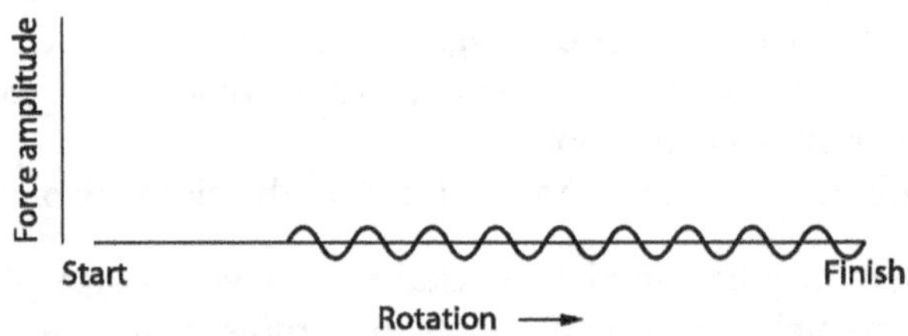

FIGURE 9.30 General forcing for nozzle wakes only.

effect of force mistuning cannot be included. This aspect is discussed in the mistuning section.

Figure 9.28 depicts a partial admission force profile experienced by a blade in one revolution including the shock-load effect and the force profile due to each nozzle opening. Figure 9.29 is the force profile due to the partial admission whereas Figure 9.30 shows the forces due to nozzle openings (nozzle-passing force). It should be realized that it assumes a uniform force from each opening. This might not be the case for a mistuned force. Thus, the response of a blade is composed of two components, one due to nozzle force and the second due to partial admission (shock load). Using the concept of a magnifier to the steady-state response due to these two situations,

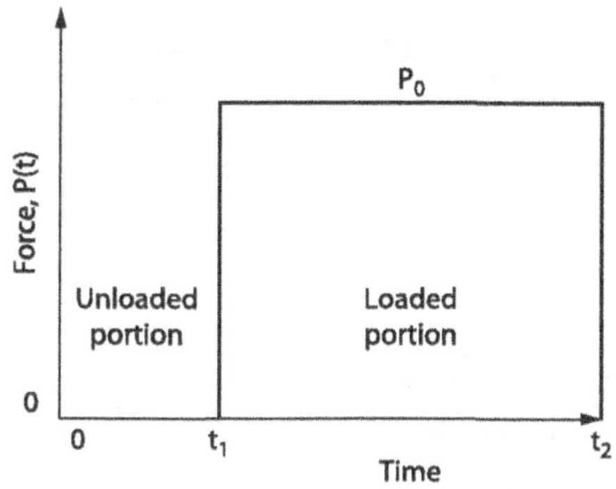

FIGURE 9.31 Simplified load pattern.

following relationship Eq. (9.14) can be used for the total response in terms of stress:

$$\sigma_{total} = \mu_1(\sigma_{steady}) + \mu_2(\sigma_{steady}) \quad (9.14)$$

where μ_1 is the magnification factors for partial admission and μ_2 is the magnification factors for nozzle wakes.

Kroon (1940) examined the response behavior of a blade by using a single spring-mass-damper system. He used a simplified load pattern shown in Figure 9.31.

In each revolution, the blade will go through successive instantaneous loading and unloading. The expression for the magnifier was approximated in successive steps ending in an expression for low damping. The three main equations are given in Eq. (9.15) through (9.17). Kroon assumed that n was the number of cycles of vibration, n_1 is the cycle of vibration in the unloaded portion, n_2 is the loaded portion, y_{osc} is the amplitude of vibration, y_{st} is the amplitude due to a steady load and δ is the logarithmic damping.

$$n = 60\omega/N$$

ω is the frequency and N is the rotor speed in RPM.

$$\mu_1 = y_{osc}/y_{st} = (0.5 + (2 + e^{-n1\delta} + e^{-n2\delta})/(2(1 - e^{-n\delta})\, e^{\delta/}2)) \quad (9.15)$$

After neglecting higher order terms, the expression becomes

$$\mu_2 = y_{osc}/y_{st} = (0.5 + (2 - n\delta/2)/(n\delta(1 + \delta/2))) \quad (9.16)$$

For low damping, terms containing $\delta/2$ and $n\delta/2$ are neglected resulting in the following equation:

$$\mu_3 = y_{osc}/y_{st} = 2/(n\delta) \quad (9.17)$$

Blade Magnifier due to Blockage	
n_1	0.73333
n_2	13.93333
n	14.66667
log decrement, δ	0.01885
Rotor speed, RPM	4500
Natural frequency, Hz	1100
Arc of admission, θ, degrees	342
Number of blades	60
# of blocked passage	3
Magnifier	
μ_1	8.20

TABLE 9.2 Magnifier Based on Kroon's Method

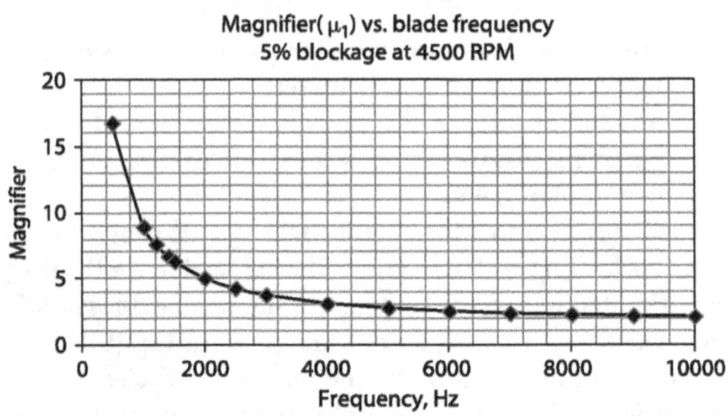

FIGURE 9.32 Magnifier vs. frequency.

The number of blades, natural frequency of blade, percent blockage, and speed of the blade are listed in Table 9.2 with the estimated value of the magnifier. An expression of magnifier of Eq. (9.15) has also been plotted for magnifier versus blade frequency (Figure 9.32). It is worthwhile to note that with this simple calculation, the possible magnitude of magnifier μ_1 can be estimated for any design at hand.

Kroon also developed an expression for a "severity factor" by a load profile (Figure 9.33) applied to a spring-mass model given below.

After solving the equation, Kroon defined the severity factor, η, as follows:

$$\eta = (T/\pi\tau)\sin(\pi\tau/T)$$

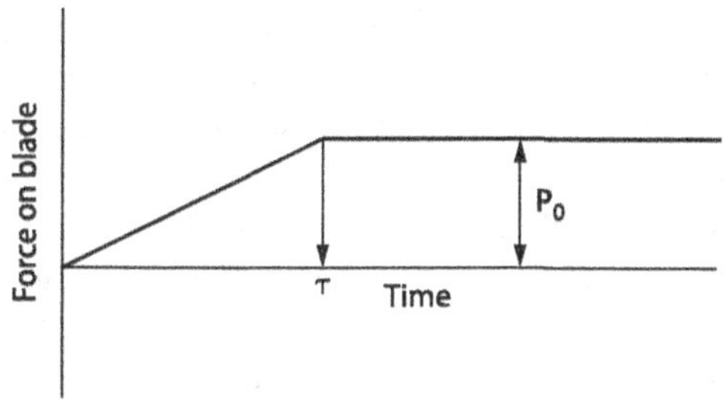

FIGURE 9.33 Simplified loading profile.

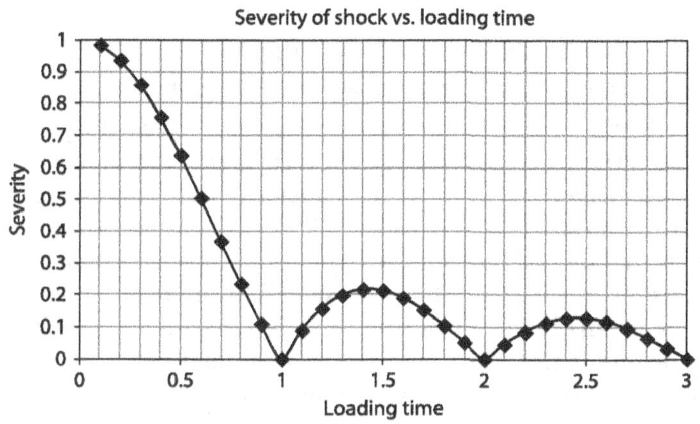

FIGURE 9.34 Severity factor vs. loading time.

where T is the period of free undamped vibration $= 2\pi/\omega$ and $\omega = (k/m)^{1/2}$

The plot of η versus force loading time, τ, is given in Figure 9.34.

The large motions are predicted when the number of vibrations per revolution is not an integer (Figure 9.34). Another important observation is that the lowest motion can be expected when the loading time, τ, is an integer multiple of the vibration period, T, of the blade.

Singh and Lucas (2011) by using SMD described a similar simplified way of analysis. After the expander is started, the blade outside the partial admission zone will respond as a free damped system. At the time it enters the partial admission zone the initial displacement and velocity are the displacement and velocity experienced at the end of the free vibration case. At the time the blade leaves the partial admission zone, the displacement and velocity become the initial conditions for free vibration analysis. This was done by solving differential equation in successive order with changing initial condition. Results of some partial type of loadings (multiple) are presented.

Damped forced vibration equations are used to estimate displacement and velocity as a blade enters and leaves the arc of admission. They noted, however, that sometimes the amplitude of vibration might not go to the maximum value. This may be due to the fact that the blade might not have entered in the loaded portion of the arc in its extreme backward position and it might not have left the loaded portion in its extreme forward position during vibration. A careful matching of those conditions may show the maximum amplitude.

9.5.2 Influence of Tolerance Stack Up in Root-Disc Attachment on Peak Stress and Natural Frequencies

Singh (1998) examined the case of tolerance stacked up due to manufacturing tolerances on a steam-turbine root disc assembly. The proposed method has the main ingredients for analyzing an expander blade root and disc assembly. It dealt with a root construction normally used in a steam turbine, but the basic behavior of an expander root will be similar. The concept of analysis and the influence of tolerance stack up for an expander blade root and disc attachment are the same. This method is described in detail in Singh and Lucas (2011). The peak stress both in the root attachment and disc slot is influenced by the stacking of machining tolerances between the root attachment and the disc slot. The assumption made during the design phase that each hook shares the load equally for multihook (lands, lobes) root construction might not be realized due to machining tolerances. It is more likely that not all hooks will be in contact at the start-up of expander. Therefore, there will be a gap between the blade root and the disc's land. Location of the gap between hooks will depend on the manufacturing tolerances at the time of assembly. As an expander starts, an existing gap at assembly may close at a speed below the operating speed or it may not at all. This situation may pose a problem if the resulting stresses are larger than estimated based on contacts at each hook, which becomes evident even for a very small gap. This point can be clarified if the assumption is made that the closure of the gap will result from pure radial extension of the root section neglecting any bending displacement of the loaded hook. Singh presented the result of a two-hook construction which was done to understand the behavior. Assume that there is a chance of a 0.002-in gap at either of the hooks of the root attachment. Distance between the lower and upper hooks is 0.50 in. A strain equal to 0.004 in/in will result to close this starting gap. If the modulus of elasticity, E, has a value of 29.0×10^6 psi, then magnitude of stress equals 116,000 psi. To close the gap an average stress equal to 11600 psi is required. Near the fillet area stress will be further magnified owing to the stress concentration. If there is sufficient centrifugal load to impose this magnitude of average stress, the gap will close before reaching the operating speed. The other hook starts to pick the load after closure of the gap and the stress in the previous hook starts to decrease.

9.6 Reliability Evaluation of Blade Design

The basic ingredient for safe operation of a blade is ensured through good design, good quality assurance, good manufacturing, and operation. If important aspects are not considered during the design phase then it invites trouble during operation; no matter how good the manufacturing process is and all protocol of quality process and careful operation within guidelines of OEM are followed. On the other hand, a good design may run into trouble if any other processes are not diligently followed. Therefore, it is incumbent on all involved to follow good practices with each required process.

The damage in the blades of the expander is almost always due to fatigue, or sometimes creep phenomenon plays a role in damage. The reliability analysis of blades consists of essentially of two parts: stress analysis and dynamic analysis of the blade or bladed disc. The strategy for a blade design is to avoid the coincidence of harmonics of running speed and nozzle-passing frequencies with the natural frequency of the blade. The concept of resonance, in which mode shape and the shape of the forcing also play a defining role, has been explained earlier.

The stresses to assess reliability of a blade design are centrifugal stress due to speed, bending stress due to gas load and dynamic stress as a result of blade excitation by fluctuating forces. The blade dynamics analysis compares the blade's natural frequencies with exciting force frequencies to determine the vibratory stresses in the blades. Fatigue damage is due to the alternating stress imposed on the blade, and it is also influenced by the magnitude of mean stress (e.g., centrifugal stress) imposed on the blades. The centrifugal stress is the mean stress at the operating speed of the expander while the alternating stress is the result of unsteady forces that exist in the turbine. The magnitude of the alternating stress is dependent on the nature and magnitude of unsteady forces, damping, and the resonance condition. Traditionally, the reliability for high-cycle fatigue of a blade design is judged based on the estimated magnitude of the factor of safety using a Goodman-type diagram.

9.6.1 Resonant Response of Blade during Vibration

Conditions of resonance for a blade and bladed disc have been discussed is the explanations of Campbell and Safe diagrams. Resonance of blade may increase the gas stress of many folds. For reliability evaluation, however, stress needs to be determined. The basic logic has been that the bending stress caused by the gas loading on the blade gets modified depending on the dynamic characteristics of the rotating blade.

A simple equation to estimate blade vibratory stress is given below:

$$\sigma_{vib} = S\,(\pi/\delta)\,(RRF)\,\sigma_{hp} \tag{9.18}$$

σ_{vib} = calculated blade vibratory stress

σ_{hp} = steady gas bending stress in the blade

S = stimulus factor; the percent of the gas force that is unsteady (ranges from 0.1 to 0.3)

π/δ = resonant amplification factor. If a critical damping ratio of 0.003 is used, then the log decrement damping (δ) is equal to $0.003(2\pi) = 0.01885$ and the resonant amplification factor is equal to $\pi/0.01835 = 166.66$

RRF = resonant response factor depending on the mode of vibration and other factors (ranges from 0 to 1). Weaver and Prohl (1956) defined the resonant response factors (RRF) that range from 0.1 to 0.9. RRF depends on the mode shape of the blade vibration and phasing of the exciting force, which is consistent with the concept of the Safe diagram. In the case of mode shape and shape of the forcing being identical, RRF will equal 1.0. Any value of RRF less than 1.0 signifies a forced response.

The two factors, S and RRF, of Eq. (9.18) are very important for good design. Acceptance criteria based on these methods is intentionally conservative to include each type of design. However, individual features of stationary vanes and rotating blades will control the magnitude of these two factors. It is safe to assume that there might be a lot of variation depending on design. This only can be ascertained by detailed analysis (e.g., FEA and CFD).

9.6.2 Running-Speed Harmonic Excitation

Running-speed harmonic excitation occurs due to any circumferential variation or interruption in the gas flow in a full admission stage in one complete revolution. Some of the sources of running-speed harmonic excitation are exhaust, diaphragm vane manufacturing or spacing variations, stationary vanes, struts, and supports in the exhaust casing. Per revolution exciting forces are generally the largest in the exhaust stages with taller blades. The other reason is that the fundamental frequency of these blades is often below 15 × RS.

The resonant response magnification factor, μ, caused by running-speed harmonic excitation can be estimated by the following equation:

$$(\mu) = \beta\,(\text{RPM})/\omega_n \tag{9.19}$$

where μ = resonant response magnification factor
RPM = turbine speed
ω_n = fundamental blade frequency (HZ)
β = a constant, typically 0.9~1.0

The above equation provides the expression for a resonant response magnification factor as a function of the blade frequency and the harmonic of running speed. This magnifier is equally valid for both shrouded and freestanding blades. For variable speed, mechanical drive expanders, the fundamental blade frequency should be designed to be above the 4 × RS exciting force. For generator drive expanders (constant speed), the blade fundamental frequency is generally tuned between the harmonics of running speed if the blade frequency is below the fourth harmonic of running speed. The calculated magnification factor is reduced by a factor of two for tuned blades. When blade frequency tuning is not possible or the vibratory stresses are large, a z-lock type construction may be used to help reduce the vibratory stresses. The calculated magnification factor can be reduced by a factor of four for a blade design with a z-lock design.

9.6.3 Reliability Evaluation during High-Cycle Fatigue

Given steady stress (centrifugal + bending) and alternating stress (equation of stimulus) and after adjusting for fatigue strength, a Goodman diagram (Figure 9.35) is created. A predetermined factor of safety defines acceptance of the design for high-cycle fatigue.

$$\begin{aligned}
\sigma_{mean} &= \sigma_{CF} + \sigma_{bending} \\
\sigma_{vib} &= S\,(\pi/\delta)\,(\text{RRF})\,\sigma_{bending} \\
1.0/\text{FS} &= \sigma_{vib}/\sigma_{fs} + \sigma_{mean}/\sigma_{ult} \\
\sigma_{fs} &= k_1, k_2, k_3, \ldots (\sigma'_{fs})
\end{aligned} \tag{9.20}$$

The factor of safety does not provide any indication of the life of the blades. To ascertain the real total operating cost the life of the blade to operate safely should be estimated. It is important to have the answer of this question so that the real cost of the stage

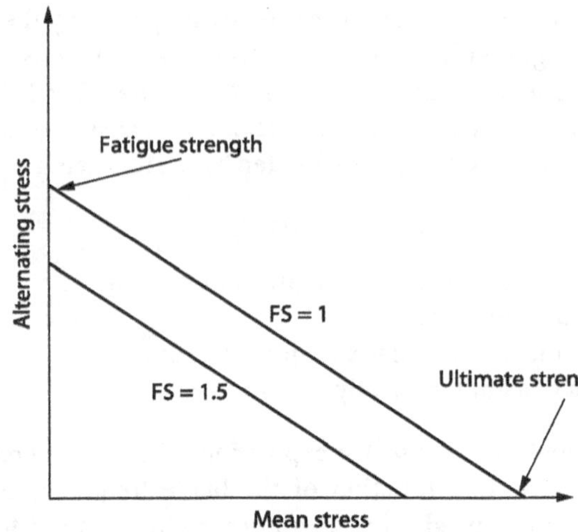

FIGURE 9.35 Basic design reliability estimation.

can be estimated. This information will also help in the repair and replacement strategy.

9.7 Mathematical Expression of Work Done by Mistuned Force on a Mistuned System

9.7.1 Expression for Work Done

The general expression for the mode shape can be written as

$$X(M) = \Sigma A_L \sin(L\theta + \phi_L) \tag{9.21}$$

where L is the half of the number of blades on the disc:

$$\begin{aligned} X(M) &= A_1 \sin(\theta + \varphi 1) + A_2 \sin(2\theta + \varphi 2) \\ &\quad + A_3 \sin(3\theta + \varphi 3) + \cdots + A_L \sin(L\theta + \varphi_L) \\ &= A_1(\sin\theta \cos\varphi 1 + \cos\theta \sin\varphi 1) \\ &\quad + A_2(\sin 2\theta \cos\varphi 2 + \cos 2\theta \sin\varphi 2) \\ &\quad + A_3(\sin 3\theta \cos\varphi 3 + \cos 3\theta \sin\varphi 3) \\ &\quad + \cdots + A_L(\sin L\theta \cos\varphi_L + \cos L\theta \sin\varphi_L) \end{aligned} \tag{9.22}$$

In the general case when forcing is also mistuned, then the shape of the force can be expressed as follows:

$$\begin{aligned} F(N) &= F_0 + F_1 \cos(\theta + \varphi 1) + F_2 \cos(2\theta + \varphi 2) \\ &\quad + F_3 \cos(3\theta + \varphi 3) + \cdots + F_L \cos(L\theta + \varphi_L) \\ &= F_0 + F_1(\cos\theta \cos\varphi 1 - \sin\theta \sin\varphi 1) \\ &\quad + F_2(\cos 2\theta \cos\varphi_2 - \sin 2\theta \sin\varphi_2) \\ &\quad + F_3(\cos 3\theta \cos\varphi 3 - \sin 3\theta \sin\varphi 3) + \cdots \\ &\quad + F_L(\cos L\theta \cos\varphi_L - \sin L\theta \sin\varphi_L) \end{aligned} \tag{9.23}$$

$$dW = F(N).d(X(M))$$

Finally, the work done, W, is given by

$$\int_0^{2\pi} dW = \int_0^{2\pi} F(N).d(X(M))$$

Using the complete integral relationship, Eq. (9.23) reduces to the following:

$$\begin{aligned} W = \pi(&A_1F_1 \cos\varphi_1 \cos\phi_1 + 2A_2F_2 \cos\varphi_2 \cos\phi_2 \\ &+ 3A_3F_3 \cos\varphi_3 \cos\phi_3 + \cdots + LA_LF_L \cos\varphi_L \cos\phi_L) \end{aligned} \tag{9.24}$$

The following conclusions can be drawn:

- The blade will respond to the harmonics of force that match with the harmonic contents of the mode shape.
- The resulting response distribution may be different from the mode shape.
- It is quite possible that there will not be a true resonance, but a forced response in such a situation.

CHAPTER 10

Damage in Material and Life Analysis

Materials resist any imposed force. Mechanical structure made of any material develops internal resistance to applied forces. This resistance manifests itself in two ways; a reaction force equal and opposite of the applied force and dimensional change in the structure. There is a limit of the force that can be safely sustained by the material before damage. When this force is larger than the capability of the material to sustain the load then result may be large distortion, development of cracks and breakage of structure in two or more pieces. Magnitude of the reaction forces and displacements may depend on the location within the structure. Theory of solid mechanics deals with the resulting quantities of reaction forces and displacements. Analytical methods deal with quantities such as stress (reaction force per unit area) and strain (displacement per unit length). The damage to material depends on the type of forces being imposed on the structure. Many theories have been propounded and successfully applied in the design of mechanical structure. These theories deal specifically with the type of force e.g., static load or time varying dynamic load. The mechanisms of damage may be rupture due to static load, distortion or rupture when the static load is applied for long time resulting in creep, high-cycle fatigue (HCF) under dynamic load below yielding of material, low-cycle fatigue (LCF) under dynamic load when material yields and the final fracture after a crack has been initiated due to load.

Basic concepts of damage in a material and methods to estimate the effect of stress and strain of mechanical components are discussed in this chapter. Finally, it leads to a possible estimation of reliability and the safe life of the component under an applied load for a specified operational cycle. The methods are equally applicable to conducting an analytical failure analysis of damaged components during or after operation.

10.1 Behavior of Material under Static Load

The basic response of a material under load is obtained through a test on a uniaxial polished specimen (Figure 10.1). Using the applied load and the measured elongation in a known length of the specimen, a curve can be plotted. This describes the relationship between load and elongation.

When a load is converted into stress (force per unit area) and displacement is converted in strain (elongation per unit length), the resulting plot is known as the Stress-Strain curve (Figure 10.2).

For example some materials return to its original position when load is removed (perfectly elastic material). Some materials behave in a perfectly plastic manner. Displacement in these types of material increases at a constant load until it breaks. Most structural materials display a combination of elastic and plastic behavior (ductile material).

Most of the terms used to describe mechanical behavior have been mentioned by Mitchell (1979).

Beyond the yield point where material goes through a plastic deformation, the total strain (Figure 10.3), ε_t, is the summation of elastic strain, ε_e, and plastic strain, ε_p, and it is given below.

$$\text{Total strain, } \varepsilon_t = \varepsilon_e + \varepsilon_p \tag{10.1}$$

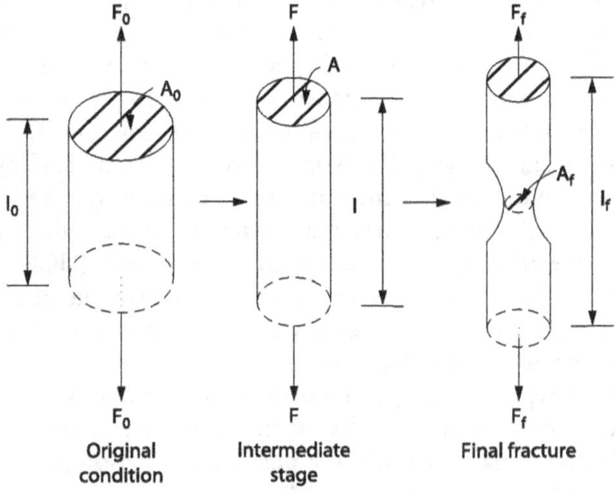

Figure 10.1 Axial sample behaviors for monotonic loading.

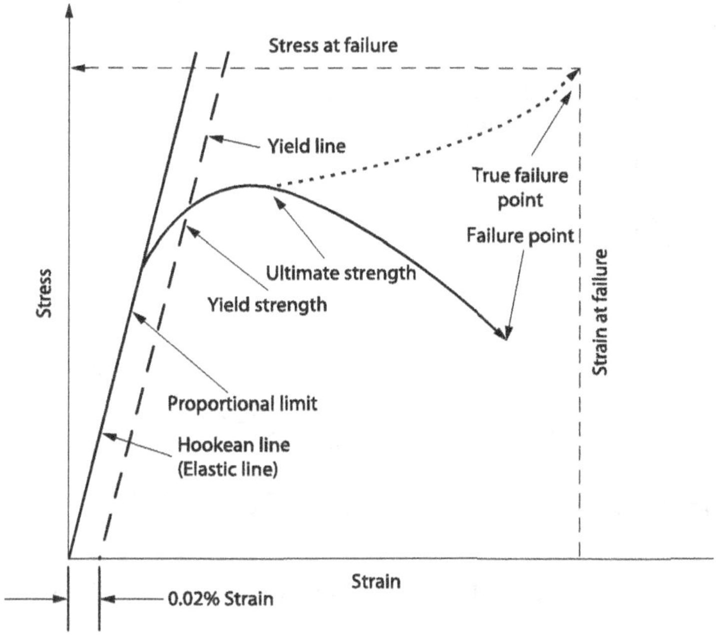

Figure 10.2 A typical engineering and true stress-strain curve.

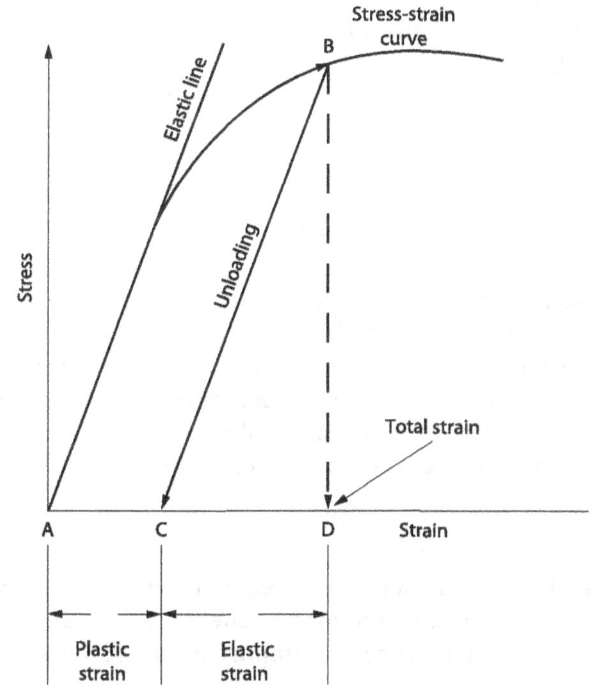

Figure 10.3 Elastic and plastic strains after unloading.

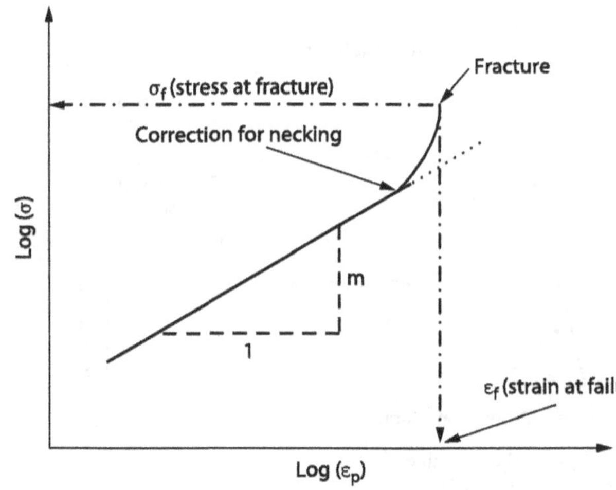

FIGURE 10.4 Log-log plot of plastic strain vs. stress.

The elastic portion of strain as given by Hooke's law is given below.

$$\varepsilon_e = \sigma/E \tag{10.2}$$

When the log (ε_p) versus the log (σ) is plotted then the resulting relation is a straight line up to the start of the necking in the material (Figure 10.4).

Stress and strain (before necking) follows a straight-line relationship Eq. (10.3):

$$\text{Log}(\sigma) = \log(K) + m\,\text{Log}(\varepsilon_p) \tag{10.3}$$

$$\varepsilon_p = (\sigma/K)^{(1/m)} \tag{10.4}$$

where m stands for a monotonic strain hardening exponent and K represents a monotonic strength coefficient.

When substitution is made of Eq. (10.2) and Eq. (10.4) into Eq. (10.1), an equation for stress-strain is obtained. This provides the relationship between stress and strain of the material and it is known as the Ramberg-Osgood relationship Eq. (10.5).

$$\varepsilon_t = (\sigma/E) + (\sigma/K)^{(1/m)} \tag{10.5}$$

For analytical assessment for reliability and to estimate the safe life of a mechanical component, coefficients of the Ramberg-Osgood relationship become very important. It will become clear in later sections.

10.2 Behavior of Material under Cyclic Loading

Material goes under progressive deterioration under application of a critical lower alternating stress. For reliability and life evaluation of the mechanical components of high-temperature turbomachinery, the phenomenon of high-cycle fatigue, low-cycle fatigue, and creep (including cycle) are of primary concern as in the case of FCC expanders. Cause of failure depends on the functional requirements, the geometrical construction of the components and the environment in which the expander operates.

The failure of metals owing to repeated loads below what was considered to be "safe" was observed. Specifically, ductile metals failed in a brittle manner. Many theories have been advanced to explain fatigue failure of material. These studies ranged from dislocation mechanisms to phenomenological material behavior to full-scale structure. It is beyond the scope of this discussion to deal with the above in detail. Mitchell (1979) provided a list of some key contributions to explain fatigue behavior in materials. Some of these are the work of Coffin (1954, 1971) and Manson (1953) (working independently, thermal cycling, low-cycle fatigue and plastic strain consideration). The Coffin-Manson law is most used to estimate life for low-cycle fatigue where plastic strain becomes the dominant factor.

10.2.1 Cyclic Stress-Strain Curve

The properties of material under cyclic loading can be very different compared to monotonic properties. Depending on the initial state of the material and test conditions, a metal may exhibit different behavior such as cyclic softening, cyclic hardening, being cyclically stable, and showing a combination of the above behavior. Figure 10.5 shows typical σ-ε curves for monotonic, cyclic softening, and cyclic hardening materials. It should be noted that (a) different stress levels are required to achieve the same strain levels, and (b) after unloading from the same strain level, they did go to different strain levels at zero stress.

If they are loaded to the same stress they reach different strains, and, after unloading the strain levels, they are vastly different (Figure 10.6).

10.2.2 Equation of Cyclic Stress-Strain Curve

Cyclic stress-strain properties of material are obtained in an axial cyclic strain-controlled test of smooth polished specimens. Hysteresis loop is the locus of each loading-unloading and loading history beyond the yield point (Figure 10.7).

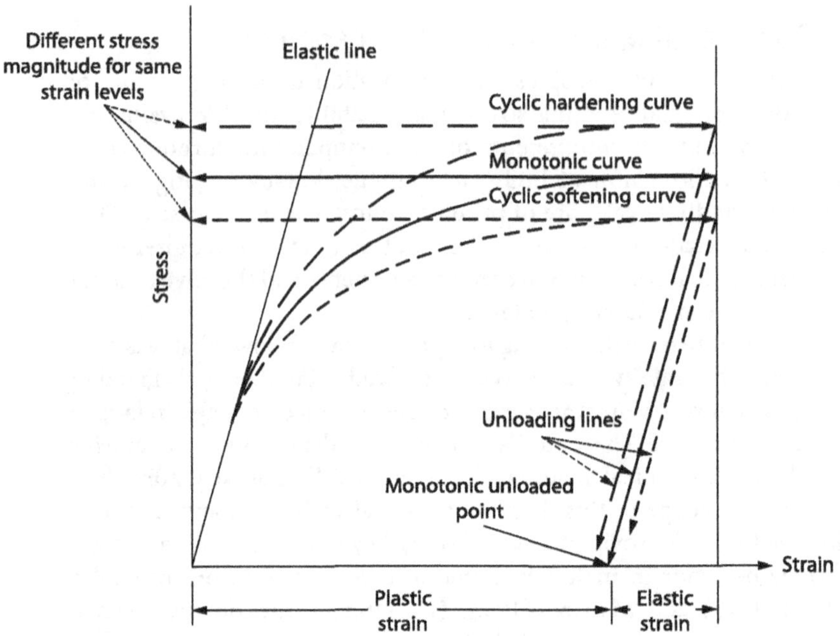

Figure 10.5 Representative stress-strain curves in monotonic and cyclic loadings.

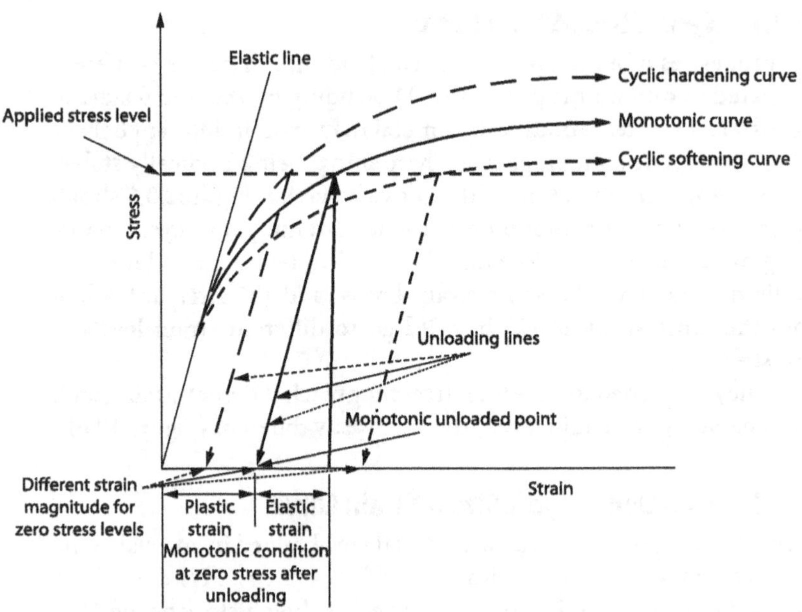

Figure 10.6 Unloading from same stress level.

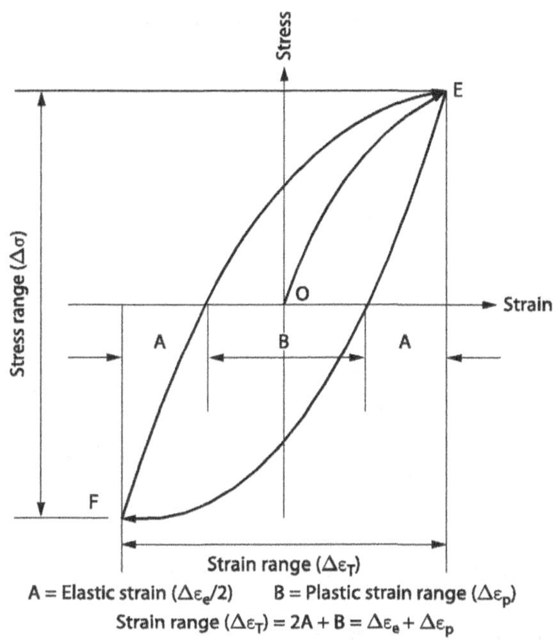

FIGURE 10.7 Typical hysteresis loop with definition of stress and strain ranges.

Stress range, $\Delta\sigma$, is the height of the hysteresis loop from tip to tip. Strain range is the width from tip to tip of the loop. Strain amplitude, $\Delta\varepsilon/2$, is one-half of the width from tip to tip. The plastic strain amplitude is the difference between elastic strain amplitude, $\Delta\varepsilon_e/2$, and the total strain amplitude Eq. (10.6).The elastic strain amplitude, $\Delta\varepsilon_e/2$, is obtained by applying Hooke's law Eq. (10.7).

$$(\Delta\varepsilon_p/2) = (\Delta\varepsilon/2) - (\Delta\varepsilon_e/2) \tag{10.6}$$

The elastic strain is given by Hooke's law.

$$(\Delta\varepsilon_e/2) = (\Delta\sigma/2E) \tag{10.7}$$

where E is the modulus of elasticity.

It has been found that the plot of the log of stress amplitude versus plastic strain amplitude displays a straight line (Figure 10.8) same as it was case for the monotonic loading.

The cyclic yield hardening exponent, m′, is the power to which the plastic strain amplitude is proportional to stress amplitude Eq. (10.8).

$$(\Delta\sigma/2) = K'(\Delta\varepsilon_p/2)^{m'} \tag{10.8}$$

where $\Delta\varepsilon_p/2$ = plastic strain amplitude.

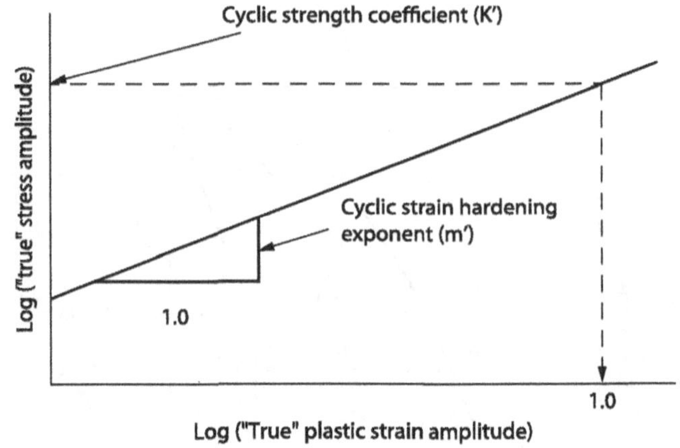

Figure 10.8 True stresses vs. true plastic strain.

Cyclic strength coefficient, K′, is the true stress to cause a plastic strain of unity.

Finally, the measured strain is converted to stress to obtain equation for cyclic σ-ε curve Eq. (10.9).

$$\begin{aligned}(\Delta\varepsilon/2) &= (\Delta\varepsilon_e/2) + (\Delta\varepsilon_p/2)\\ &= (\Delta\sigma/2E) + (\Delta\varepsilon_p/2)\\ &= (\Delta\sigma/2E) + (\Delta\sigma/2K')^{(1/m')} \qquad (10.9)\end{aligned}$$

Eq. (10.9) is the Ramberg-Osgood equation for cyclic σ-ε curve.

10.3 Definition of Force

It is important to revisit some theories of physics to lay the foundation for two situations that are encountered in the design and analysis of a mechanical system. First, when mass of the mechanical system changes during operation, for example, depositions on blades of an expander then change in forces and stress along with any change in natural frequencies should be accounted for. Second, the basis of existence of localized stress in the presence of discontinuity should be understood.

The theory of physics based on Newton's laws of motion is called Newtonian mechanics. Newton considered mass of the structure and energy of the system to be separate from each other and mass and energy to be conservative, in contrast to their treatment in relativistic mechanics. Newton's laws of motion form the basis of analyzing mechanical structure to estimate response under a specified force. The definition of force and its effects on the structure are embedded in Newton's laws.

Newton proposed three laws. The first law states "whether a body will be at rest or move with constant velocity or it will move with accelerating velocity" when acted upon by a force. The second law provides a "definition of a force" such that a mathematical relation can be written for force. The third law defines "the magnitude and direction of the reaction of the material to the applied force."

The first law states that a body maintains its state (whether of rest or a state of constant motion) unless it is forced to change. In other words, a body will acquire a velocity that will vary with time under action of a resultant force.

The definition of a force is embedded in the second law. Newton stated that force acting on a body is proportional to the rate of change of momentum (m(dx/dt)) [Eq. (10.10)]:

$$\begin{aligned} \text{Force, } P &\propto d(m(dx/dt))/dt \\ &\propto (dm/dt)\cdot(dx/dt) + m(d^2x/dt^2) \end{aligned} \tag{10.10}$$

During operation, the mass of blades of an expander may change due to deposition, corrosion, and erosion. The first term of Eq. (10.10) must also be included to estimate magnitude of the force if there is change in the mass of the structure. Force changes at the rate of change of mass. Natural frequency will also change due to change in the mass of the blade.

When mass of the structure does not change, then $dm/dt = 0$

$$\text{Force, } P = K\, m(d^2x/dt^2)$$

K is a constant.

When units of force, mass, and acceleration are chosen, such that $K = 1$ and rate of change of velocity (d/dt(dx/dt)) is equal to a (acceleration), then

$$P = ma \tag{10.11}$$

The third law states that any action has an equal and opposite reaction. This means that the force acting on the structure must balance. This leads to the equation of motion for dynamics, as well as, to steady-state stress evaluation.

It also means that either

- if the force is as defined by Newton, then energy must conserve
- if energy conserves then the force must be as defined by Newton's laws

10.3.1 Stress at a Point

In almost all practical mechanical structure stresses and strains are not uniform at each point. Near any kind of geometrical discontinuity, stress is many times larger than the other places. This phenomenon is called stress concentration. It is important to estimate peak stress and peak strain because damage to a structure initiates from this location. A notch or a discontinuity makes the distribution of forces nonuniform. Damage to a structure starts from the location of peak stress even though the magnitude of average stress across a cross section might be acceptable. It has been understood that damage starts from the location of the largest stress and to appreciate the implication of stress concentration, stress, and strain at a point need to be explained. It is important to consider a distinction among scalar, vector, and tensor quantities. When a load is applied, three important parameters are considered: magnitude, direction, and point of application, which results in stress and strain in the structure. A scalar quantity is described completely by its magnitude (e.g., mass, density, and temperature). For a description of a vector quantity, direction is specified in addition to magnitude (e.g., velocity, force, acceleration). On the other hand, to completely define stress and strain, quantities (e.g. magnitude, direction, and a point of application) are required.

The average stress (S) at a point as in Figure 10.9 is stated as:

$$S = (\Delta F/\Delta A)$$

The local stress at the point using the reaction force δR is as follows:

$$\sigma = \lim_{\delta A \to 0} (\delta F/\delta A) = (dF/dA) \tag{10.12}$$

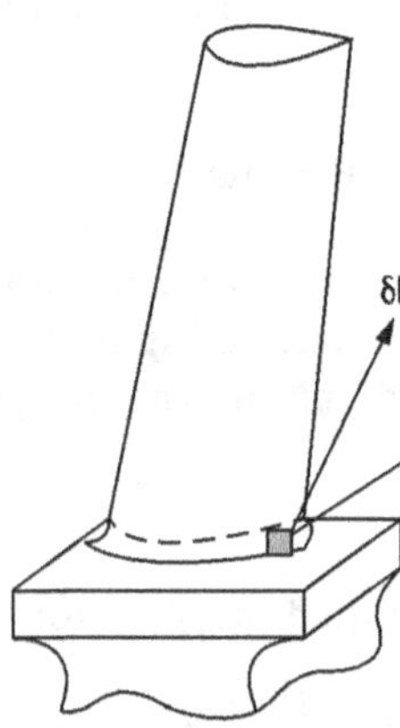

Figure 10.9 Force and stress on a small area.

10.3.2 Stress Concentration

In the presence of a notch or any type of discontinuity the applied load is not uniformly distributed, for example, as shown in the Figure 10.10. In such a situation the reaction forces needed for equilibrium will also be nonuniform. To match exactly to the applied force at each point, the reaction forces will be equal and opposite. This results in the stress at each point across the cross section to be different, thus it will not be uniformly distributed. The definition of stress at a particular point is required to define local stresses and to estimate the magnitude of stress concentration.

Eq. (10.25) provides a way to estimate stress concentration if the force at each point of the structure is known. Localized stress can be estimated by multiplying the average cross-section stress by a theoretical stress-concentration factor in the absence of any plastic deformation, that is, when the localized stress is less than the yield strength of material. Many mechanical components of expanders have fillets and notches as in Figure 10.11. Tables and equations of many geometrical and loading conditions are available to estimate the stress concentration, K_T, called theoretical stress concentration. By multiplying K_T to the average stress, the peak stress is estimated by assuming that material follows Hooke's law. Finite element analysis may be used to estimate actual localized stress. In the situation of a highly localized

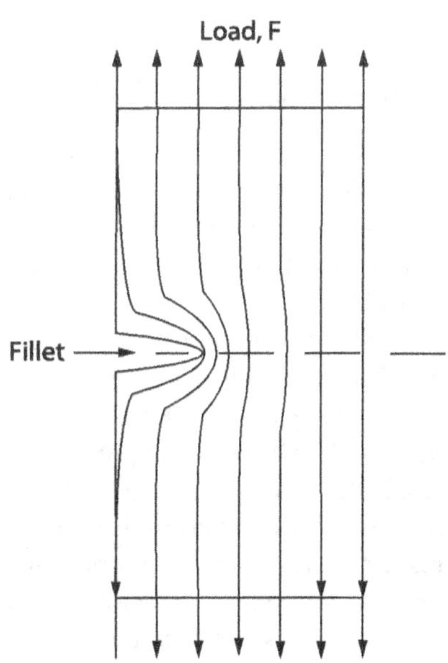

Figure 10.10 Stress contours showing concentration of stress near fillet.

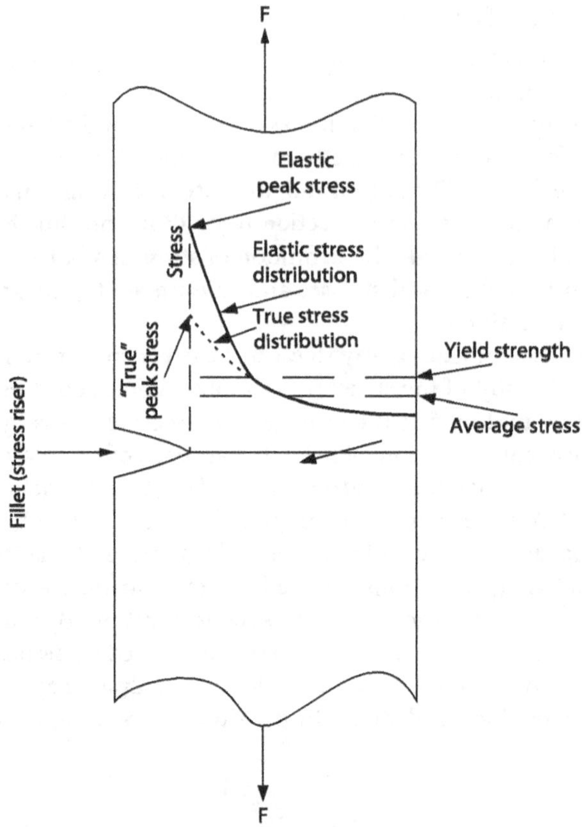

FIGURE 10.11 Localized stress in the vicinity of a fillet.

stress, a quick estimation is often made by some simple rules without spending valuable resources on finite element analysis. A method called Neuber's rule to estimate the peak stress is very popular and is explained by Figure 10.12.

Therefore, the stress concentration, K_T, is known as "elastic" stress concentration. The resulting peak stress is expected to lie on the straight line representing the Hookean line. It is a straight line drawn from the origin of the stress-strain plot with a slope representing Young's modulus. If the peak stress is greater than the elastic limit of the material then a plastic deformation is expected. In such a situation the actual stress and strain of the material must be estimated by its properly locating this state on the stress-strain curve of the material. Special care must be exercised when the resulting stress is greater than the yield strength of the material. Post-yielding or beyond the elastic limit the correct state is determined by the stress-strain curve.

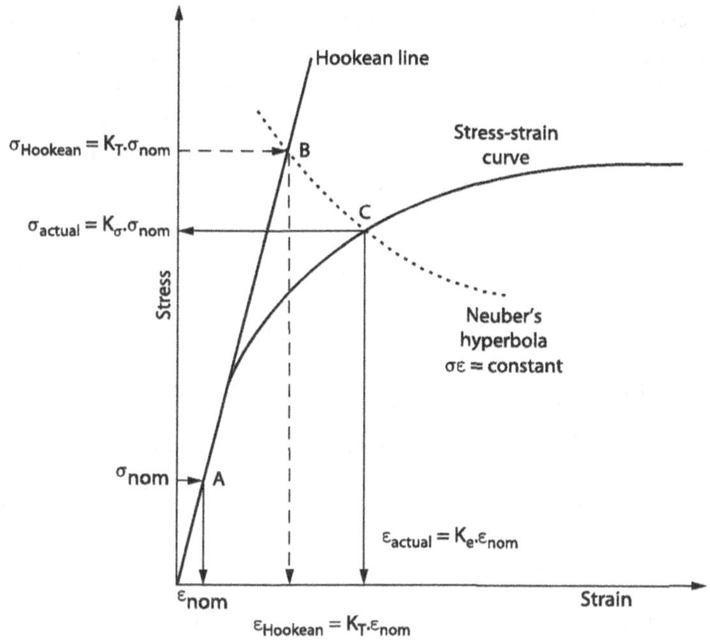

Figure 10.12 Neuber's hyperbola.

10.3.3 Neuber's Method to Estimate Localized Stress and Strain

Figure 10.12 shows an idealized stress-strain curve along with the Hookean line and Neuber's hyperbola. The average stress (σ_{nom}) is at point A, and when it is increased by a factor of "theoretical stress concentration, K_T," the resulting peak stress follows the Hookean line. The resulting peak stress, σ_H, is denoted by point B. This stress is larger than the elastic limit and the yield stress of the material. It should be recognized that any stress and strain condition of the material must fall on the stress-strain curve. Stress at point A is the nominal stress, σ_{nom}, whereas stress at point B is magnified in the vicinity of the fillet. The actual stress and actual strain should follow the σ-ε curve. Neuber's equation given in Eq. (10.13) allows estimation of the location of point C on the stress-strain curve by using the elastic peak stress of point B. A review of Figure 10.12 shows that if the real stress-strain curve is considered then actual stress beyond the elastic limit will be smaller than that estimated by the elastic stress concentration. However, the actual strain will be greater than estimated by elastic considerations. Actual stress concentration, K_σ, will be less than K_T but actual strain concentration, K_ε, will be greater than K_T. Below the elastic limit these three concentrations ($K_T = K_\sigma = K_\varepsilon$) are equal. Many finite element codes are capable of elastic-plastic analysis, which is expected to provide a better state of stress and strain at the location of peak stress. It may be time consuming or even costly to warrant this type of analysis in most cases. Depending on the situation, a quick and an approximate solution might be desirable.

Neuber rule can be expressed by the Eq. (10.13):

$$K_\sigma \cdot K_\varepsilon = K_T^2 \tag{10.13}$$

where $K\sigma = \sigma_{actual}/\sigma_{nom}$ is "actual" stress concentration (10.14)

$K\varepsilon = \varepsilon_{actual}/\varepsilon_{nom}$ is "actual" strain concentration (10.15)

$K_T = S_H/\sigma_{nom}$ is "elastic" stress concentration (10.16)

S_H is stress from a linear analysis

E is Young's modulus

Substituting Eqs. (10.14) to (10.16) into Eq. (10.13) yields the following:

$$(\sigma_{actual}/\sigma_{nom}) \cdot (\varepsilon_{actual}/\varepsilon_{nom}) = (S_H/\sigma_{nom})^2$$

$$\sigma_{actual} \cdot \varepsilon_{actual} = (S_H/\sigma_{nom})^2 \cdot (\sigma_{nom} \cdot \varepsilon_{nom}) \tag{10.17}$$

$$\sigma_{actual} \cdot \varepsilon_{actual} = (S_H)^2 \cdot (\varepsilon_{nom}/\sigma_{nom}) = (S_H)^2/E = C, \text{ Constant} \tag{10.18}$$

Ramberg-Osgood's equation for the stress-strain curve is given by

$$\varepsilon_{actual} = \sigma_{actual}/E + (\sigma_{actual}/K)^{1/m} \tag{10.19}$$

Solution for σ_{actual} and ε_{actual} can be obtained either using a numerical technique or using a graphical method by drawing Neuber's hyperbola superposed on the stress-strain curve as represented by Eqs. (10.18) and (10.19).

10.4 Damage Mechanisms in Material

Some reasons for the degradation and damage to a material, but not all, are described below. Under applied force, deformation damage may occur owing to a change in the physical dimensions or shape of a component. This might indicate failure if it is sufficient for components' function to be lost or impaired. If a component is cracked and it is separated into two or more pieces, it is termed *fracture*. In case of material removed from the surface due to abrasion or sticking between solid surfaces due to relative motion, it is called wear. Erosion is also wear, but it is caused by action of fluid (gas or liquid). Corrosion is loss of material due to a chemical action. In general terms, the damage can be classified into three main categories, as shown below.

1. Deformation
   a. Time dependent (creep)
2. Degradation (environment)
   a. Corrosion
   b. Erosion
   c. Wear
3. Fracture
   a. Cyclic loading
      High-cycle fatigue
      Low-cycle fatigue
      Crack and crack growth
      Fretting fatigue
      Corrosion fatigue
   b. Static loading
      Brittle fracture
      Ductile fracture
      Creep rupture

An explanation of the response of a material under an applied force and the resulting system of stress and strain is required to understand the basis of damage phenomenon.

10.4.1 Failure Theories for Sustained Static Loading (Creep)

Creep is a time-dependent permanent deformation of a material under applied stress constantly maintained. Due to a sustained loading for an extended period of time, a permanent deformation may occur. Even at a relatively low stress, deformation may occur in relatively less time owing to high temperature. A crack may initiate due to creep near the area of the localized stress in a shorter period of time. Creep damage is traditionally considered in stationary components of turbomachinery. It becomes important for blades, for example, in the fillet area, at the base of the airfoil or in the root area where high localized stress may exist.

Sometimes the temperature of a stage may rise due to flow conditions or change of pressure of the stage.

Creep deformation, which accumulates with time, may become so large that the component fails to perform its intended function. It might eventually rupture if the duration is large. Virtually any material will creep upon approaching melting temperature.

Creep properties are determined through a test under a constant load or stress on a specimen. The resulting strain is plotted as a function of time as shown in Figure 10.13. After initial strain, a decelerating strain stage called primary creep is reached, which is followed by a steady minimum creep rate called secondary creep. Finally, an accelerating stage, tertiary creep, is reached ending in fracture.

Creep deformation is a thermally activated process. The rate of activation process may be governed by the Arrhenius rate Eq. (10.20), which is represented below:

$$d\varepsilon/dt = A\, e^{-Q/RT} \tag{10.20}$$

Q is a physical constant called "activation energy." For the molecular motion to occur this energy must be overcome. A method most widely used in the mechanical design is briefly described next.

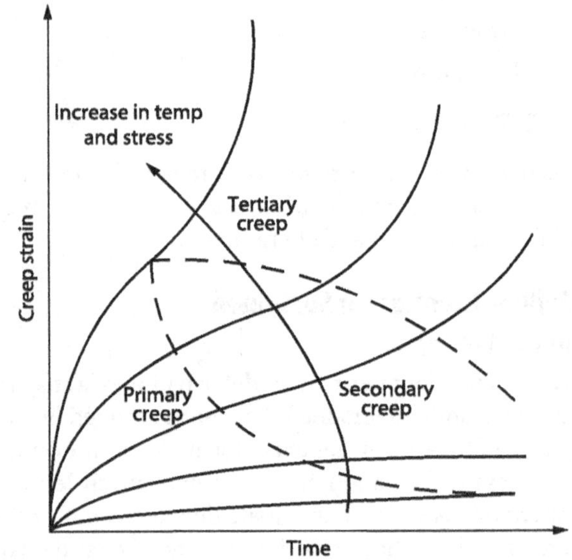

Figure 10.13 Schematic basic creep behavior of a material.

Larson-Miller Parameter

The Larson-Miller parameter (LMP) method was described by Larson and Miller (1952). It is widely used in design. Larson and Miller rearranged Eq. (10.21) as follows:

$$0.217Q = T\,(\log 10\ t_r + C)$$

where $C = -\log 10\ \theta_r$

$$LMP = (T + 273)\,(\log 10\ t_r + C) \tag{10.21}$$

The value of C is the intercept on a plot of log 10 t_r versus 1/T for a given stress as shown in Figure 10.14. Values of the constant C for rupture of various steels are often near 20. This value of C may be used as an estimate if more specific information is not available.

It has been found that the plot of stress versus LMP results in a single curve. The LMP curves for many materials are available in literature. A typical representative LMP curve for IN 738 superalloy is shown in Figure 10.15.

Evaluation of creep life of blades becomes important either during the design phase or when a failure occurs. Utilization of LMP to estimate time to rupture will be explained later in the section to estimate life with an example.

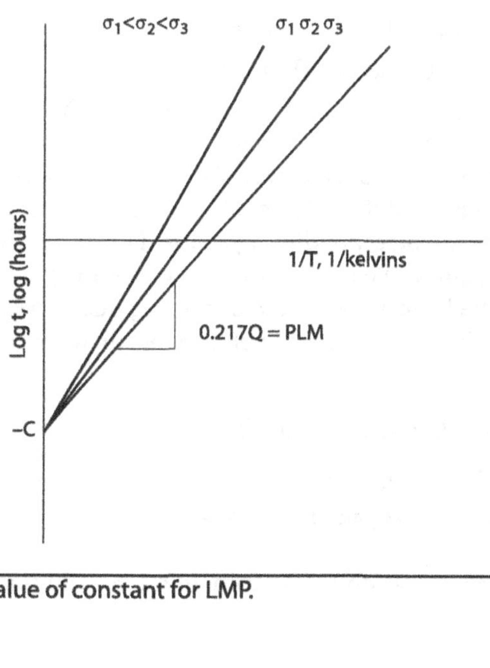

FIGURE 10.14 Value of constant for LMP.

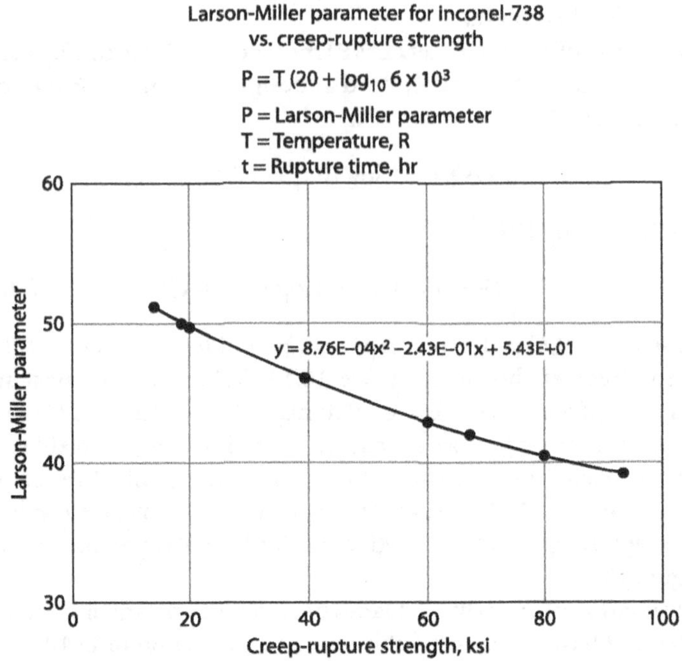

FIGURE 10.15 Typical Larson-Miller curves for IN 738.

10.4.2 High-Cycle Fatigue Loading and Evolution of Goodman Diagram

The fatigue resistance of a material is defined as the number of constant amplitude stress or strain reversals that causes failure. For life estimation, material properties are determined by a number of test samples to failure at different stress levels. The plot of stresses on the ordinate and the corresponding cycles to failure on the abscissa is called the S-N curve (Figure 10.16). An estimation of cycles to failure under an applied stress level is done by using this curve.

The total life of a component can be divided in three parts (Figure 10.17):

1. Life spent to nucleate a crack
2. Life spent to initiate a crack
3. Life spent in propagating a crack

Total Life = Nucleation Life + Initiation Life + Propagation Life

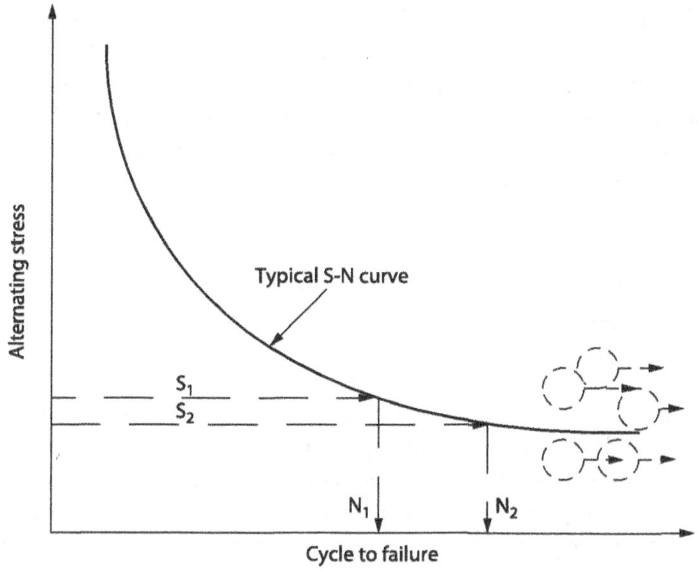

Figure 10.16 S-N curves.

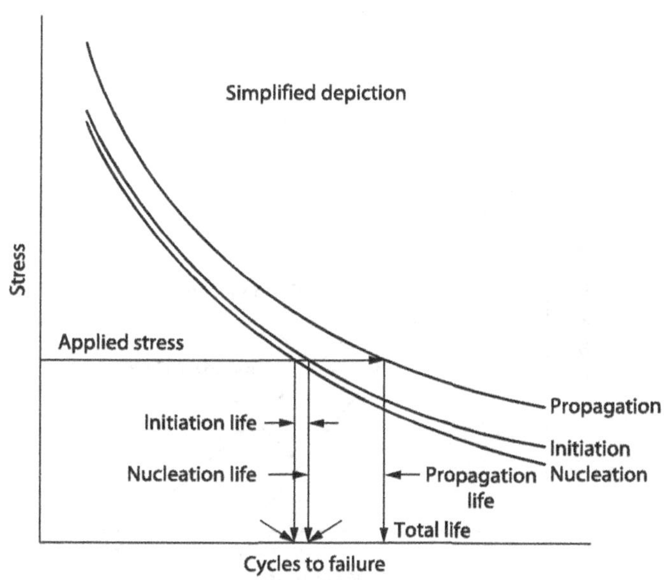

Figure 10.17 Curves for nucleation, initiation and propagation of a crack.

Traditionally, in terms of life, the life above 10^6 cycles is considered to be high-cycle fatigue. Mechanistically, damage under negligible plastic strain can be considered high-cycle fatigue. Influence of mean stress on fatigue is generally shown on a constant life diagram. Locus of a constant finite life to failure is drawn for different combinations of alternating and mean stresses. Mean stress is the average between the maximum and minimum stress of the fatigue-loading cycle. Alternating stress is the amplitude of the cyclic stress, or half the difference between the maximum and minimum stress of the cycle. Margin in design is estimated through use of a term "factor of safety." Many different diagrams have been used (e.g., Soderberg diagram and Goodman diagram). The alternating stress is plotted on the vertical axis and the mean stress on the horizontal axis. These methods are used to estimate reliability of the blades by way of establishing an acceptable factor of safety under combined loading of steady stress and alternating stress. A factor of safety above a certain predetermined number provides an acceptable design. Equations describing two popular methods, namely Soderberg and Goodman methods, are given below (Figure 10.18).

The Soderberg line is represented by line a. The equation of this line is given below.

$$\sigma_a/\sigma_f + S_m/\sigma_y = 1$$

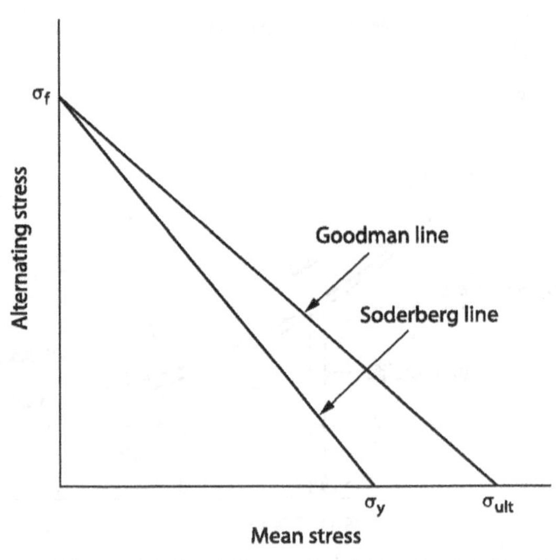

FIGURE 10.18 Effect of mean stress on fatigue strength (Goodman and Soderberg lines).

The Goodman line is represented by line b and the equation of this line is given below:

$$\sigma_a/\sigma_f + S_m/\sigma_{ult} = 1$$

where S_m = combined steady stresses in the blade
σ_a = alternating stress in the blade
σ_{ult} = ultimate strength of the blade material
σ_f = adjusted fatigue strength of the blade material
σ_y = yield strength of material

The Goodman line is drawn from fatigue strength to the ultimate strength. This is beyond the yield strength of the material. In fact, it allows small localized yielding until it reaches ultimate strength. It might be proper to consider strain in place of stress for criteria even in the case of high-cycle fatigue if yielding is at a very low level as is the case for low loads.

10.4.3 Factor of Safety Calculation for High-Cycle Fatigue

Reliability of a turbine blade is expressed in terms of safety factor. The factor of safety may be estimated by the use of Goodman equation (Eq. (10.22)). The failure line (when the factor of safety equals 1.00) on a Goodman diagram is the straight line that connects the fatigue strength with the ultimate strength of the blade material (Figure 10.19):

$$1.0/FS = \sigma_a/\sigma_f + S_m/\sigma_{ult} \tag{10.22}$$

FS = factor of safety
S_m = combined steady stresses
σ_a = alternating stress
σ_{ult} = ultimate strength of material
σ_f = adjusted fatigue strength of material

Any design is considered safe for operation when it achieves an established acceptable minimum value for the factor of safety. It should be realized that the minimum value of safety factor depends on many factors, for example, accuracy and method of stress estimation, accuracy of material properties, effect of environment on material properties, and experience with this type of design in the desired operation. If a change in design is required and material is not changed, decision is dictated by whether the steady stress or

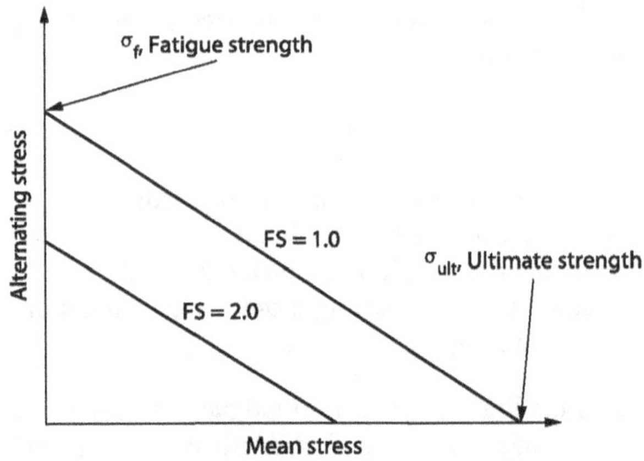

FIGURE 10.19 Goodman diagram representing factor of safety.

alternating stress component is contributing most to the low factor of safety. A geometrical change may be required. If alternating stress is the result of vibration of structure such as blade, then change in the natural frequency of the blade or the exciting frequency is often possible to avoid resonance resulting in high vibratory stresses. It is not possible to avoid all possible resonance in variable speed machine, though. Thus, the largest responses are avoided and the blade is made strong enough to operate at the remaining resonances within the operating speed range. Another possibility may be to increase damping in the design to lower the vibratory stress level.

Adjustment of Fatigue Strength for Mechanical Components

The fatigue strength of a material depends on many factors. Therefore, it should be adjusted to reflect the influence of those factors for a real application. Fatigue properties are evaluated by tests on polished specimens machined from the material to be used in the mechanical components. To evaluate the properties of the material of construction, these properties should be characterized using materials machined from actual manufactured parts. However, it is not always practical for two reasons. First, at the design stage parts may not be available and it may be very costly for some parts to be destroyed for this test. Second, it may not be possible to conduct the test in the actual operating environment.

A method to estimate properties of the parts is to modify the fatigue properties obtained through polished specimen test. For many applications this is proven to be adequate. It is done by using modifying factors to the specimen's properties. These modifying factors have been determined after testing on materials. These, however, should be used as approximation. If the fatigue strength estimated from test sample, σ_f, then the fatigue strength, σ_f' of the part can be approximated by the Eq. (10.23). Lipson and Sheth (1973) list many of these factors:

$$\sigma_f' = K_1 \cdot K_2 \cdot K_3 \ldots K_n \cdot \sigma_f \qquad (10.23)$$

The K_ns depend on the influences of parameters being considered including, but not limited to, the types of loading:

- Surface finish
- Reliability factor
- Strength reduction due to notch
- Size effect
- Surface treatment
- Environment

σ_f is the fatigue strength at a given life and σ_f is the adjusted fatigue strength. It has been customary to use the term, σ_e, known as the endurance strength of the material in place of s_f. When alternating stress is less than σ_e, the high-cycle fatigue life is approximated to be infinite. Many times test data is not available to support this assertion at least in the environment other than inert environment. Care is taken by using a factor in a Goodman-type diagram called the factor of safety. This does not provide any idea about life. For example, if the estimated life is 10^{10} cycles and the frequency of the alternating stress is 2000 Hz (blade frequency), then the structure will fail in about 16 days for a factor of safety equal to 1.00.

10.4.4 Goodman Diagram with Lines Representing Cycles to Failure

From the argument given above about the need for estimating cycles to failure, there is another way to draw the radial lines on a Goodman diagram. These radial lines represent different lives. Radial lines are drawn from ultimate strength to fatigue strength representing cycles to failure. This is read from S-N curve for a given alternating stress representing influence of appropriate mean stress on fatigue property. Figure 10.20 shows such a diagram. This becomes valuable to estimate life during high-cycle fatigue. This topic will be dealt in detail in section for life.

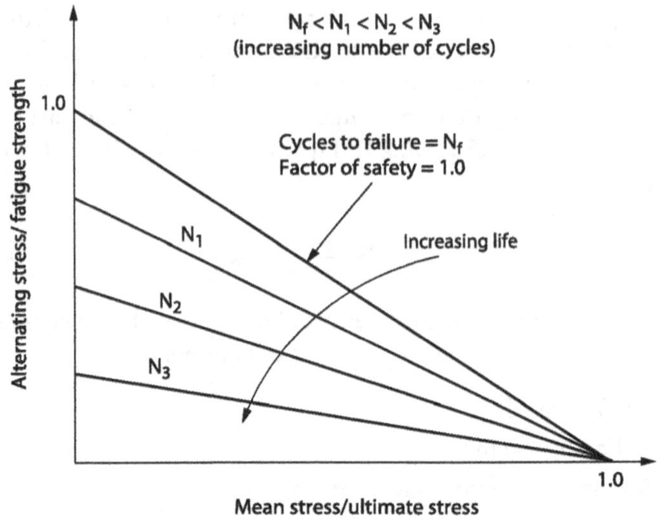

FIGURE 10.20 Alternate Goodman diagram with lines representing life.

10.5 Estimation of Component Life

Reliability estimation leads to find out safe life of operation of a mechanical component such as blade, disc, or any mechanical component of expander. This information helps the buyer to make decisions, the operator to plan an overhaul and repair, and most importantly, it will help the designer and manufacturer to deliver product that will support the customers' needs. Failure due to static loads for a long period of time at a high temperature is called creep, and failure due to a cyclic load is called high-cycle fatigue (HCF) or low-cycle fatigue (LCF) are of main concern. Creep and fatigue phenomena of damage are discussed in this section. HCF seems to occur when the resulting alternating stress is low enough so that there is no permanent deformation at any location of the structure, whereas LCF occurs when there is a plastic deformation in the structure. Theory and properties of materials are available to make the estimation possible. Moreover, these types of method are being used for safe design and also being used to explain unexpected failures of components.

10.5.1 Life due to Creep

It has been found that the plot of stress versus LMP for most materials results in a single curve. The LMP curves for many materials are available in literature. A typical representative LMP curve for IN 738 superalloy is shown in Figure 10.21.

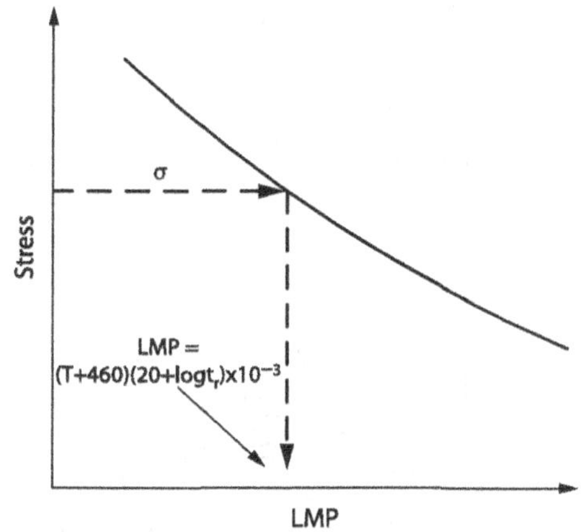

Figure 10.21 Rupture in the case of constant stress across crass section.

Evaluation of a creep life of blades becomes important either during the design phase or when a failure occurs. The following examples will explain this point:

Example: An expander blade made of IN 738 is subjected to creep under simple tension at a stress of 39 ksi at 1268°F and the measured rupture life is 323 years. Estimate creep rupture life of the blade is continuously operated at a temperature of 1450°F:

Larson-Miller Method

$$\begin{aligned} t_{r(1268)} &= 323 \times 365 \times 24 \text{ hours(h)} = 2{,}829{,}000 \text{ h} \\ \text{LMP} &= (\text{T} + 460)\,(\log_{10} t_r + \text{C}) \\ &= (1268 + 460)\,(\log_{10}(2829000) + 20) = 45{,}708 \\ \log_{10} t_{r(1450)} &= \text{LMP}/(\text{T} + 460) - \text{C} \\ &= 45708/(1450 + 460) - 20 \\ &= 3.93 \\ t_{r(1450)} &= 10^{3.93}\,\text{h} = 8529\,\text{h} = 0.97 \text{ years} \end{aligned}$$

Estimation of creep rupture life is done by using LMP curve of the material by using average stress across a cross section. The horizontal line is drawn from the stress value on the vertical axis to touch the LMP curve. From this point the corresponding value of LMP is determined (Figure 10.21).

This value of LMP is used to estimate the time to rupture (t_r) at the temperature, T, of the component Eq. (10.55). In case of an

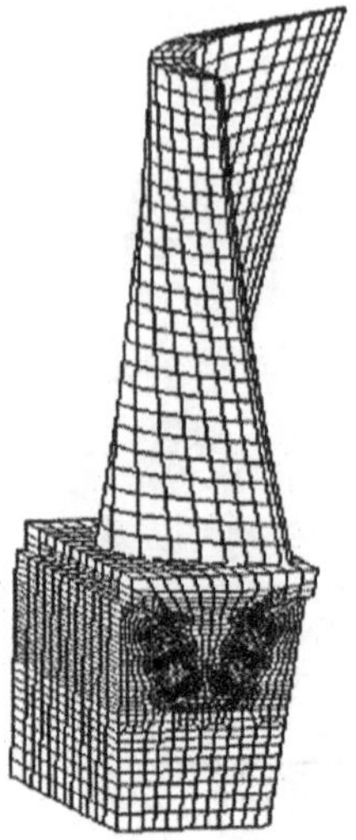

Creep Life Calculation

	Height, %	Stress, psi Centri-fugal	Gas Bending	Total = Cent. + 0.5 × GB
BASE	0	33.00	12.00	39.00
	10	30.97	6.58	34.26
	20	28.67	3.76	30.54
	30	26.08	2.53	27.34
	40	23.21	2.10	24.26
	50	20.05	1.90	21.00
	60	16.62	1.57	17.41
	70	12.91	0.98	13.40
	80	8.91	0.21	9.02
	90	4.64	0.10	4.69
TIP	100	0.00	0.05	0.03

Table 10.1 The Stress at Different Blade Height

expander blade, stress and temperature vary along the length. Therefore, it is expected that creep rupture life will be different at each height of the airfoil. Larson-Miller parameter was used to evaluate the creep rupture life of an expander blade. Average stress of the blade cross section owing to rotation from the base to tip of the blade was calculated. The gas-bending stress as a function of height was also evaluated. For the purpose of creep life calculation 50 of bending stress was added to the average stress due to rotation. Stresses are listed in Table 10.1. Temperature variation along the length of the blade is given in Table 10.2.

Creep Life Calculation

	Height, %	Temp, F	LMP	Time/1.0E05, hrs	Time, hrs/4
BASE	0	1269	46.155396	28.3	7.1
	10	1276	47.00282773	66.5	16.6
	20	1284	47.69507762	126.9	31.7
	30	1291	48.31084893	217.7	54.4
	40	1299	48.92089648	369.2	92.3
	50	1306	49.58235586	666.1	166.5
	60	1313	50.33561979	1346.0	336.5
	70	1321	51.20182632	3125.0	781.3
	80	1328	52.1797665	8309.7	2077.4
	90	1336	53.18027242	22557.4	5639.3
TIP	100	1343	54.29392555	70089.2	17522.3

Table 10.2 Creep Rupture Life Estimation at Different Blade Height

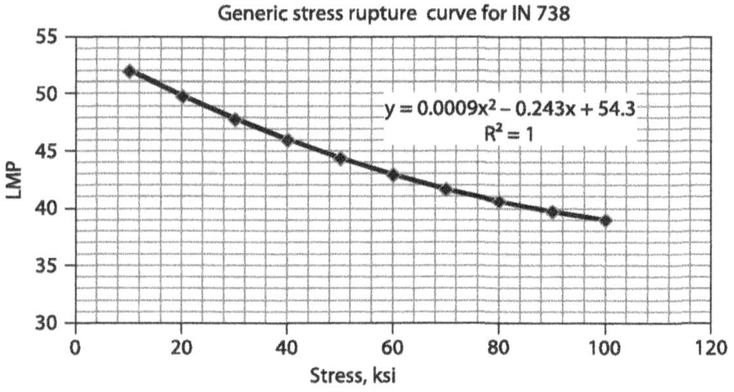

FIGURE 10.22 Typical Larson-Miller curves for IN 738.

The Larson-Miller parameter curve for IN 738 shown in Figure 10.22 has been used. The LMP equation has been used to estimate creep rupture life at each blade height; it is listed in Table 10.2.

As stress and temperature are the largest at the base of the blade for the above example, the creep rupture life at the base is the lowest.

Creep Crack Initiation and Time to Rupture in Case of Localized Stress

A notch or a discontinuity makes the distribution of forces nonuniform. Damage to a structure starts from the location of peak stress even though the magnitude of average stress across a cross section might be acceptable. This means that the crack initiates at the location where the combined effect of stress and temperature is the worst. Stress varies from a high value to a lower value through thickness in the vicinity of a stress riser. Then, depending on the condition of stress and temperature beyond this point, the initiated crack might grow to some distance through the material or may continue to final rupture. The estimated peak stress may be greater than the yield strength of the material. It has been discussed earlier that in this situation the actual stress may be lower than estimated. The actual peak stress is so localized that it can be estimated by a simple rule called Neuber's rule, which has been described in detail earlier. For completeness, Figure 10.23 is presented graphically show the Neuber's method.

Creep damage occurs when the localized stress is large enough to initiate a crack in a very small time through a very small depth. Once the crack is initiated it may grow under the remaining stress through the thickness. This is known as "creep crack growth" or is due to fatigue if the cyclic strain is large.

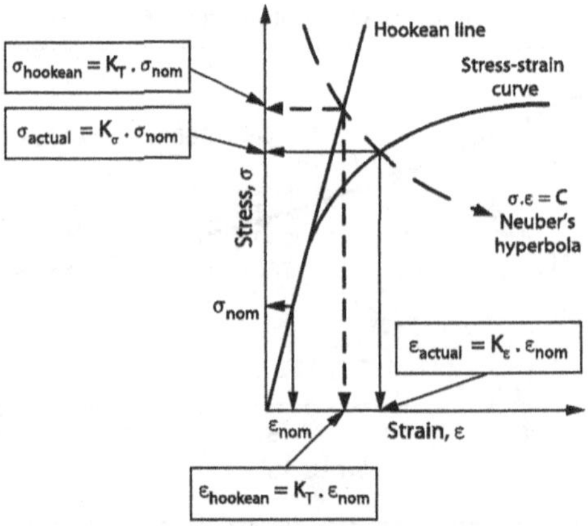

FIGURE 10.23 Graphical demonstration of Neuber's rule.

If the stress has been estimated based on elastic consideration, that is, the behavior of material has been assumed to follow Hooke's law and the peak stress is greater than the yield strength of material, then the actual stress must be estimated. In such a case one may use Neuber's method for estimating the actual stress. The estimated actual peak stress using the elastic stress as a function of depth of the cross section is shown in Figure 10.24.

Distribution of elastic stress (S) is converted to the actual stress (σ) at a different distance along the cross section using Neuber's method. Elastic stress, S_1, is estimated to actual stress (σ_1), S_2 is estimated to actual stress (σ_2), and S_3 is estimated to actual stress (σ_3), and so on.

On the other hand, if actual stress is estimated, then distribution of the actual stress and temperature variation through the thickness are as shown in Figure 10.25.

The thickness (a) is divided into small segments as shown in Figure 10.25. Stress and temperature are estimated from the curve at these locations as shown below.

$$\sigma_1, \sigma_2, \sigma_3, \cdots \sigma_n$$
$$T_1, T_2, T_3, \cdots T_n$$
$$a_1, a_2, a_3, \cdots a_n$$

There is a corresponding LMP for each set of σ and T (e.g., σ_1 and T_1 yield an LMP_1, σ_2 and T_2 yield an LMP_2, σ_3 and T_3 yield an LMP_3, and so on). When these LMPs are used in conjunction with temperature, T, these yield corresponding t_r's.

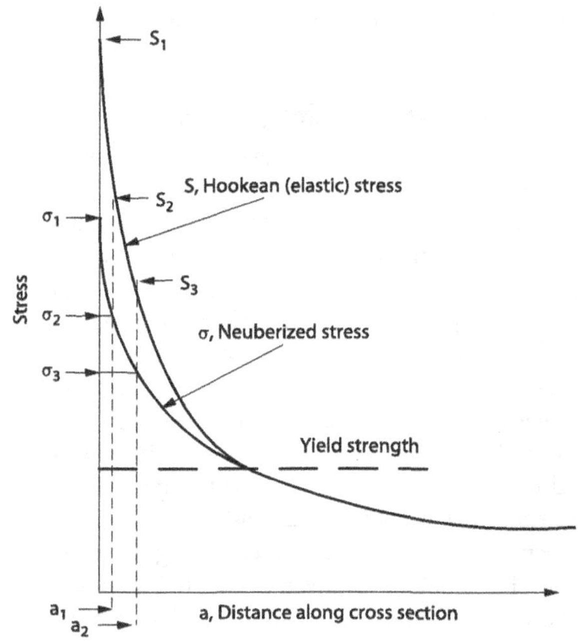

FIGURE 10.24 Estimation of actual stress from elastic stress.

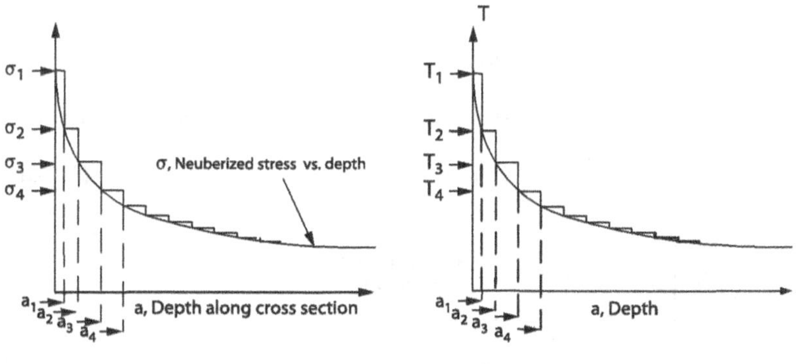

FIGURE 10.25 Neuberized stress and temperature through thickness.

These are shown below.

$$\sigma_1, T_1 \; LMP_1 \rightarrow t_{r1} \rightarrow$$
$$\sigma_2, T_2 \; LMP_2 \rightarrow t_{r2} \rightarrow$$
$$\cdots\cdots\cdots\cdots$$
$$\sigma_n, T_n \; LMP_n \rightarrow t_{rn} \rightarrow$$

The first rupture time, t_{r1} is considered as the crack initiation of crack length, a_1.

10.5.2 Life under Cyclic Loading (Fatigue)

Energy is stored in the material during loading of a component. This stored energy is the area under the stress-strain curve. During unloading, energy is completely released if loaded below the elastic limit of the material. However, upon unloading, some energy is stored in the material if loaded beyond elastic limit. Thus, the released energy is the elastic portion of energy. The remaining stored energy is due to the plastic deformation of the component. The area enclosed by the hysteresis loop is the stored energy. For life estimation, it is easier to consider strain is in place of stress when plastic deformation occurs. A brief description of the strain-based life estimation method is discussed next. The equation developed is applicable for low-cycle fatigue situation as well as high-cycle fatigue.

It has been observed that the plot of the log of "true" stress amplitude versus cycles to failure results in a straight line (Figure 10.26). The equation of the straight line representing the relationship between stress and cycle can be written as follows:

$$\text{Log}(\Delta\sigma/2) = \log(\sigma_f'/2) + b\,\text{Log}(2N_f) \tag{10.24}$$

or

$$(\Delta\sigma/2) = (\sigma_f'/2)\,(2N_f)^b \tag{10.25}$$

Finally, in terms of elastic strain, amplitude equation reduces to

$$\Delta\varepsilon_e/2 = (\sigma_f'/2E)\,(2N_f)^b \tag{10.26}$$

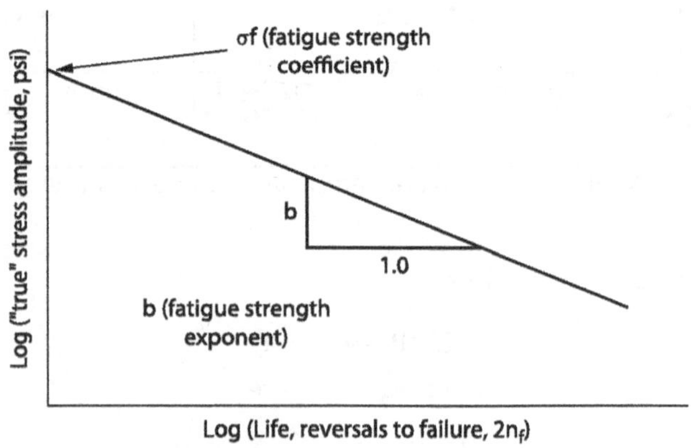

FIGURE 10.26 Stress amplitude versus cycles to failure.

Also, the plot of the log of "true" plastic strain amplitude versus amplitude of cycles to failure displays a straight-line behavior (Figure 10.27).

The relationship between plastic strain and cycles to failure can be written as follows:

$$\text{Log}\,(\Delta\varepsilon_p/2) = \log\,(\varepsilon_f') + c\,\text{Log}(2N_f) \tag{10.27}$$

$$\Delta\varepsilon_p/2 = \varepsilon_f'\,(2N_f)^c \tag{10.28}$$

Finally, the total strain amplitude is obtained by adding the elastic strain amplitude and the plastic strain amplitude.

$$(\Delta\varepsilon_t/2) = (\Delta\varepsilon_e/2) + (\Delta\varepsilon_p/2)$$

$$(\Delta\varepsilon_t/2) = (\sigma_f'/2E)\,(2N_f)^b + \varepsilon_f'(2N_f)^c \tag{10.29}$$

Eq. (10.29) is known as the Coffin-Manson equation for fatigue. If during high-cycle fatigue there is negligible permanent set or plastic strain, then the first part of the equation covers high-cycle fatigue. The second part is used for low-cycle fatigue where plastic strain is dominant.

Plots of the expressions represented by Eqs. (10.26), (10.28), and (10.29) are shown in Figure 10.28. Log-log plot of the strain amplitudes versus cycles to failure are straight lines for both elastic and plastic components of the strain. The total strain curve versus number of cycles to failure is also shown.

As load is increased, the plastic strain amplitude becomes larger in comparison to the elastic strain amplitude. Lower life seems to be controlled mainly by the plastic strain. Larger life is mainly controlled

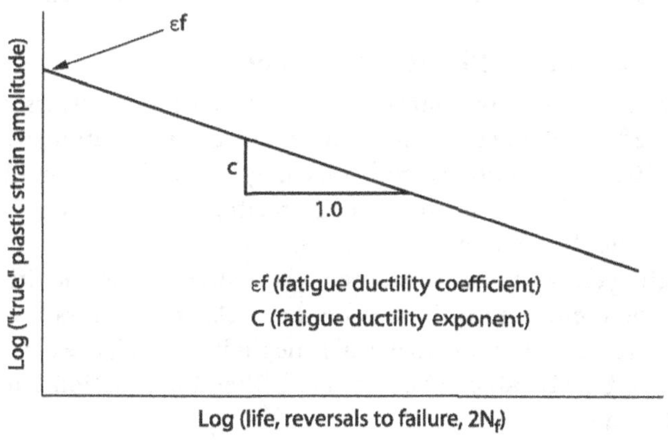

Figure 10.27 Plastic strain amplitude versus cycles to failure.

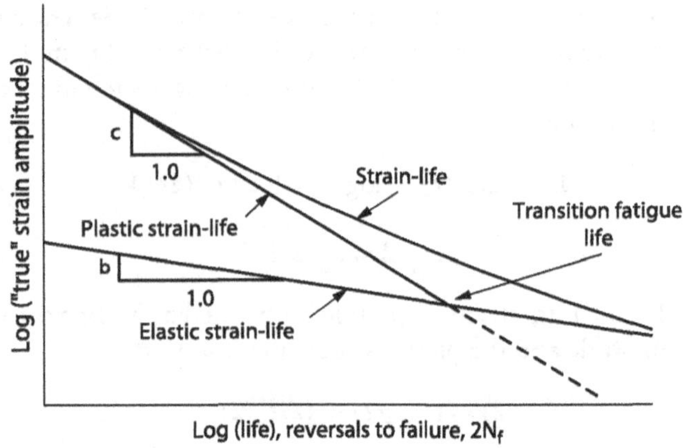

FIGURE 10.28 Strain amplitude versus cycles to failure.

by the elastic strain. It is also known that the fatigue property of a material depends on a combination of the alternating stress and the mean stress. Morrow (in Graham et al., 1968), by observing test data on many materials, modified Eq. (10.29) to include the effect of mean stress, σ_0, which is expressed in Eq. (10.30).

$$(\Delta\varepsilon/2) = ((\sigma_f' - \sigma_m)/2E)\,(2N_f)^b + \varepsilon_f'(2N_f)^c \tag{10.30}$$

Inclusion of the mean stress into the elastic portion of Eq. (10.63) indicates that Morrow assumed (or found) that only the high-cycle fatigue life is affected by the mean stress. The effect of mean stress on low-cycle fatigue life is negligible. Based on Morrow's equation, curves are plotted for different tensile mean stress. Figure 10.29 graphically demonstrates the implication of Eq. (10.67) for tensile mean stress.

10.5.3 Life due to High-Cycle Fatigue

The fatigue strength of material is a function of mean stress. Curves for the effect of mean stress shown are clearly demonstrated in Figure 10.29. The endurance limits may be different for different mean stress. Morrow's modified life relationship has been used to demonstrate the effect of mean stress.

High-cycle fatigue is considered where stress is low and life is very large (where endurance limit is defined). At these stress levels, the plastic deformation in the material is negligible. Thus, the plastic portion of the Eq. (10.30) can be neglected. Then the equation reduces to the following:

$$(\Delta\varepsilon/2) = ((\sigma_f' - \sigma_m)/2E)\,(2N_f)^b \tag{10.31}$$

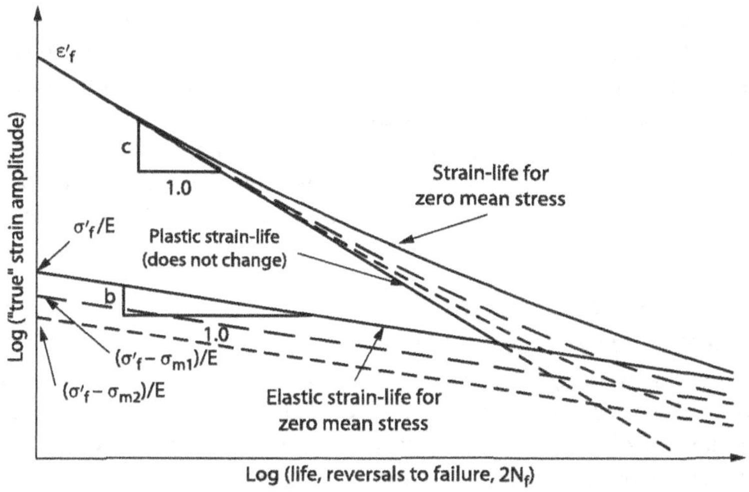

FIGURE 10.29 Effect of mean stress on fatigue properties.

After substituting for alternating strain amplitude ($\Delta\varepsilon/2 = 2\sigma_a/2E$), the equation reduces to the following:

$$(2\sigma_a/2E) = ((\sigma_f' - \sigma_m)/2E)\,(2N_f)^b$$

$$\sigma_a = (\sigma_f' - \sigma_m)\,(2N_f)^b \tag{10.32}$$

The magnitude of b is always negative.

$$2N_f = (\sigma_a/(\sigma_f' - \sigma_m))^{1/b} \tag{10.33}$$

When mean stress is zero, the above equation reduces to

$$2N_f = (\sigma_a/\sigma_f')^{1/b} \tag{10.34}$$

The number of cycles is expected to be large under a cyclic load with relatively small stress. For an expander blade, the magnitude of the cyclic stress due to the load arising from flow distortion is typically low, but it can be magnified many folds due to resonant condition. The major design consideration in this situation revolves around the avoidance of resonance. This is graphically depicted on a Campbell diagram or in a Safe diagram. A most widely used design criterion in a high-cycle fatigue situation is called Goodman criteria, and it is depicted on a Goodman diagram. The cyclic material properties used is obtained from stress (S) versus cycles to failure (N) curve generated from laboratory test on the material. The slope of this curve for most ductile materials is almost horizontal after about 10^6 cycles. After 10^6 cycles, a small reduction in stress produces a large increase in life. Reliability is estimated by a factor of safety by Goodman equation. It should be noted that the knowledge of the factor of safety does not provide any clue about the expected life of the component. Normally, factor of safety, mean stress, and some time-alternating stress are available. A method has been proposed by Singh (2001) to calculate a factor of safety based on life.

Assessment of the Useful Life of a Blade in the Presence of High-Cycle Fatigue

Failure due to fatigue is the major reason of the vast majority of component failures. Most expander blade failures are the result of the application of a cyclic load. Thus, attention in design is given to deal with and avoid/alleviate fatigue problems. However, this method does not yield a quantitative estimate of safe life, that is, how many cycles are required for fatigue damage.

Almost all the time, a Goodman diagram and the factor of safety for the component are available without any estimation of life. The number of cycles to failure during high-cycle fatigue can be estimated by combining the Goodman equation and Morrow's equation for life. This method provides an estimate of life by taking advantage of stresses used in constructing a Goodman diagram (Singh, 2001 Singh and Lucas, 2011).

An alternative way to plot a Goodman diagram is to draw failure limit lines representing different lives. Figure 10.30 is the plot of the alternating stress on the vertical axis corresponding to any desired

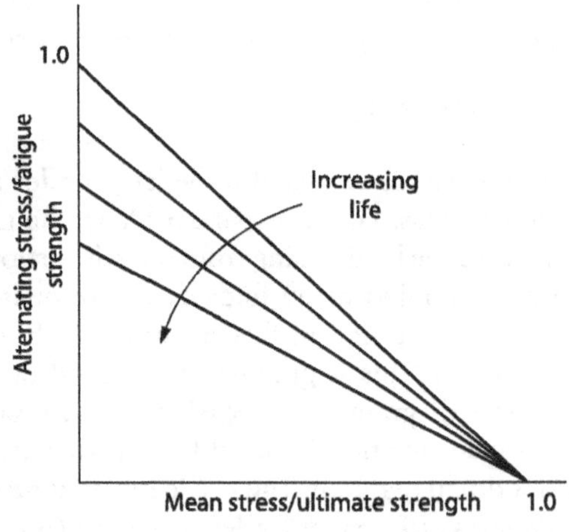

Figure 10.30 Goodman diagram with constant life lines.

life. This point is connected to the ultimate stress on the horizontal axis by a straight line. This line represents a life that is expected, if the applied alternating stress is equal to the point on the vertical axis for zero mean stress.

Traditional use of a Goodman diagram is shown in Figure 10.31. A radial line corresponding to the fatigue strength represents a factor of safety equal to unity. A point is located on this diagram corresponding to alternating stress and mean stress is estimated for the design in question. From this design point, a parallel line to the original radial line is drawn to estimate a factor of safety.

For estimating life for HCF can be done by using Eq. (10.32) developed earlier, and it is given below.

In place of reversal to failure ($2N_f$) is replaced with N_f to be taken as cycles to failure.

$$\sigma_a = (\sigma_f' - \sigma_m)N_f^b \tag{10.35}$$

Singh (2001) showed that when mean stress is zero that is, $\sigma_m = 0$, then $\sigma_a = \sigma_e$. The resulting relationship between cycles to failure and stress is given by Eq. (10.34).

$$N_f = (\sigma_e/\sigma_f')^{1/b}$$

Substituting Eq. (10.32) into Eq. (10.33) and rearranging yields the following equation:

$$\sigma_a/\sigma_e + \sigma_m/\sigma_f' = 1 \tag{10.36}$$

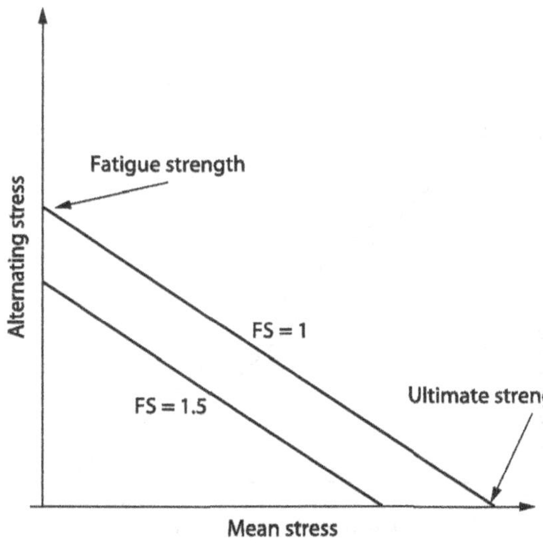

Figure 10.31 Goodman diagram with factor of safety line.

The relationship between (σ_a/σ_e) and (σ_m/σ_f') is a straight line for a constant life as shown by Eq. (10.34). The straight line is defined by the x-intercept of (σ_m/σ_f') and the y-intercept of (σ_a/σ_e). This equation has the same form as the Goodman equation, which contains the quantity σ_{ult} in place of σ_f'. Straight lines representing Eq. (10.34) and Goodman Eq. (10.35) and are shown in Figure 10.32. Morrow's equation in essence estimates a longer life for zero-alternating stress than that given by the Goodman radial line and σ_f' (Point A) as the failure point compared to σ_{ult} that is used in the Goodman diagram. σ_f' is the stress at final failure as can be estimated from the cyclic stress-strain curve (as discussed in an earlier section).

The Goodman equation is given below.

$$\sigma_a/\sigma_e + \sigma_m/\sigma_{ult} = 1/FS \tag{10.37}$$

As Goodman equation is used to estimate the factor of safety and for estimation of life, σ_f' is substituted with σ_{ult} to be consistent, and it is recognized that it should provide conservatism. By combining the Goodman equation and the life equation, an expression for the cycles to failure, N_f, for a specified FS is obtained. It can be expressed either in terms of mean stress or alternating stress for a given factor of safety. These expressions are given as Eqs. (10.36) and (10.37).

The expression for N_f when FS and mean stress are known is given below.

$$N_f = ((1/F.S - \sigma_m/\sigma_{ult})\,(\sigma_e/(\sigma_{ult} - \sigma_m)))^{(1./b)} \tag{10.38}$$

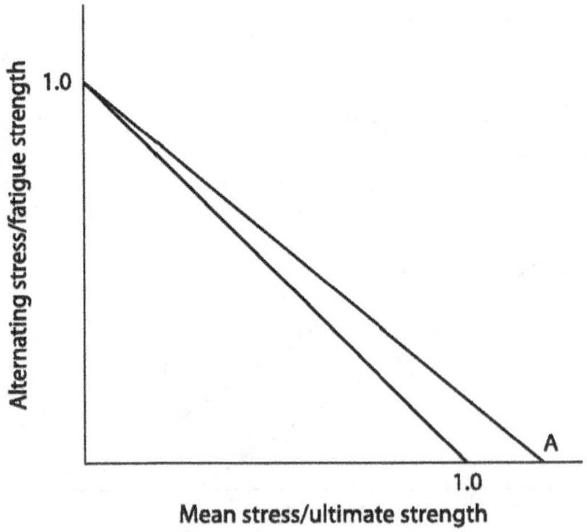

Figure 10.32 Comparison of Goodman and Morrow equations.

The expression for N_f when FS and alternating stress are known is given below.

$$N_f = ((\sigma_{ult} - \sigma_{ult}(1/F.S - \sigma_a/\sigma_e))/\sigma_a)^{(-1./b)} \qquad (10.39)$$

The implication of Eqs. (10.36) and (10.37) regarding the life of a blade design can be observed in Figure 10.33. This shows a superimposition of Figure 10.19 and 10.20. The radial lines for N_f, N_1, N_2, and N_3 represent constant life in increasing order. A Goodman-based factor of safety line as defined by Eq. (10.35) is also drawn. The intersecting points A, B, and C, although they represent the same factor of safety, lie on lines denoting different lives. The implication of this observation is that designs having the same factor of safety may yield different lives.

Singh and Lucas (2011) provided an example of eight different designs of similar components with identical materials. Damage in these designs is considered to be because of high-cycle fatigue. It was found that design having a lower factor of safety may last longer. The important information that should be sought is the safe operating life of the blades. The factor of safety in itself does not provide this crucial information. The HCF life of each design and the ranking based on life are estimated. The surprising fact was that the worst design based on the factor of safety ranked the best if ranking was done based on life. This is important information because the main concern must be safe operating life.

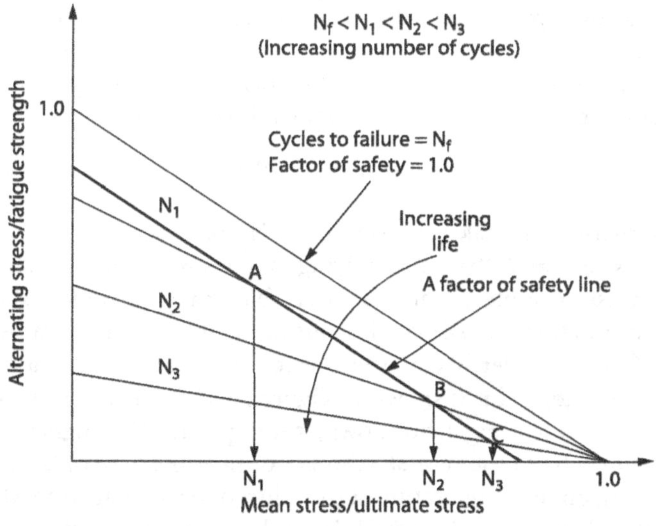

Figure 10.33 Superposed Goodman lines.

The result of the example demonstrated the following points:

1. It is important to estimate the life of the design together with the safety factor.
2. If the stresses that have been used to estimate the factor of safety are known, then the life estimation can be made.
3. It might be possible for a design with a lower safety factor to yield higher life than one that has a higher safety factor.
4. It seems that longer blades that may have a higher mean stress are designed for the same factor of safety as for smaller blades having lower mean stress.
5. Many times, difficulties arise in ranking designs for technical superiority and the techniques shown so far may be used to gain insight into the reliability of various alternative designs.

10.5.4 Life Estimation due to Low-Cycle Fatigue

The equation of Coffin-Manson as modified by Morrow to include the effect of mean stress is given below for completeness.

$$(\Delta\varepsilon/2) = ((\sigma_f' - \sigma_m)/2E)\,(2N_f)^b + \varepsilon_f'(2N_f)^c$$

The first part on the right side of the equation pertains to high-cycle fatigue while the second part is used to evaluate life for low-cycle fatigue (LCF). It should be worth noting that this equation shows the effect of mean stress only for high-cycle fatigue life. On the other hand, the low-cycle fatigue life entirely depends on the plastic portion of strain.

Therefore, the equation to estimate LCF life of a component after neglecting the first term reduces to the following Eq. (10.38):

$$(\Delta\varepsilon/2) = \varepsilon_f'\,(2N_f)^c \qquad (10.40)$$

Estimation of peak stress or strain is done either by estimating average stress and then multiplying it by stress concentration or through use of finite element analysis. Normally, in both instances, elastic peak stress is estimated. Consider a start-up and shut-down cycle of an expander blade. The elastic (Hookean) stress varies linearly with time, and it is shown in Figure 10.34 for loading (up to full speed) and unloading (shut-down, zero speed). The highest stress location may be in the root attachment or near the base of the airfoil. Depending on the temperature at these locations, fatigue properties of the material might be different. It is prudent to analyze different locations to find the worst stress and temperature combination. This will

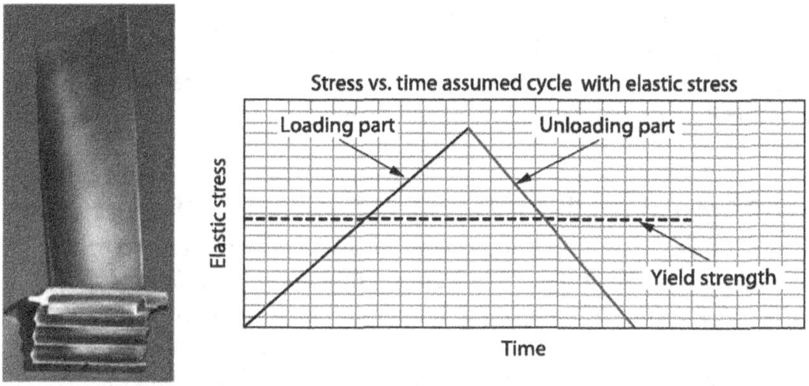

Figure 10.34 Elastic (Hookean) stress varies linearly with time.

pinpoint the limiting location as far as LCF is concerned. It is assumed that the peak stress is larger than the yield strength of the material of construction, material is expected to yield at the location of this stress level as depicted.

The stress cycle needs to be converted to follow the actual σ-ε curve. As discussed earlier for a monotonic loading case, Neuber's method may be used to estimate actual stress and strain beyond yield strength of material. A similar Neuber method may be used for cyclic loading. A computer program is used to convert two elastic stress cycles to yield the desired hysteresis loop (Figure 10.35). The magnitude of strain is estimated from this analysis.

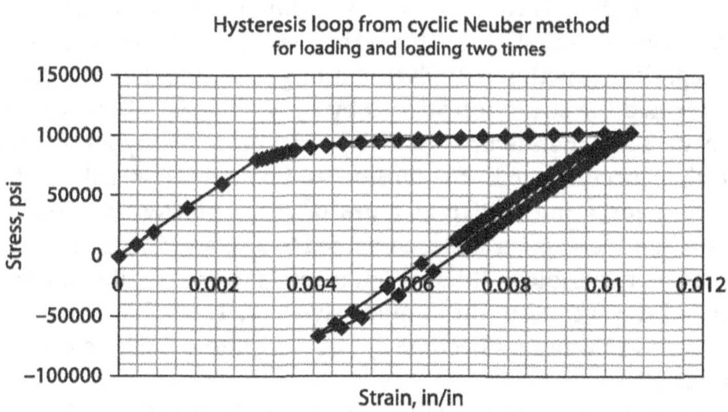

Figure 10.35 Two elastic stress cycles is converted to yield hysteresis loops.

The next information needed is the material property for LCF. Generally, a strain-controlled fatigue test is conducted to obtain this data. The coefficients for the Coffin-Manson equation are estimated from this data or can be read from the graph.

10.6 Life Evaluation during Creep Fatigue Interaction

During long operation of components at high temperatures, creep of material may occur. If it is cycled during operation, then fatigue may also occur. Fatigue and creep interaction may occur when a component is subjected to a cyclic load with a possibility of creep deformation in the material during cycling. The analysis of damage due to fatigue and creep is complex. Many methods of treating this phenomenon have been proposed, a large number of which are similar. There is a basic link in the framework of the method to one or more of the mechanisms that is relevant to the high-temperature problems.

Manson and colleagues (1971) proposed the method of strain range partitioning (SRP). Many materials have been tested using the SRP framework. While some limitations were recognized, the general usefulness of the approach was verified. Thus, the basic concepts behind SRP (both a scientific explanation of material behavior and a technologically viable method) have been examined and verified. The major portion for the following description of the SRP method has been taken from the work of Halford et al. (1973).

10.6.1 Types of Strains during Cyclic and Creep Loading

The SRP method defines two types of deformation that can occur in the creep range for some materials. These strains are introduced by grain boundary (g.b.) sliding and slip plane (s.p.) sliding. If the loading is rapid, there is no time for g.b. sliding to occur, and all strains are absorbed as s.p. sliding. This is the case of plastic deformation by only s.p. sliding without involvement of g.b. sliding. For this reason, they termed the strain introduced by s.p sliding as "plasticity," even if it occurs at a high temperature. By combining two types of strains with their occurrence in the tension and compression halves of the hysteresis loop, four different types of strain ranges were defined. "Plastic" and "creep" strains are referred to as P and C. Strain range is defined by two indices, the first letter refers to the tensile part of the loop and the second to the compressive half. Thus, the four permutations of strain ranges become ε_{pp}, ε_{cc}, ε_{cp}, and ε_{pc}. These are listed in Table 10.3.

The four permutations of loading termed as PP, CC, CP, and PC cycles lead to different micro-mechanistic effects (Figure 10.36).

Hysteresis Loop		Symbol of Strain	Designation of Cycles
Tension Portion	Compression Portion		
P	P	ε_{PP}	PP
P	C	ε_{PC}	PC
C	P	ε_{CP}	CP
C	C	ε_{CC}	CC

TABLE 10.3 Four Permutations of Strain Ranges Become ε_{pp}, ε_{cc}, ε_{cp}, and ε_{pc}

For the ε_{cp} strain range (Figure 10.36a), the tensile strain form B to C is achieved by rapidly increasing the stress to a value in the vicinity of the yield point. Stress is held at a constant value for creep deformation to occur along CD. For the compression half of the loop, load is decreased rapidly so that creep effect is minimized. The plastic deformation takes place along DEA. The micro-mechanistic effect of this type of cycle is shown in Figure 10.36(b) to (d). "Creep" implies g.b. sliding together with slip plane sliding and plasticity implies only slip plane sliding, then it can be seen that during CD there will be displacement of one grain relative to other along the grain boundary GH, while the strain reversal DEA will occur along slip plane EF. After one loop, a small notch remains on the surface

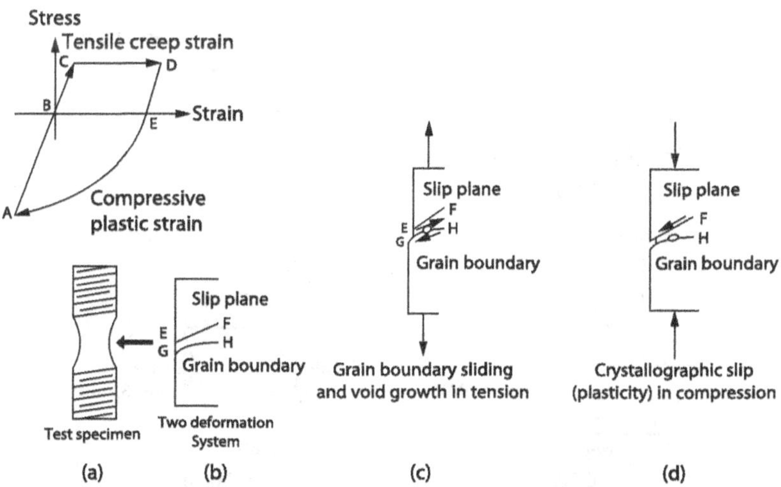

FIGURE 10.36 Schematic illustration of creep-fatigue interaction.

at an Inter intersection with grain boundary. Not only is oxidation more probable at the freshly created surface, but cavitations along the grain boundary weakens it and should make it a preferred site for oxidation. Additional cycles of this type of loading exacerbate the damage, each cycle adding more monotonic creep strain along the grain boundary and plasticity along the slip plane. Final fracture, for materials subject to grain boundary sliding, is frequently intercrystalline, and it occurs after relatively few cycles compared to other ways of imposing the same strain range.

The other types of strain range ε_{pp}, ε_{cc}, and ε_{pc} involve different types of combined slip plane and grain boundary sliding, producing different deformation effects and life relationships. An overall view of the four types of strain range, the idealized associated hysteresis loops, and the manner in which the two types of sliding interact are shown in Figure 10.37 (a and b).

Each of the four generic types of strain ranges involves different micro-mechanisms of crack development. Therefore, it is expected that fatigue lives should differ among them for the same applied strain ranges. Many materials do indeed satisfy this expectation.

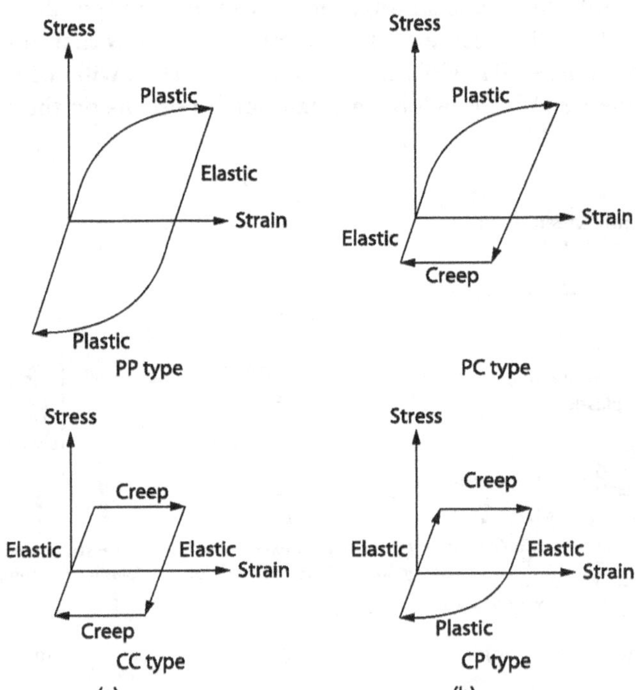

FIGURE 10.37A Simple cyclic deformation models.

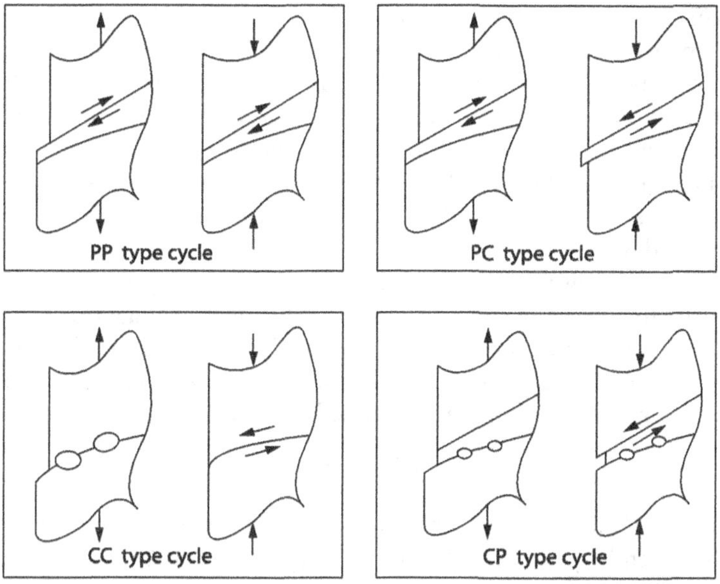

FIGURE 10.37B Simple cyclic deformation models.

10.6.2 Hysteresis Loop Containing More Than One Type of Strain Range

Such pure generic strain range loops rarely occur in real operation of a machine. Manson and coworkers considered a loop as shown in Figure 10.38. It is composed of three loops each of 1 percent, but having generic types PP, CC, and CP together with where the three independent 1 percent strain range loops are drawn.

Figure 10.38 shows a still-idealized but more realistic loop that contains PP, CC, and CP strain ranges. The tensile strain consists of 1 percent plasticity and 2 percent creep, whereas the compressive

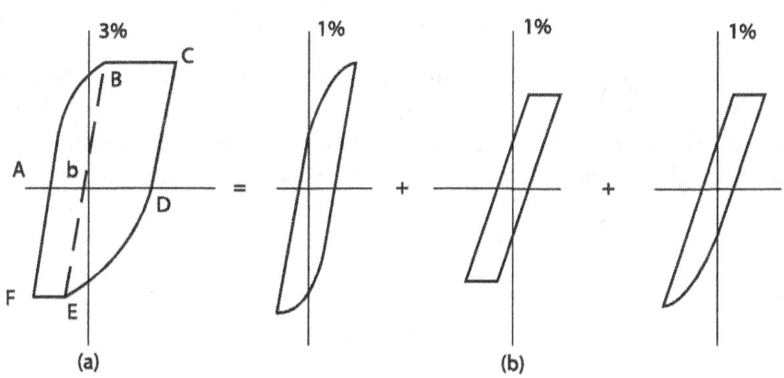

FIGURE 10.38 Analysis of complex loop by linear damage rule.

strain consists of 1 percent creep and 2 percent plasticity. The PP component generated by the tensile plasticity Ab is balanced by half of the compressive plasticity DE. The CC component to be generated by half the tensile creep bD is balanced by the compressive creep bA. The CP component to be generated by half the tensile creep bD is balanced by half the compressive plasticity DE. For simplicity, they assumed the strains PP = CC = CP = 1 percent each. Similarly, strain range magnitudes of any hysteresis loop can be partitioned into idealized loops containing PP, CC, CP, and PC types. They proposed a method to estimate life for any generic loop called "linear damage rule."

They reasoned that in one loop of ABCDEF would use up $1/N_f$ of life. Since one loop of the PP, CC, and CP, each of 1 percent, the life used up by applying the three generic loops would be

$$1/N_{pp} + 1/N_{cc} + 1/N_{cp}$$

where N_{pp} is life for 1 percent strain range on the PP curve, and similarly N_{cc} and N_{cp} are for CC and CP curves. Thus, the life due to this loop will be given as follows:

$$1/N_f = 1/N_{pp} + 1/N_{cc} + 1/N_{cp} \tag{10.41}$$

This equation was satisfactory for materials such as 316 SS, but when they considered materials of other character, it did not work out. When they considered MAR-M200 for which all the life lines are coincident, the life estimation did not make sense at all. They discarded this method.

In an effort to align damage rule for combining the effects of concurrent strain ranges, they proposed that these life relations must be modified. They reasoned that there is an "interaction" among the concurrent strain ranges which influences all the life curves.

They proposed a damage rule called the Interaction Damage Rule. Thus, life calculation became according to the relation:

$$1/N_f = f_{pp}/N'_{pp} + f_{cc}/N'_{cc} + f_{cp}/N'_{cp} \tag{10.42}$$

where $f_{pp} = \varepsilon_{pp}/\varepsilon_{in}$, $f_{cc} = \varepsilon_{cc}/\varepsilon_{in}$ and $f_{cp} = \varepsilon_{cp}/\varepsilon_{in}$. This formulation changes the N_{pp}, N_{cc}, and N_{cp} values of Eq. (10.39) to N'_{pp}, N'_{cc}, and N'_{cp} those are the life values from the generic life relations at the total strain range of the original hysteresis loop (Figure 10.39). In this case, then

$$N'_{pp} = N'_{cc} = N'_{cp} = 1000 \text{ cycles}$$

Since

$$f_{pp} = f_{cc} = f_{cp} = 1/3$$

The calculated value of N_f becomes 1000, as would be expected.

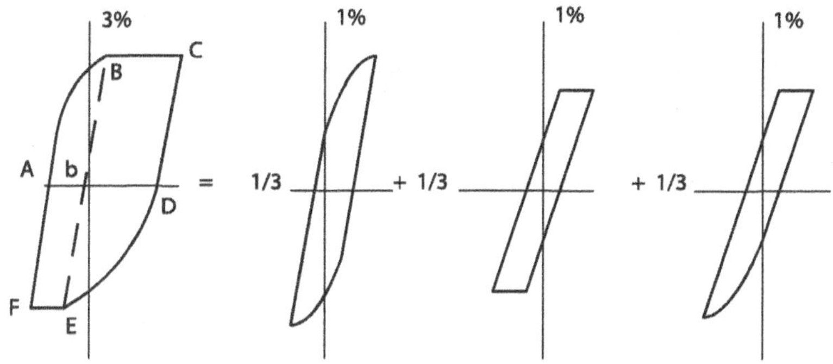

FIGURE 10.39 Hysteresis loop summation by interaction damage rule.

10.7 Linear Elastic Fracture Mechanics

A crack or cracklike discontinuity may weaken a component. A crack might grow to failure so that fracturing it will be into two or more pieces. This can occur even at stresses below the material's yield strength. Singh and colleagues (2004) described salient aspects of fracture mechanics. It is described as follows. Emphasis is placed on the basic principles with the fracture mechanics model that can be used in making decision about reliability of components with reportable discontinuities.

In the fracture mechanics analysis, the stress fields near the crack tips are classified as three basic types, shown in Figure 10.40. Mode I

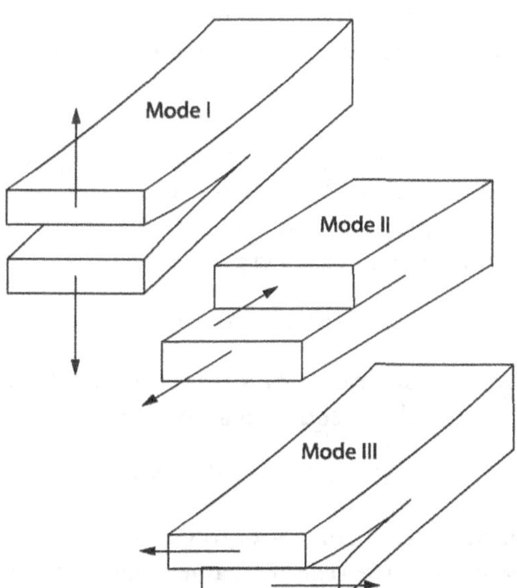

FIGURE 10.40 Modes of fracture.

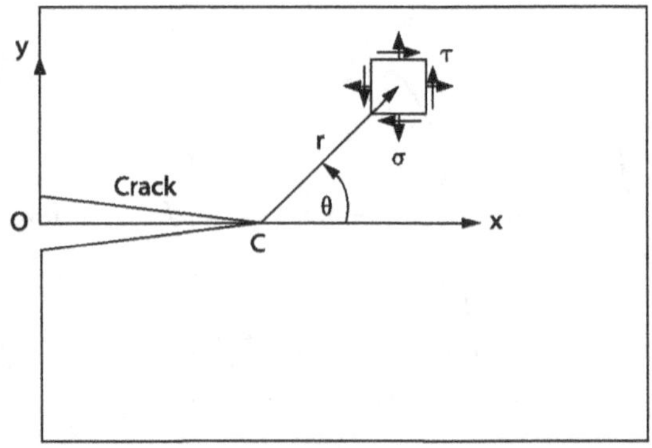

FIGURE 10.41 Stress analysis at the crack tip.

is opening mode, Mode II is sliding mode, and Mode III is tearing mode. Of the three modes, Mode I is the most pertinent for brittle crack propagation.

The theory of elasticity provides the vehicle to estimate stress distribution near the crack. To simplify the analysis, approximation is made that the crack is like a sharp slit (Figure 10.41). The tip of the crack is assumed to be perfectly sharp in the unstressed condition. The surfaces generated by the growing cracks are considered to be free of forces at all stages of loading. The stress field at the crack tip takes a simple solution of stress field near the tip (the distances from the tip are small compared to the characteristic dimensions of the crack system).

The characteristics of the stress field and its strength at the crack tip are defined by the stress intensity factors (K_I, K_{II}, K_{III}). It depends only on the loading and crack geometry. The stress intensity factor is the mathematical expression that describes the redistribution of load paths for transmitting load past a crack.

$$K_I = 1.12(\pi a)^{1/2}(\sigma) \tag{10.43}$$

Foe fatigue analysis, range of stress intensity factor and stress range are used in the above equation as follows:

$$\Delta K_I = 1.12(\pi a)^{1/2}(\Delta\sigma) \tag{10.44}$$

A plot of experimental data of crack growth versus ΔK (Figure 10.42) shows that there are three regions, namely very slow growth, stable slow growth, and fast fracture.

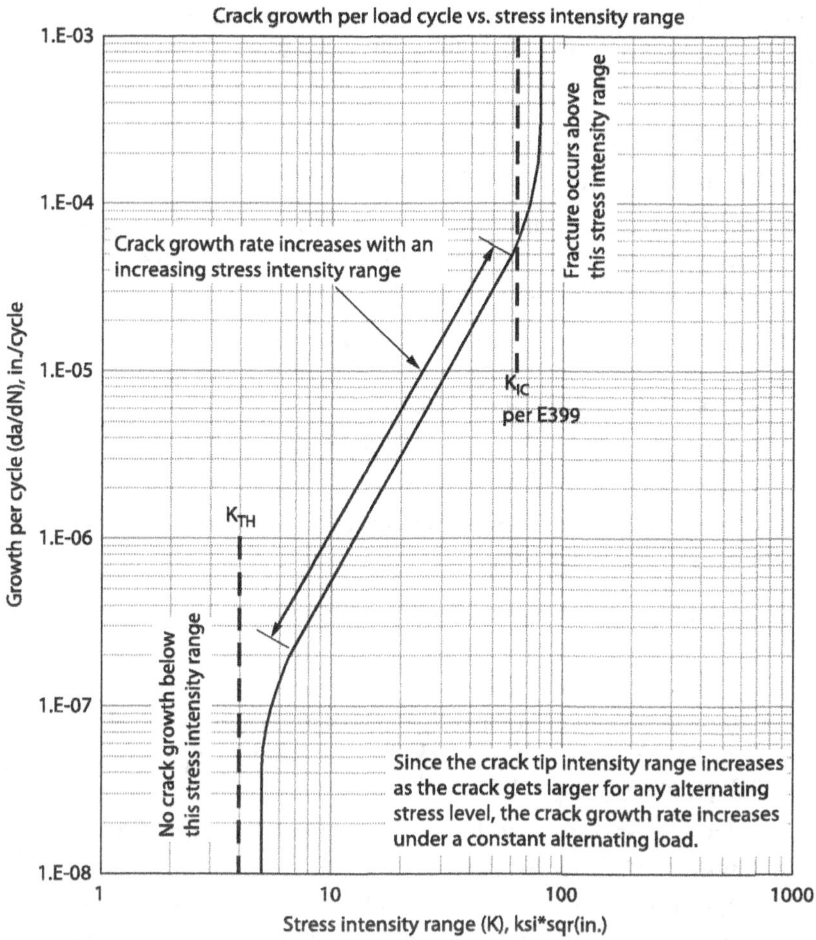

FIGURE 10.42 Crack growth behavior of material.

An expression for the crack growth rate (called stable slow crack growth regime) versus alternating stress intensity factor (ΔK) is expressed as follows:

$$da/dN = C\,(\Delta K)^m \tag{10.45}$$

or

$$\log(da/dN) = \log(C) + m\log(\Delta K) \tag{10.46}$$

The plot of the above expression on a log-log plot is a straight line with m as the slope of the straight line. C and m are regarded as crack growth material's properties and these are obtained by testing. Eq. (10.90) is known as Paris law. Reliability evaluation of a mechanical component can be accomplished with the following information:

1. Location, size, and orientation of indication
2. Far field stress near the indication
3. Analytical expression of the stress intensity factor at the crack tip, e.g. the crack tip stress intensity factor of a surface crack with a circular cross section is expressed as:

$$K_I = 1.12\,(\pi a)^{1/2}\,(\sigma) \tag{10.47}$$

   where a is the radius of the circle
4. Fracture mechanics material properties, namely fracture toughness, threshold, and growth data (C and m for Paris law).

The following analyses can be performed to assess reliability of a component with an indication.

a. Will the indication become unstable in the first application of load?

   A stress intensity factor (K_I) is estimated for the given indication and component geometry, indication size (a) and stress (σ). For a surface crack the expression of K_I as given in Eq. (10.45) is $K_I = 1.12(\pi a)^{1/2}\,(\sigma)$

   The stress intensity factor is compared to the fracture toughness (K_{IC}) $\cdot$ K_I < or = or > K_{IC}

   If K_I is less than K_{IC} then the indication will not be unstable in the first application of load; on the other hand, if the reverse is true, then it is expected to fracture in the first application of load.

b. Will this indication grow if it is stable after the first load application?

   An estimate stress intensity factor range uses the expression, $\Delta K_I = 1.12(\pi a)^{1/2}\,(\Delta\sigma)$

   The stress intensity factor range is compared to the threshold (ΔK_{TH}) $\cdot$ Δ_{KI} < or = or > ΔK_{TH}

   If ΔK_I is less than ΔK_{TH} then the indication will not grow after the load is applied; on the other hand, if the reverse is true then it is expected to grow in a stable manner after load application. Paris Law ($da/dN = C(\Delta K)^m$) is used for the stable growth estimation.

c. Estimate the critical crack size (a_c).

Critical size is the one at which the crack becomes unstable. It is estimated from the equation,

$$K_{IC} = 1.12(\pi a_c)^{1/2} (\sigma) \quad (10.48)$$

d. What is the safe number of load cycles that can be applied before unstable fracture?

Estimate the number of cycles required for the initial crack (a_i) to reach the critical crack size (a_c). Paris law is used to determine the expression of the number of cycles.
Many times a closed-form expression cannot be obtained. In such a situation, a numerical approach is adopted. For this example, the closed form solution is

$$N = (a_f^p - a_i^p)/(1.12\pi)^{0.5m}\, pC(\Delta\sigma)^m)$$

where $p = (2.0 - m)/2.0$.

e. Estimate the final crack length (a_f) after application of N cycles with initial crack length of a_i. The closed-form solution for the expression of K_I as given earlier is given by

$$a_f = (a_i^p + (1.12\pi)^{0.5m}\, pC(\Delta\sigma)^m N)^{1/p} \quad (10.49)$$

10.8 Failure of Cracked Components Fracture Mechanics Approach

The presence of a crack in a component may weaken it so that it fails by fracturing into two or more pieces. This can occur at stresses below the material's yield strength. The fracture mechanics approach is discussed in the following example. It is to demonstrate the utility of fracture mechanics to evaluate the reliability of a component. This can be applied to evaluate an expander disc if a discontinuity is detected. Indications of cracklike discontinuity were discovered near the blade slot (fir tree) of an expander disc.

LEFM was used to determine the suitability of its use. Stress levels were estimated using a detailed finite element model. The material properties were taken from those provided in Singh et al. (2005).

For example, the following disc material properties for Waspaloy at 800°F were used:

Yield strength:	$\sigma_{YS} = 106.0$ ksi
Fracture toughness:	$K_{IC} = 75.0$ ksi in$^{1/2}$
Threshold:	$\Delta K_{TH} = 5.0$ ksi in$^{1/2}$

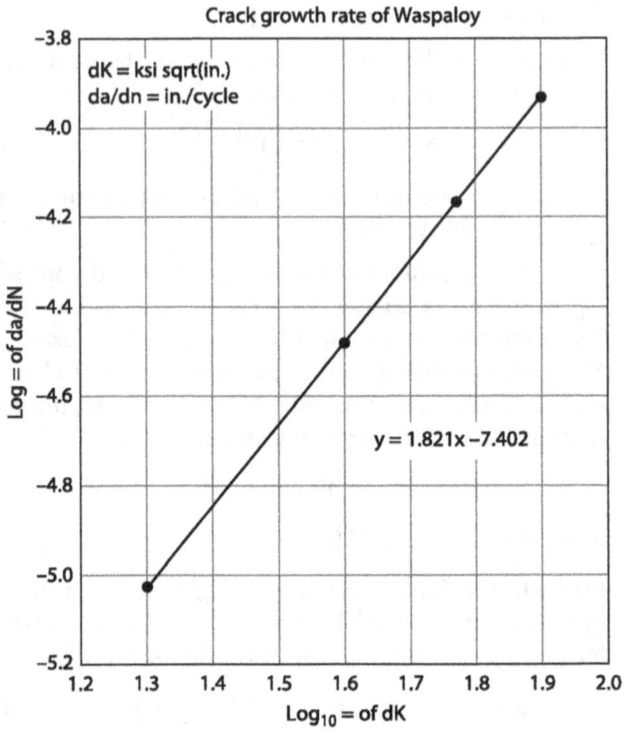

FIGURE 10.43 Plot of log(ΔK) vs. log(da/dN) for Waspaloy at 800°F.

The crack growth rate is represented by the equation below (Figure 10.43).

$$\log_{10}(da/dN) = 1.821x\log_{10}(\Delta K) - 7.402 \qquad (10.50)$$

$$\log_{10}(da/dN) = \log_{10}((\Delta K)^{1.821}) + \log_{10}(3.96 \times 10^{-8})$$

$$(da/dN) = 3.96 \times 10^{-8}(\Delta K)^{1.821}$$

Thus, the coefficients of Paris law are

$$C = 3.96 \times 10^{-8}$$

$$m = 1.821$$

where da/dN = crack growth rate, in/cycle

ΔK = stress intensity factor range at the crack tip point, ksi in$^{1/2}$

A finite element analysis was performed to calculate stress levels. Stress levels in the area of the indications were taken from a solid (3-D) finite element stress analysis of a disc attachment under centrifugal loading at the rated speed of the expander. The centrifugal loading was assumed to be equal to the longest blade that can be used.

The first principal stress at the indication location was the hoop stress (σ) at 30.1 ksi.

Following fracture mechanics evaluations helped in the disposition of the disc.

Will the Indication Become Unstable in the First Application of Load?

The stress intensity factor is estimated using a far-field hoop stress, and it is compared to the fracture toughness of the material. The diameter of the reported indication was 0.266 in. To account for the inaccuracy in the measurement of the area of the indication, area of reported indication was multiplied by 10 to simulate a flaw of 10 times the area (assuming a circular indication). Then the applicable diameter of the indication is calculated.

Reported area of the indication $= \pi\ (0.266)^2/4 = 0.0556\ \text{in}^2$

Adjusted area of the indication $= 10 \times 0.0556 = 0.556\ \text{in}^2$

Diameter of the adjusted indication $= 0.8412$ in

Radius of approximate circular indication $= 0.42$ in

Therefore, for the evaluation, the initial flaw size (half of the diameter) has been taken as 0.42 in.

The following expression was used to estimate the stress intensity at the crack tip:

$$K_I = \sigma\ (1.21)(\pi a)^{1/2} = (30.1 \times 1.12)\ (\pi 0.42)^{1/2} = 38.71\ \text{ksi in}^{1/2}$$

Since, K_I (38.71) $< K_{IC}$ (75), the indication will not become unstable in the first application of load.

Will This Indication Grow if It Is Stable after the First Load Application?

$$\Delta K_{TH} = 5.0\ \text{ksi in}^{1/2}$$

Since, ΔK_{TH} (5.0) $< \Delta K$ (38.71), the indication will grow.

What Is the Critical Crack Size (a_c)?

It is estimated from the equation, $K_{IC} = 1.12(\pi a_c)^{1/2}\ (\sigma)$

$$a_c = (K_{IC}/(1.12\sigma))^2/3.14 = 1.577\ \text{in}$$

What Is the Safe Number of Load Cycles That Can Be Applied before Unstable Fracture?

To be conservative, it was assumed that the indication spanned the bottom of the attachment for the entire thickness of the disc (about 4 in). It was further assumed that the flaw spans to the attachment bottom. This was done to add even more conservatism. It should be noted that the flaw being evaluated is not the size of the indication as reported, but it was a 0.6-in deep over the entire thickness of the disc (about 4 in).

The following expression was used to estimate the stress intensity at the crack tip:

$$N = (a_c^p - a_i^p)/((1.21\pi)^{0.5m}\, pC(\Delta\sigma)^m)$$

where $p = (2.0-m)/2.0$

$$p = (2.0-1.821)/2.0 = 0.0895$$

$$N = (1.57^{0.0895}-0.42^{.0895})/((1.21 \times \pi)^{(0.5x1.821)} \times 0.0895 \times 3.96 \times 10^{-8} \times (30.1)^{1.821}) = 21{,}100 \text{ cycles}$$

The calculated start–stop cycles for crack to grow from 0.42 in to 1.57 in (critical crack size) is equal to 21100 cycles.

Estimate the Final Crack Length (a_f) after Application of N Cycles with Initial Crack Length of a_i

Let us assume that operation requires start–stop cycles once a month, and it is required to monitor the crack size for 21,100 cycles or 1758 years. The following formula can be used to estimate the crack size.

$$a_f = (a_i^{\,p} + (1.21\pi)^{0.5m}\, pC(\Delta\sigma)^m\, N)^{1/p}$$

$$a_f = (0.422^{0.0895} + (1.21 \times \pi)^{(0.5\times 1.821)} \times 0.0895 \times 3.96 \times 10^{-8} \times (30.1)^{1.821} \times 21100)^{1/0.0895}$$

$$a_f = 1.57 \text{ in}$$

It is equal to the critical crack size as estimated earlier.

Figure 10.44 show an expanded estimation of life in years for different number of start–stop cycles per month.

Another interesting and useful curve can be drawn for the given material that will show the critical crack size as a function of stress for a given geometry. The formula used earlier for the geometry has been used for the calculation. Figure 10.45 shows stress versus

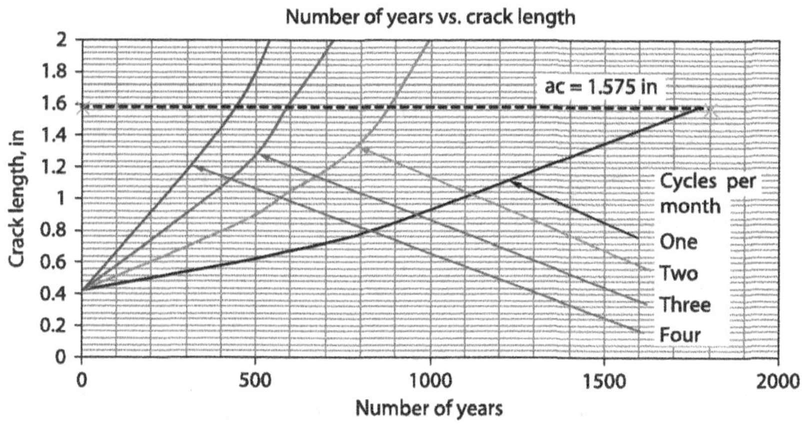

Figure 10.44 Plot of crack length after years of operation for different number of start–stop cycles per month.

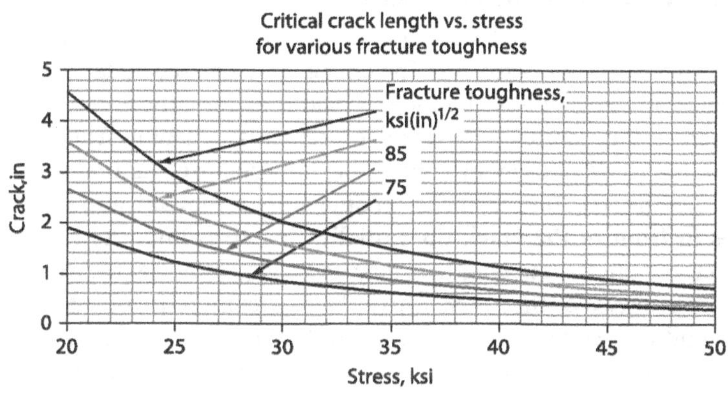

Figure 10.45 Stresses versus critical crack length for different values of fracture toughness (K_{IC}).

critical crack length for different values of fracture toughness (K_{IC}) for Eq. (10.46).

Another fruitful way of looking at the influence of material property and crack length can be visualized from a plot. Figure 10.46 is a plot of stress versus crack length as a function of fracture toughness and threshold values of the material. A crack should not grow for any combination of stress and crack length below the threshold curve. Between the threshold curve and the fracture toughness curve, a crack should grow. Beyond the fracture toughness curve the crack is expected to rupture.

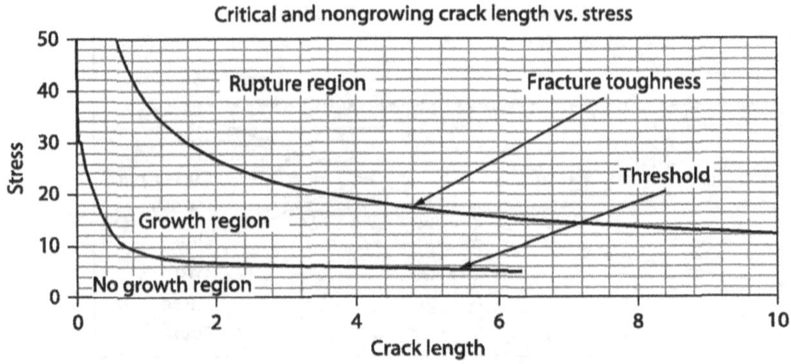

Figure 10.46 Stress versus crack length for fracture toughness (K_{IC}) and stress intensity factor threshold (ΔK_{TH}).

10.9 Crack Growth during Thermal Cycle

It has been observed that during thermal cycles if a crack grows then it slows down with growth. The examples discussed so far had a mechanical loading imposed on it and load remains constant. Either stress across the ligament increases or remains constant. Thermal stress is due to displacement control of the structure. As the crack grows, then the structure seems to become more compliant, and stress reduces in front of the crack tip. Initially, the stress intensity factor increases, but after some length, it may decrease, resulting in slowing down the progress of the crack. This subject gets more involved, and it is beyond the scope of this book. Interested readers should refer to literature on this subject for more detailed information. A brief explanation of this phenomenon is provided with the help of Figure 10.47.

Three different scenarios have been shown in Figure 10.47. A stress intensity factor is a function of stress and crack length. For a pure mechanical cycle, the stress intensity factor increases as a crack grows. Crack growth starts when the stress intensity factor is larger than the threshold stress intensity factor represented by point A. It is true whether stress remains constant or increases as the remaining cross-section area decreases. The curve labeled ABC represents a mechanical-loading situation. As soon as the stress intensity factor reaches the value of fracture toughness (K_{IC}), point B, the velocity of crack growth increases and the component goes to rupture. For a thermal-cycle situation, curves for two different stress intensity factors versus crack length are shown, namely ADE and AHI. For each case, stress intensity factor reaches a maximum value and then it decreases as the crack grows. The maximum value of the stress intensity factor

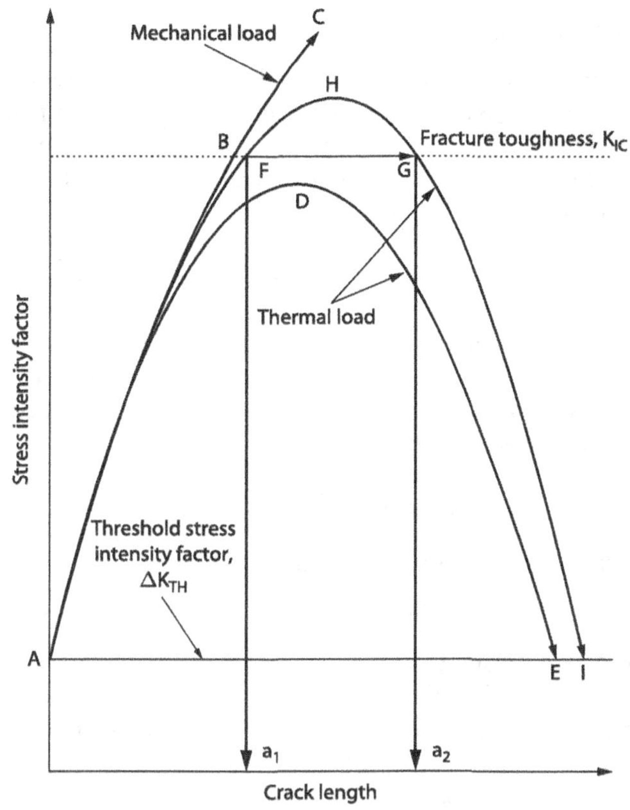

FIGURE 10.47 Stress intensity factor versus crack length.

for the curve ADE does not reach the value of fracture toughness. After this point, the crack starts slowing down and it might reach to point E and stop. On the other hand, the maximum value exceeds the fracture toughness represented by the curve AHI at the point represented by F or the crack length a_1. At this point the crack velocity increases and jumps to point G, represented by crack length a_2. After this crack starts to slow down to point I, its stops its growth. In both situations of thermal cycling. as long as there is enough remaining area, the crack will arrest. This is a common observation of damage during thermal cycling. This brief description of LEFM discussed is useful to estimate preliminary reliability of a component. There are limitations like closed form mathematical expression of stress intensity factor that might not be available, care should be taken for any plasticity occurring at the crack tip, crack when grows might not follow a straight line path etc. Detailed analysis of crack growth or stress intensity factor for the component geometry being considered should be done.

10.10 Creep Crack Growth

The creep rupture due to localized stress beyond yield stress of the material has been discussed in an earlier section. The first rupture time (t_{r1}) is considered as the crack initiation of crack length, a_1. Once t_r's are estimated for each combination of stress and temperature, the creep crack growth is estimated by use of the linear elastic fracture mechanics (LEFM). The theory of LEFM (to be discussed later) calculates a quantity called stress intensity factor, K_I, which has been shown to be a function of stress, crack length, geometry, and type of loading. To demonstrate this calculation method, Singh and Lucas (2011) considered an airfoil root attachment slot of a disc as shown in Figure 10.48.

The expression of K_I for the above case can be approximated for blade root and is given below:

$$K_I = \sigma\,(\pi a)^{1/2}\,F(a/b) \tag{10.51}$$

$$F(a/b) = (A_0 + A_1(a/b) + A_2(a/b)^2 + A_3(a/b)^3 + A_4(a/b)^4)/(1 - (a/b))^{1/2} \tag{10.52}$$

A_0 through A_4 are constant.

Using Eq. (10.49), the values of K_I corresponding to each combination of σ and a are estimated starting from σ_2 and a_2 as shown below.

$\sigma_2, a_2 \ldots K_{I2}$

$\sigma_3, a_3 \ldots. K_{I3}$

$\sigma_4, a_4 \ldots K_{I4}$

...

....

$\sigma_n, a_n \ldots K_{In}$

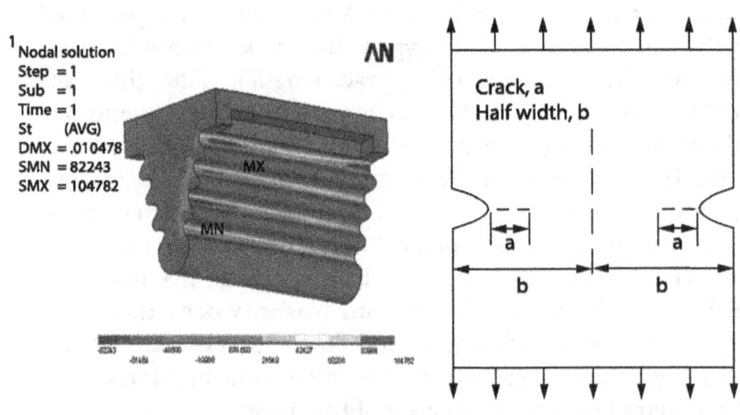

FIGURE 10.48 Stress localization in a root attachment.

To estimate the crack growth with time, the following expression for da_c/dt (refer to BS 7910:1999) or similar can be used.

$$da_c/dt = 0.014((K_I)^2/(\sigma.t_r))^{0.85} \quad (10.53)$$

Finally, the total time, t, to final fracture is calculated by using the following expression:

$$t = t_{r1} + \Sigma_i ((a_i - a_{i-1})/(0.014((K_{Ii})^2/(\sigma_i t_{ri}))^{0.85})) \quad (10.54)$$

where $i = 2,3, \ldots n$

10.11 Cumulative Damage

More often different fatigue cycles with different stress magnitudes are imposed on a structure in a given order. The theory presented so far deals with single fatigue cycle. There is a need to estimate damage due to many cycles. The following sections present a method to analyze fatigue damage in such a situation. It is termed cumulative damage analysis.

10.11.1 Miner's Method

A linear damage rule was put into common use based on a method published by Miner (1945). This assumes that total life per part can be estimated by just adding up the percentage of life consumed by each cycle. This is a linear cumulative damage method.

$$n_1/N_1 + n_2/N_2 + n_3/N_3 + \cdots + n_N/N_N = 1.0$$

or

$$\Sigma n_i/N_i = 1.0, i = 1, 2, 3 \ldots N$$

where n's are the applied cycles and Ns are the corresponding cycles to failure.

Miner's original test data showed the value of $\Sigma n_i/N_i$ between 0.62 and 1.45. Though very popular and simple to use, it suffers on a basic point that any damage is irreversible process and damage is path dependent.

10.11.2 Damage Curve Concept

The interaction damage rule considers only one type of hysteresis loop containing different strain range types applied to structure. The situations where many hysteresis loops of different types to each other are imposed on a given structure cannot be handled by interaction damage rule. A cumulative damage analysis is required to account for the fatigue damage done by each type of loop with different loading history. Practical cases have been encountered where high-cycle fatigue and low-cycle fatigue are superposed with a large difference in amplitude and frequencies. Manson and Halford (1967, 1981) developed the damage curve approach (DCA) to analyze such a situation. In a situation, the DCA lends to clean calculation strategy when many different types of load history are imposed on a structure. This method is suitable for automation on computer.

Damage to any structure has two important attributes. Namely, damage is an irreversible process, and it is path dependent. The cumulative rule proposed by Miner, which is essentially a path-independent rule, is used in many situations. Essentially, it is a linear damage rule. Figure 10.49 shows the plot of the accumulation of "damage" as a function of cycle ratio for various life values. The linear damage rule requires that the curves be coincident for all life level. It is their separateness that produces the loading order effect. All curves start at the origin which represents the initial condition of the material where the damage state is zero, and terminate at failure F where D = 1.0.

The DCA concept is that damage accumulation proceeds along the curve associated with the life level at which a cycle ratio is applied. For example, if a cycle ratio n_1/N_1 is first applied at the life level N_1,

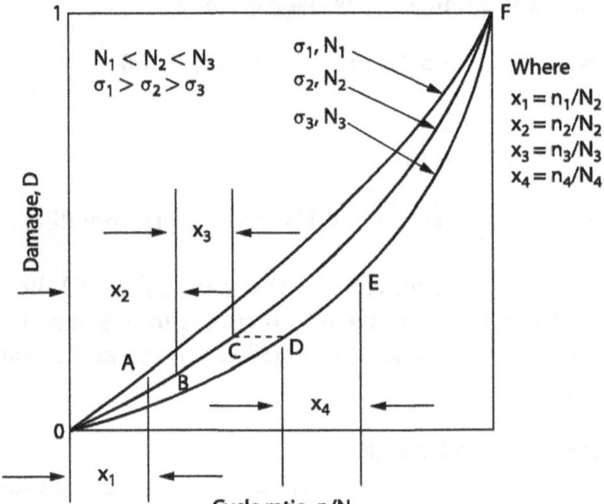

FIGURE 10.49 Damage curve concept for cumulative damage analysis for complex loading.

the damage will go from zero to A. If at this point a new loading level is introduced, the life of which is N_2 and the damage curve is OBF.

The point B on this curve has the same damage as point A on curve OAF. If n_2 cycle is then applied at the N_2 life level, the point C is located as shown. Similarly, if n_3/N_3 is applied at the N_3 life level, point D is located at the same level as C and increment DE by the cycle ratio n_3/N_3 and so on for an arbitrary sequence of loading.

They defined damage (D) as

$$\left(a_0 + \left(a_f - a_0\right)\left(n / N_f\right)^{(2/3)N_f^{0.4}}\right) \tag{10.55}$$

Applying Eq. (10.53) for the first increment of loading n_1/N_1 along OA in Figure 10.49, one obtains the value of D at A and B

$$D_A = \left(1/a_f\right)\left(a_0 + \left(a_f - a_0\right)\left(n_1 / N_1\right)^{(2/3)N_1^{0.4}}\right)$$

If one proceeds to point B along path OB, the cycle ratio required would be n_2/N_2 and the damage

$$D_B = \left(1/a_f\right)\left(a_0 + \left(a_f - a_0\right)\left(n_2' / N_2\right)^{(2/3)N_2^{0.4}}\right)$$

$$\text{As } D_A = D_B \tag{10.56}$$

$$n_2'/N_2 = \left(n_1 / N_1\right)^{(N_1/N_2)^{0.4}}$$

When added n_2/N_2 cycle ratio is applied along BC, the total cycle ratio at the point C is

$$n_2'/N_2 + n_2 / N_2 = \left(\left(n_1 / N_1\right)^{(N_1/N_2)^{0.4}} + n_2 / N_2\right) \tag{10.57}$$

Proceeding in the same manner to find the total cycle ratio at point E it becomes

$$\left(\left(n_1 / N_1\right)^{(N_1/N_2)^{0.4}} + n_2 / N_2\right)^{(N_2/N_3)^{0.4}} + n_3 / N_3 \tag{10.58}$$

If point E coincided with point F, Eq. (10.56) would be set to unity, implying that failure occurs. In a more general case when K loadings are applied before failure occurs, the equation for DCA becomes

$$(((((n_1/N_1)^{(N_1/N_2)^{0.4}} + n_2/N_2)^{(N_2/N_3)^{0.4}} + n_3/N_3)^{(N_3/N_4)^{0.4}} + \cdots + n_{k-1}/N_{k-1})^{(N_{k-1}/N_k)^{0.4}} + n_k/N_k) = 1 \quad (10.59)$$

Sequence of the loadings is denoted by subscripts 1, 2, 3, . . . K. As it happens that the only constant that remains in the final Eq. (10.57) is the exponent 0.4. Values in the range of 0.3 to 0.5 have been examined and they decided on the value of 0.4 to yield better results.

SRP method has been used to estimate life for fatigue creep interaction by the following steps:

1. Test material to generate strain range versus life data for the basic pure generic strain types, for example, PP, PC, CC, and CP.
2. Operational load histories need to be analyzed to estimate the resulting hysteresis loops.
3. The interactive damage rule is used to estimate the number of cycles to failure for each load history.
4. A damage curve analysis method is then used to assess the life of the component by combining different load histories.
5. This procedure is also applied in case of pure LCF (PP) type of loadings.

Example: The following example contains five different types of sequential operational cycles for an expander. Table 10.4 lists the sequence of loadings, estimated cycles to failure, and applied number of cycles. Table 10.5 lists the ratios of cycles as needed to apply Eq. (10.57) to estimate damage by DCA.

Damage due to the loadings has been estimated by Miner's rule and DCA for comparison.

Damage, D, by Miner's rule:

$$D = 2.37 \times 10^{10}/2.11 \times 10^{12} + 20/8040 + 10/2160 + 5/86 + 5/6740 = 0.0772$$

Load #	Load Type	Cycles to Failure	Applied Cycle
1	hcf	2.11E+12	2.37E+10
2	Normal startup	8040	20
3	Transient	2160	10
4	Alternating transient	86	5
5	Afterburn	6740	5

Table 10.4 Sequence of Loadings with Cycles to Failure and Applied Operational Cycles

Ratio of Load #	Applied Cycles/Cycles to Failure (n_i/N_i)	Ratio of Load #	Ratio of Cycles to Failure (N_i/N_{i+1})
1/1	1.12E-02	1/2	1.18E+09
2/2	2.49E-03	2/3	2
3/3	4.63E-03	3/4	2
4/4	5.81E-02	4/5	1
5/5	7.42E-04		

Table 10.5 Ratio of Cycles

Damage by DCA using Eq. (10.57):

$$D = (((((1.12E-02)^{(1.18E+09)^{0.4}} + 2.49E-03)^{(2)^{0.4}} + 4.63E-03)^{(2)^{0.4}} + 5.81E-02)^{(1)^{0.4}} + 7.42E-04)$$

$$D = 0.0598$$

There is a difference in the final damage assessment between Miner's estimation and DCA estimation. It is clear that the sequence of loading has an effect on damage.

Next to further demonstrate the effect of the sequence of loading; the sequence of loading has been arbitrarily changed. This is listed in Table 10.6. Table 10.7 lists the ratio of cycles as needed to apply Eq. (10.57) to estimate damage by DCA.

Damage due to the loadings has been estimated by Miner's rule and DCA for comparison.

Damage, D by Miner's rule:

$$D = 10/2160 + 20/8040 + 5/86 + 5/6740 + 2.37 \times 10^{10}/2.11 \times 10^{12} = 0.0772$$

Load #	Load Type	Cycles to Failure	Applied Cycle
1	Transient	2160	10
2	Normal startup	8040	20
3	Alternating transient	86	5
4	Afterburn	6740	5
5	hcf	2.11E+12	2.37E+10

Table 10.6 Assumed Sequence of Loadings

Ratio of Load #	Applied Cycles/Cycles to Failure (n_i/N_i)	Ratio of Load #	Ratio of Cycles to Failure (N_i/N_{i+1})
1/1	4.63E-03	1/2	2.69E-01
2/2	2.49E-03	2/3	9.35E+01
3/3	5.81E-02	3/4	1.28E-02
4/4	7.42E-04	4/5	3.20E-09
5/5	1.12E-02		

Table 10.7 Ratio of Cycles

Damage by DCA:

$$D = (((((4.63E-03)^{(2.69E-01)^{0.4}} + 2.49E-03)^{(9.35E+01)^{0.4}} + 5.81E-02)^{(1.28E-02)^{0.4}} + 7.42E-04)^{(3.20E-09)^{0.4}} + 1.12E-02)$$

$$D = 1.0110$$

Table 10.8 lists the estimated damages for both sequences estimated by Miner's rule and DCA. It is important to note that damage estimated by Miner's rule is independent of the sequence, but DCA captures the effect of the sequence demonstrating that damage is path dependent.

10.12 Evaluation of Damage after Service

Many theories and methods described so far are considered together with experience of particular OEM in the design process of mechanical component. Some or all may be used depending on the operational conditions. Components that are considered safe during design even based on past experience have experienced damage during operation. In this type of situation a more detailed look is warranted.

Sequence of Load #	Damage by DCA	Damage by Miner's
1	0.059800327	0.077222473
2	1.011025631	0.077222473

Table 10.8 Comparison of Damage by DCA and Miner's Method

It is also very useful to reexamine the design based on operational experience. Methods described in this chapter can be used to do a detailed analysis. An example of expander blade damage is presented to demonstrate the applicability of methods to determine the possible cause of damage.

Let us consider an assumed blade damage of an expander that was discovered owing to high vibration after about three years of service. It happened during start-up of the expander just after three years of service. After examining the disassembled disc and blades, cracks of about half-inch long in the blade attachment were found. A root cause failure analysis can be performed to find the possible cause of damage.

The construction details and operational history are listed below.

No. of blades	90
No. of nozzles	40
Temperature	1000°F
Material – IN 738	

Operational history

Operational speed	Design	6240
	Max	6940
	Min	5675

Time to damage	3 years
No. of start–shut down cycles	Once per quarter

Detailed Analysis

Metallurgical examination of the cracked surface found that it was a fatigue failure. The crack was progressing in the tangential direction to about half an inch with many tightly spaced beach marks.

Analytical method should start with some basic calculation such as Safe or Campbell diagrams, steady-state stress, alternating stress to assess the reliability.

Steady bending stress = 3600 psi
Average centrifugal stress = 50,000 psi

As the failure was estimated due to fatigue, modal solution of the bladed disc should be obtained to find out if it is due to any type of resonance problem. A finite element analysis should be used to yield a Safe diagram or a Campbell diagram. Preference should be for Safe diagram because it yields information about the mode shape of concern if any. A finite element analysis was conducted to yield a

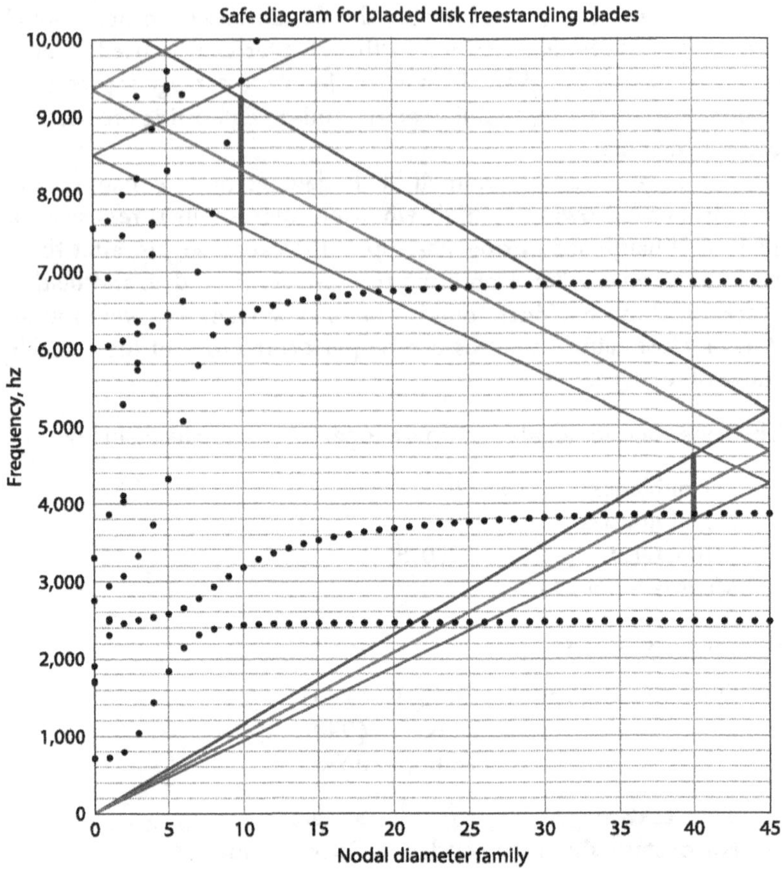

FIGURE 10.50 Safe diagram for an assumed expander disc.

Safe diagram of this (assumed) construction and it is shown in Figure 10.50. It shows that there is no concern about resonance at the design speed of the expander due to 1xNPF excitation due to 40 nozzles. However, there is a possibility of forced excitation near the minimum speed of the expander. A forced response type of analysis near the lower speed was performed to estimate the alternating stress if an expander ran at this speed. This seems to be the only place where damage might occur during start-up of the machine, if it did.

The basic reliability type of analyses are conducted that included Goodman analysis and creep rupture analysis.

Even though the maximum steady-bending stress was 3600 psi, the alternating stress estimated from force response analysis is 18000 psi.

Goodman Factor of Safety

Alternating stress = 18000.0 psi
Steady stress = 50000.0 psi
Factor of safety, FS = $1/(\sigma_a/\sigma_e + \sigma_m/\sigma_u)$

$$\begin{aligned} FS &= 1/(18/60 + 50/135) \\ &= 1/(0.3 + 0.37) \\ &= 1/0.67 \\ &= 1.49 \end{aligned}$$

Goodman safety factor seems to work for design, but, as was discussed earlier, it does not provide any information for how long it might take to initiate a crack if it is possible. The next analysis was performed to estimate HCF life.

Cycles to Crack Initiation due to HCF

Material fatigue properties for life estimation are given below.

$$\sigma_f' = 183280 \text{ ksi}$$
$$b = -0.062$$

Life equation in case of HCF-type loading was discussed earlier.

$$\begin{aligned} \text{Number of cycles to initiate HCF crack} &= (\sigma_a/(\sigma_f' - \sigma_m))^{1/b} \\ &= (18/(183.28 - 50))^{(1/-0.062)} \\ &= 1.06E + 14 \text{ cycles} \end{aligned}$$

The total time required to initiate the crack is estimated by using the natural frequency of the blade which is about 3866 Hz

$$\begin{aligned} \text{Total time} &= 1.06E + 14/(3866 \times 3600 \times 24 \times 365) \text{ years} \\ &= 2.08E + 4 \text{ years} \end{aligned}$$

The estimated time is too large for HCF to initiate the crack.

Creep Rupture

Next analysis is to check the rupture time due to the applied steady centrifugal stress of 50000.00 psi. A typical generic curve for IN 738 is given in Figure 10.51.

Curve fit of the data provides an equation between LMP and stress, which is used to estimate LMP.

$$\begin{aligned} LMP &= 0.0009(\sigma)^2 - 0.243\sigma + 54.3 \\ &= 0.0009(50)2 - 0.243(50) + 54.3 \\ &= 44.40 \end{aligned}$$

$$
\begin{aligned}
LMP &= T(20 + \log t) \times 10^{-3} \\
\log t &= (44.40 \times 1000)/(1000 + 273) - 20 \\
t &= 10^{14.87824}\ h \\
&= 7.5551E + 14\ h \\
&= 8.62E + 10\ years
\end{aligned}
$$

Again, it is too long a time for creep rupture damage when the expander ran only for about three years. Moreover, it confirms the metallurgical evaluation that damage was due to fatigue.

Creep Crack Initiation due to Localized Stress

Even though creep rupture of the blade based on the average cross-section stress is not feasible, the question needs to be answered about the cause of initiation. The peak stress at the site of initiation is not 50 ksi but much larger than that due to stress concentration. Theoretical stress concentration near the initiation site was estimated to be 4.6.

Average cross-section stress = 50 ksi
Theoretical stress concentration, $K_T = 4.6$
Elastic peak stress = 4.6 × 50 ksi = 230 ksi

As explained earlier that crack might initiate due to creep at the site of localized stress in less time. The estimated localized peak stress of 230 ksi is termed *elastic*. There is a need to estimate actual stress and it has been done using Neuber's method. Stress-strain curve or Ramberg-Osgood equation for stress-strain curve was not available. The bilinear approximation of the curve was used to estimate actual stress level.

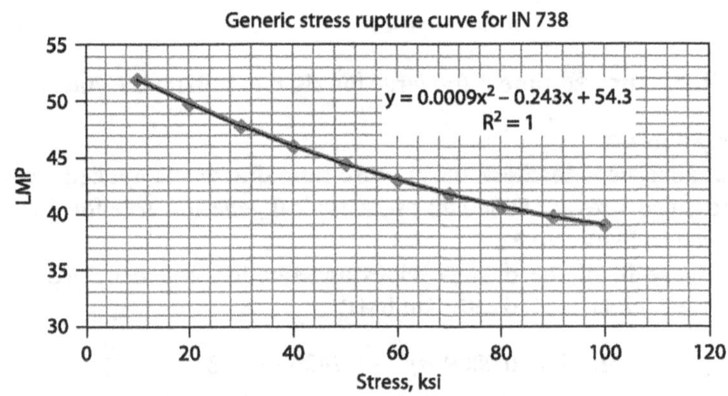

FIGURE 10.51 Typical stress rupture curve for IN 738.

Estimated Neuberized actual stress = 126 ksi

Analysis follows the same way as in case of stress rupture, but using the actual stress. It is assumed that the stress is so localized that if there is a crack rupture due to localized stress the crack size is very small. This crack size can be taken as initiated crack length.

$$\begin{aligned}
\text{LMP} &= 0.0009(\sigma)^2 - 0.243\sigma + 54.3 \\
&= 0.0009(126)2 - 0.243(126) + 54.3 \\
&= 37.97
\end{aligned}$$

$$\begin{aligned}
\text{LMP} &= \text{T}(20 + \log \text{t}) \times 10^{-3} \\
\log \text{t} &= (37.97 \times 1000)/\,(1000 + 273) - 20 \\
\text{t} &= 10^{9.827494} \\
&= 6.72\text{E} + 09 \text{ h} \\
&= 7.67\text{E} + 05 \text{ years}
\end{aligned}$$

Again, the time is too large and confirms the finding that initiation may be due to fatigue.

As it is determined so far that crack initiation is not feasible either due to HCF or localized creep then next attention is focused on the start-up and shut-down cycle. The estimated actual stress is 126 ksi that corresponds to a strain of about 0.009 in/in. As the estimated stress and strain is larger than the yield strength crack initiation due to LCF might be possible.

Crack Initiation due to Start–Shut-Down Cycle

Actual strain corresponding to actual stress of 126.00 ksi = 0.009 in/in

Coffin-Manson equation for LCF initiation cycles, $N = (\varepsilon/\varepsilon_f')^{1/c}$

$$\begin{aligned}
\varepsilon_f' &= 0.032 \\
c &= -0.5 \\
N &= (0.009/0.032^{(1/-0.5)} \\
&= 12.6 \text{ cycles}
\end{aligned}$$

Number of start–shut down cycles to initiate a crack = 12.6

The estimated number is very close to the total number of start-up and shut-down cycles of 12 (3 years × 4 per quarter). Usually this type of estimation does not yield such close results. However, it should be taken as a possibility of crack initiation in 12 starts and shut-down cycles.

Once it is accepted that there is a possibility of crack initiation due to start-up and shut-down cycle, the next analysis to be conducted is the mechanism of crack growth. Fatigue data from test that is obtained on a sample of 1/8 in in diameter and the number of cycles obtained is to complete separation. It was discussed in the earlier section that due to this reasoning the crack length at initiation is assumed to be 1/8 in. Linear elastic fracture mechanics may be used to determine what it will take for this crack of 0.125 in long to grow and what might be the duration for this growth. During the next start-up, before the expander reaches the design speed, the expander experiences high vibration. The focus goes to the alternating stress (18 ksi) developed near the natural frequency, which may be the driving stress for growth.

Crack Growth Analysis

$$\text{Frequency} = 3866 \text{ Hz}$$
$$\Delta\sigma = 18.00 \text{ ksi}$$
$$\text{Initial crack size, } a_i = 0.125$$
$$\text{Final crack size, } a_f = 0.5$$

Is There a Possibility That the Initial Crack Will Grow?

$$\Delta K_{TH} = 5 \text{ ksi in}^{1/2}$$
$$K_I = 1.12\,\sigma\,(\pi a)^{1/2} = 1.12 \times 18.0\,(\pi \times 0.125)^{1/2} = 12.63 \text{ ksi in}^{1/2}$$
$$K_I > \Delta K_{TH}$$

The applied stress intensity factor due to stress (18 ksi) is far greater than the threshold value; therefore, the crack is likely to grow.

To find out if the initial crack will go to rupture after the first application of stress following estimation is performed.

Estimation of Critical Crack Size for Rupture

$$\text{Fracture toughness, } K_{IC} = 75 \text{ ksi(in)}^{0.5}$$

Following equation for critical size will be used.

$$a_c = (K_{IC}/(1.12\,\sigma))^2/\pi$$
$$a_c = (75/(1.12 \times 18.00))^2/\pi = 4.41 \text{ in}$$

The initial crack size of 0.125 in is too small compared to critical crack size of 4.41 in. It is inferred that with the first application of stress equal to 18 ksi, the crack will not go to rupture.

Now it is established that initial crack will grow, but it will not go to rupture, then analysis is conducted to estimate number of cycles for crack to reach the reported length of 0.5 in.

Cycles for Crack to Grow from 0.125 in to 0.5 in

The expression for number of cycles required for crack to grow from initial crack size (a_i) to the final crack size (a_f) is given below.

$$N = (a_f^p - a_i^p)/((1.12\pi)^{0.5m}\, pC(\Delta\sigma)^m)$$

where $p = (2.0 - m)/2.0$

Paris constant for the material is listed below.

$$C = 3.96E - 08$$

$$m = 1.821$$

$$p = (2.0 - 1.821)/2 = 0.0895$$

$$N = (0.5^{0.0895} - 0.125^{0.0895})/((1.12\pi)^{(0.5x1.821)} \times 0.0895 \times 3.96E - 08(18)^{1.821})$$

$$N = 5.14 \times 10^4 \text{ cycles}$$

Time in seconds = N/Blade frequency = $5.14 \times 10^4/3866 = 13.2$ s

The estimation of time for growth from 0.125 in to 0.50 in is small and it is possible that damage may have occurred during this start-up cycle.

10.13 Summary

The methods presented in this chapter may help to set up design method(s) for mechanical components. It has been demonstrated through many examples. It is also demonstrated that with the help of these methods an analytical protocol can be outlined to conduct failure analysis of mechanical components.

CHAPTER 11

Probabilistic Concept and Risk Assessment

Any decision to purchase equipment is not made only on its initial price but also on its likelihood of failure during operational cycle. Premature failure has its consequences. Decision is made on the risk that a business entity may be willing to accept by balancing profit and loss.

The likelihood of equipment failure for any kind of operation depends on its design and manufacture. The consequences of failure depend on many factors, including the monetary loss a business has to incur.

Chapter 7 laid down the basic equations available to deal with many types of damage to mechanisms, for example, creep and fatigue. Assumptions are made that knowledge of all the parameters involved in these equations is identically known. But this assumption is hardly realized in practice. In reality, nearly each parameter has scatter, and this is best defined through statistical means.

This chapter introduces the basic probability theory. Many examples are given to estimate the risk associated with engineering decisions. This is done to show the basic process that can be followed to estimate the risk of failure or the likelihood of failure. The probability theory helps to estimate engineering risk of failure (likelihood of failure).

Next, we see how risk of failure or likelihood of failure in conjunction with the consequences of failure are used to estimate business risk. To demonstrate the process we include an example of the failure of an expander blade.

11.1 Brief Probabilistic Theory Applied in the Design Process

No two parts produced are exactly the same, no matter how carefully they are made. The differences may be small; nevertheless, they exist. This, in turn, may have an important bearing on the reliability of design. The variability, or scatter, complicates the task of proper designing. In the absence of scatter, that is, when all parts are exactly the same, the decisions for design are easy. Unfortunately, however, the parts are not the same. It is a common practice to make decisions based on the results, which in turn are based on deterministic concepts, that is, there is no variability. Design decisions are often made with insufficient information and, as a consequence, the decisions have risk associated with them. Sufficient margin in the design is made to ensure safe operation; however, the risk is not known. If a decision is made based on probability/reliability/risk assessment, then this will bring manufacturers and operators closer in considering variations in the loads, geometry, and material properties. This may enable designers and decision makers to conduct a probabilistic type of analysis that might provide more realistic estimate of reliability or risk.

The importance of this concept is underscored by Siddall (1983) and also by Singh and Drosjack (2008).

Siddall stated, "Probabilistic design is the codification of risk judgment; and *is not* the statistical analysis of the design problem." Singh and Drosjack in relation to the design process stated, "There is a need for combining physics-based theory of material behavior with probabilistic methods that account for uncertainty (variability) in data in order to achieve a robust design and to estimate the risk in accepting the resulting design."

If design is based on deterministic principle, that is, a single value of any variables is used neglecting any variability among parts (e.g., material properties, operational loads) then the resulting design does not have a 100% guarantee, or, absolutely no potential for failure. There is always a finite risk, however small it may be.

Drosjack et al. (2005) emphasized the need to consider risk as a component in each design and stressed the need to quantify risk. Total economic risk is the product of the likelihood of an event to occur (engineering risk) times the consequence of the occurrence (business risk) of the event. For example, the probability of imposed stress on a blade or disc to be larger than an acceptable magnitude is an engineering risk. In the deterministic approach to design, various parameters are considered as single valued. To accommodate any variation, a design margin is employed during analysis to cover the influence of the uncertainty. It is called "factor of safety" (FS). The basic fact is that if complete knowledge of the events and behavior of the component are available or can be predicted, then there is no need for this margin. It is thus appropriate to term it a "factor of ignorance." Higher value of margin may reflect higher uncertainty either in presumed operation or values of parameters. It can be assumed that the larger safety factors reflect more uncertainty on the part of the designer. If the margin is set to a high number, it might force an overdesign in many instances. Failures verify this fear where the factor of safety didn't actually cover the variations in critical parameters.

Even in developing equations to estimate engineering quantities many assumptions are made to track the solution of given conditions or loadings. Most modeling protocols involve linearization of nonlinear phenomena, for example, bearing stiffness characteristics used in rotor dynamics. The laws of physics may not be precise either. Thus, no matter what engineering design decisions are made, there is a risk associated with the calculated results.

The theories of statistics and probability have evolved and grown. These have been successfully used in many branches of science. It has been driven by needs and has been aided by technical developments. It was accelerated by the advances in computers and computational capabilities. As a result, many more comprehensive analysis capabilities are available today than in past years.

The approach to be discussed assumes that parameters and properties may be multivalued. All values of parameters have an inherent variability or uncertainty. Thus, the resulting solutions may not be single-valued—rather they are best represented as probabilistic quantity. The likelihood of each answer can be described as having a technical risk associated with it. Then the designer can select an answer based on the acceptable level of risk. Thus, design decisions do not need to incorporate factors of safety (factors of ignorance) to accommodate the uncertainty. Rather, the risk may be quantified and design decisions made on the acceptable level of risk.

Once the risk is estimated, it can be managed. Due to enhancements and capabilities in analytical tools, it is possible to estimate risk once any uncertainties are defined. The need to estimate risk associated with design and operation is real, but it requires more application to grow and flourish. This method seems to promise to push designs into new frontiers from the basis of size, loads, speeds, pressures, materials, costs, and delivery times.

The dominant characteristic of engineering design is uncertainty—uncertainty about, among other things, operational loads, manufacturing dimensions, material properties, and measured data. The estimation of the behavior and/or response of a structure based on measured data will also exhibit uncertainty (randomness). There may be uncertainty about the validity of a hypothesis or theory that is to be used to predict the performance of the design. The concern is whether or not a model, which was valid in certain circumstances, is applicable to a new situation in which the circumstances are uncertain. As engineering knowledge and tools become more sophisticated, it is certain that the "factor of ignorance" approach will become unsatisfactory.

A probabilistic type of analysis may offer some of the following benefits for an acceptable design:

- It may include uncertainties with appropriate physics-based law(s).
- It may provide estimates of "risk" for the design.
- It may allow a given level of risk for any desired finite operational life with economic realities to be acceptable.

Reasons for adopting a probabilistic approach during the design phase is summarized by Haugen (1980) in the following three items:

1. *Theoretical validity*. Many phenomena encountered in a design with multivalued quantities are a theoretical reality. The correspondence between predicted and actual performance depends on variability. Many statistical assumptions use strength and stress having admissible ranges of values from 0 to infinity. This assumption results in abnormally large stresses and abnormally low strengths that rarely occur. There are functions whose admissible ranges are finite, for example, among others, the diameter of a shaft, the thickness of a shell, temperature variation, and pressure variation.
2. *Reliability*. An achievable reliability goal can be decided before designing. Better yet, it should be incorporated into the design methodology with defined levels of performance and defined levels of reliability.
3. *Economic benefits*. If an improved design is expected to minimize costly testing programs, reduced total time from design to test, optimized design based on a reliability target and other requirements then economic benefits may be realized. It will then result in appropriate weight, size, volume, and more. A rational approach to design should focus on total life cost rather than a policy of replacements and warranties. It lends itself to overdesign and unachievable manufacturing tolerances without the estimation of reliability and a goal of reliability in design. It directly relates to higher cost without any idea of risk.

To obtain material properties many identical samples are tested under identical conditions and the resulting data shows variations. The test samples are chosen at random; therefore, the average value of the property should also be considered random. This number is named as a random variable. For example, the life of a component under a load may be considered to be random. Life of a part cannot be less than zero but a finite life. Life ranges from zero to a finite life.

A probability density function (pdf) is defined as a function that will yield probability of the variable it takes at any admissible values. Depending on the nature of data, many different pdfs have been defined through mathematical expressions, as for example, normal distribution, log normal distribution, Weibull, and more.

$$\text{Mean} = \sum_{i=1}^{n} x^i / \text{n} \tag{11.1}$$

$$\text{Standard deviation} = (\Sigma(\text{x}_i - \mu)^2/(\text{n} - 1))^{1/2} \tag{11.2}$$

The function of normal distribution is given below:

$$\text{f(x)} = (1/\sigma(2\pi)^{1/2})\,\exp(-(\text{x} - \mu)^2/2\sigma^2)$$
$$-\infty \leq x \leq \infty \tag{11.3}$$

Table 11.1 lists test frequencies of 21 manufactured blades. The mean of all values and standard deviation has been estimated using Eqs. (11.1) and (11.2).

Eq. (11.3) is used to calculate pdf, and it is plotted in Figure 11.1.

A cumulative distribution function (CDF) for a continuous variable is given by the following equation:

$$\text{F}(\text{x} \leq \text{a}) = \text{F(a)} = \int_{-\infty}^{a} f(x)dx = \text{P}(\text{x} \leq \text{a}) \tag{11.4}$$

where f(x) is the probability of obtaining the value of x. $P(x \leq a)$ is given by the area under the pdf curve from $-\infty$ to $x = a$.

The above expression is a complex one, and the integral is not readily available. Tabulation is available to estimate the integral for any pair of x. A transformation of variable is used to reduce the calculated value in a single table.

Transformation is achieved by equating $(x - \mu)/\sigma = z$

Then $dx = \sigma\, dz$

The function f(x) is transformed as f(z) and it is given below.

$$\text{f(z)} = (1/(2\pi)^{1/2})\,\exp(-\text{z}^2/2) \tag{11.5}$$

Number of Tested Blades	Measured Frequency	Freq - Average	Square of Freq - Average	
1	1200	−5.5238	30.512	
2	1208	2.4762	6.132	
3	1205	−0.5238	0.274	
4	1200	−5.5238	30.512	
5	1198	−7.5238	56.608	
6	1197	−8.5238	72.655	
7	1202	−3.5238	12.417	
8	1206	0.4762	0.227	
9	1205	−0.5238	0.274	
10	1210	4.4762	20.036	
11	1212	6.4762	41.941	
12	1207	1.4762	2.179	
13	1202	−3.5238	12.417	
14	1197	−8.5238	72.655	
15	1192	−13.5238	182.893	
16	1201	−4.5238	20.465	
17	1210	4.4762	20.036	
18	1219	13.4762	181.608	
19	1217	11.4762	131.703	
20	1215	9.4762	89.798	
21	1213	7.4762	55.893	
Total	25,316		1041.238	
Average	1205.5238		7.2154	STD

TABLE 11.1 List of Measured Frequencies, Mean Value and Standard Deviation

z has a normal distribution with $\mu = 0$ and $\sigma = 1$. This is known as standardized normal distribution.

CDF of standard z distribution is plotted in Figure 11.2.

The probability of the variable x (between a and b) is estimated by calculating corresponding values of z_α as follows:

$$z_{\alpha 1} = (a - \mu)/\sigma$$
$$z_{\alpha 2} = (b - \mu)/\sigma$$

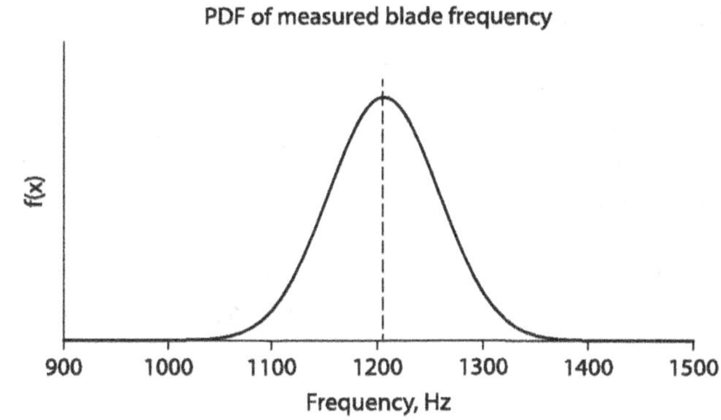

FIGURE 11.1 PDF of measured blade frequency.

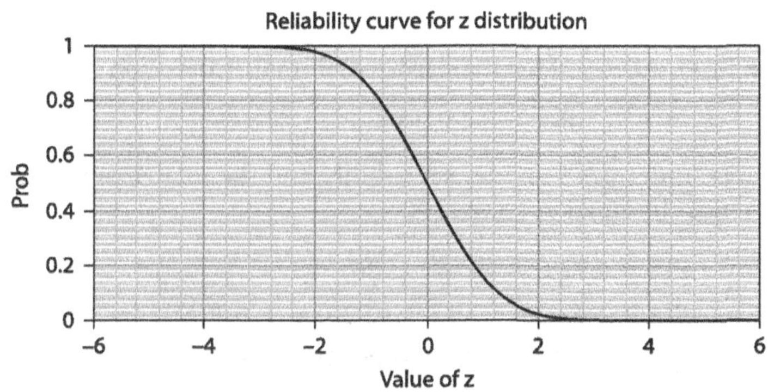

FIGURE 11.2 CDF of standardized normal distribution.

Then,

$$P(a \leq x \leq b) = P(z_{\alpha 1} \leq z \leq z_{\alpha 2}) = \int_{z\alpha 1}^{z\alpha 2} f(z)dz = \alpha \qquad (11.6)$$

A table of the value of α is given in every statistics book.

The above function enables designers to estimate the probability for frequency to be greater and less than a specified magnitude for the complete lot of blades.

It may be desirable for reliability purposes to find the probability of a blade to have frequency between a certain range. The nominal value of the frequency of the blade is 1200 Hz and the acceptable design is for the frequency to be within ±2.5% of 1200 Hz. Therefore, the acceptable range of frequency is 1170 to 1230 Hz and may be used in the design. Based on the manufactured lot of all blades, it is informative to estimate the probability of achieving that range of frequency.

Table 11.1 shows scatter in the values of measured blade frequencies with an average value of measured frequency equal to 1205.52 Hz and standard deviation equals to 7.21 Hz.

The estimated values of $z_{\alpha's}$ are:

$$z_{\alpha 1} = (a - \mu)/\sigma = (1170 - 1205.52)/7.21 = -4.926$$

$$z_{\alpha 2} = (b - \mu)/\sigma = (1230 - 1205.52)/7.21 = 3.395$$

Now

$$P(z_{\alpha 1} \leq z \leq z_{\alpha 2}) = \int_{z_{\alpha 1}}^{z_{\alpha 2}} f(z)dz = P(z \geq z_{\alpha 1}) - P(z \geq z_{\alpha 2}) = \alpha$$

$$P(z_{\alpha 1} \leq z \leq z_{\alpha 2}) = P(z \geq -4.926) - P(z \geq 3.395) = \alpha$$

Invoking symmetry of the standardized normal distribution curve (pdf)

$$P(z \geq -4.926) = P(z \leq 4.926)$$

The area under pdf provides probability and the total area under the curve must be equal to 1 (i.e., 100%).

$$P(z \leq 4.926) + P(z \geq 4.926) = 1$$
$$P(z \geq -4.926) = P(z \leq 4.926) = 1 - P(z \geq 4.926)$$
$$\alpha = 1 - P(z \geq 4.926) - P(z \geq 3.395)$$

Values of $P(z \geq 4.926)$ and $P(z \geq 3.395)$ are estimated from the z table:

$$P(z \geq 4.926) = 0.0000481$$
$$P(z \geq 3.395) = 0.000337$$

Finally,

$$\alpha = 1 - 0.0000481 - 0.000337$$
$$= 0.9996 = 99.96\%$$

This means that about 99.96% of the manufactured blades will have frequencies between the specified ranges.

The manufacturing method and the geometrical tolerances are very good.

The method is very important to control quality of design and manufacturing for single items. It is also the basis to handle the physical process with many variables, which is when mathematics gets complicated. The theory of probability has evolved to handle such problems.

11.2 Probabilistic Aspect Leading to Estimation of Likelihood of Failure

Material properties are obtained through laboratory testing on multiple samples. Resulting data shows scatter in the properties. Variation in the data will depend on the number of samples tested. When such data as yield strength of the material are plotted, it will clearly have a statistical distribution.

How the structure reacts to the applied force or imposed stress depends on the interaction between the imposed stress, S, and the component's resistance, R. As recognized, there are uncertainties in the values of both stress and material properties. These are represented in statistical terms by a pdf.

The deterministic method defines the margin as follows:

$$\text{Margin} = (\text{Material's strength, R})/(\text{Stress, S}) \tag{11.7}$$

Average values of stress and strength are given below.

$$\text{Stress, S} = 60 \text{ ksi}$$
$$\text{Material's strength, R} = 120 \text{ ksi}$$

A graphical description of stress and strength is shown in Figure 11.3.

$$\text{Margin or FS} = 120/60 = 2.0$$

After considering the scatter in the imposed stress and material's strength, the probability of success (P_s) is expressed as the following:

$$P_s = P(R > \text{or} = S) \tag{11.8}$$

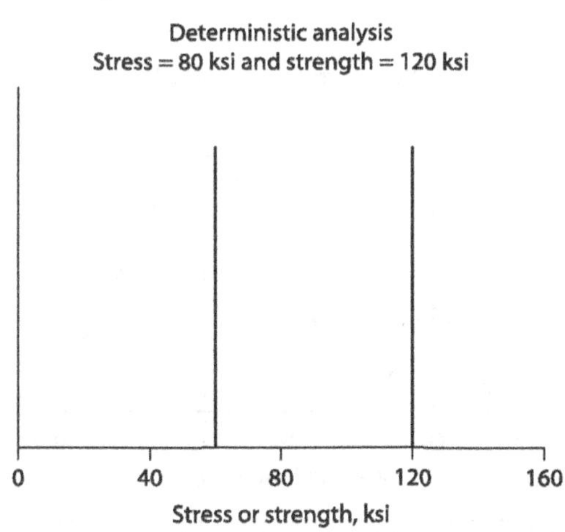

Figure 11.3 Deterministic method of estimating margin.

A general form of the above equation takes the following form:

$$P_s = P(G(R, S) > \text{or} = 0) \tag{11.9}$$

where G(R, S) is a limit function and defines the relationship between R and S. For example, if R is the yield strength of the material and S is the imposed stress, F might be expressed as follows:

$$G(R, S) = R - S \tag{11.10}$$

The above relationship states that the function, G, is equal to the yield strength less the imposed stress. For reliable operation, G must be positive, a deterministic statement.

The probabilistic of success of the problem can be stated as follows:

$$P_s = P((R - S) > \text{or} = 0) \tag{11.11}$$

or

likelihood of failure

$$P_f = 1 - P_s \tag{11.12}$$

Graphical representation of the probability density functions for R and S is illustrated in Figure 11.4. The average magnitude of S (applied stress) is less than the average strength, which satisfies the condition for safe design when a deterministic concept of design is considered. A factor of safety is equal to 2.0. However, when the possible uncertainties in R and S are considered, it becomes clear that there is still a possibility of S being larger than R in the area in the middle of the

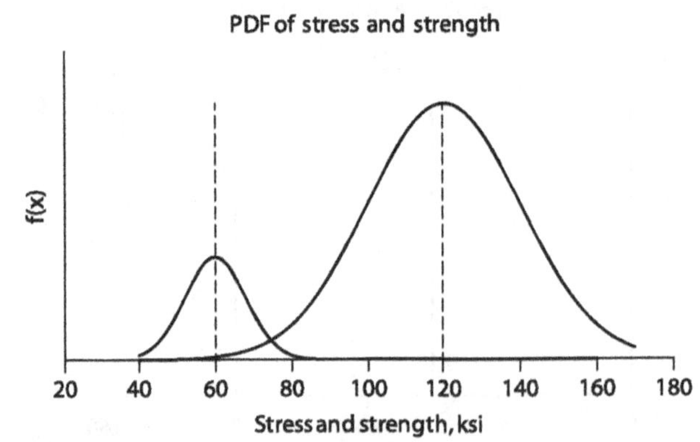

Figure 11.4 Description of probability of failure is the area common to two pdf's.

graph where the stress and strength curves cross over one another. Hence, there is a chance of failure because of the scatter in R and S, while the deterministic approach indicates success.

11.3 Basic Probabilistic Theory Using Limit Function

Let us assume that a process is defined by a function (f) as shown below:

$$f(x) = f(x_1, x_2, x_3, \ldots, x_n) \tag{11.13}$$

where $x_1, x_2, x_3, \ldots, x_n$ are variables. Some x's may be deterministic and some may have variations defined by a statistical distribution.

f(x) can be any physical relationships, such as factor of safety, natural frequency of a blade, HCF life of a component, and more.

A limit function G can be written as

$$G(f_1, x_1, x_2, x_3, \ldots, x_n) = f(x_1, x_2, x_3, \ldots, x_n) - f_{target} \tag{11.14}$$

where f_{target} is a desired value for f(x).

The probabilistic mathematical statement can be stated as follows:

$$P_s = P(G(f_{target}, x_1, x_2, x_3, \ldots, x_n) > \text{or} = 0) \tag{11.15}$$

P_s represents the probability of occurrence for a desired value of f_{target}. $P_f = 1 - P_s$, represents the risk of not achieving P_s (likelihood of failure).

The magnitude of P_s (likelihood of failure) for each value of f_{target} can be estimated for function, f(x) for a target value of f_{target}.

Thus, Eq. (11.15) can be considered as the basic formulation that can be used to conduct probabilistic analysis for any defined function, f(x). This provides the probability of occurrence.

11.4 Examples Pertaining to Expander Design

The importance of probabilistic type of analysis wherever possible can be realized through examples. Results of such exercise should point out that it is possible to make design decisions different from that based on a deterministic-type of analysis. This section includes a variety of simple examples to demonstrate application of probabilistic analysis leading to an estimation of the likelihood of failure. Most of the example pertains to blade of expander for simplicity, but theory and process are equally applicable to any component as long as the physics of behavior under load of the component is available. Probabilistic type of analysis is within reach and it yields the important information about technical risk that will help in estimating business risk.

11.4.1 Probabilistic Treatment of Bending Stress due to Gas Load

Simple beam bending theory is used to design a beamlike structure (e.g., expander blade) and a margin of resulting bending stress from desired material property is required for safe application. To demonstrate the importance of probabilistic analysis, consider the following simple example of a beam under a bending load (Figure 11.5). This load simulates the bending on the blade owing to gas load.

A deterministic-type analysis is traditionally done (neglecting any variation) based on average values.

$$\text{Desired margin is calculated in terms of factor of safety} = \sigma_y/\sigma_b \tag{11.16}$$

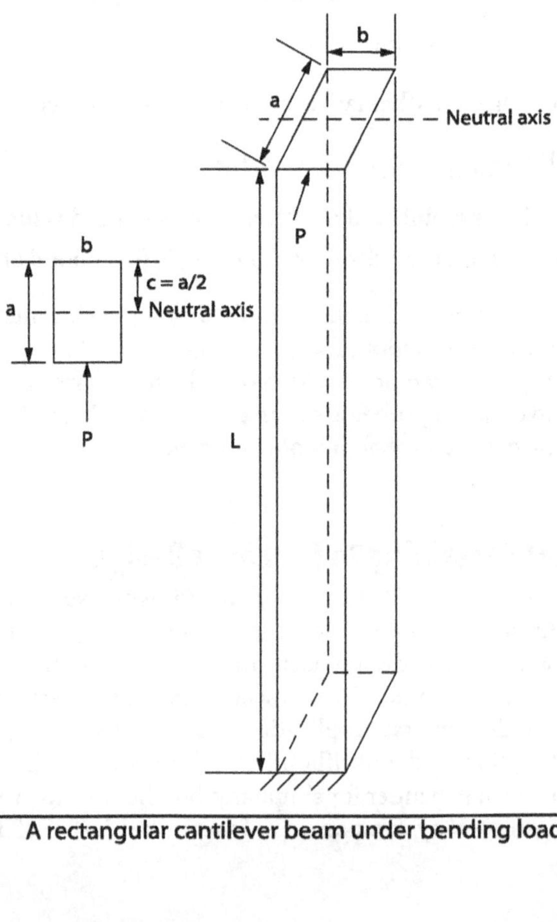

Figure 11.5 A rectangular cantilever beam under bending load.

where

$$\sigma_b = Mc/I, \text{ bending stress} \tag{11.17}$$

M is the applied moment at the cross section

σ_y = yield strength
c = distance of outer fiber from neutral axis
I = moment of inertia of the cross section = $ba^3/12$
a and b = dimension of cross section of a rectangular beam
M = PL
P = applied load
L = length of the beam
FS = average value of σ_y/average value of σ

Table 11.2 lists the dimensions of the beam, magnitude of the point load, and yield strength of the material along with calculated values of c, M, I, and bending stress at the bottom of the beam:

$$FS = \sigma_y/\sigma_b = 1.59$$

If the dimensional tolerances, variation in load, and scatter in the material data for yield strength are taken into account, then a probabilistic-type analysis is suitable. A Monte Carlo–type analysis was performed to estimate probability on FS. Table 11.3 lists the assumed variation in the variables. Standard deviation is estimated based on the total spread equal to 12 sigma (6 sigma on each side of the mean). This ensures about 99.9999998% probability of data to fall within the pdf.

$$FS = \sigma_y/\sigma_b = f(\sigma_y, \sigma_b)$$

Deterministic Estimation of FS	
P, lb	700
L, in	6
a, in	1
b, in	0.5
c = a/2, in	0.5
M = PL, lbin	4200
I = ba^3/12, in^4	0.04167
Stress bending, psi	50400
Yield Strength, psi	80000
FS	1.59

Table 11.2 Dimension of Beam, Magnitude of Load, and Yield Strength

Variables		Variation Plus Minus	S TD Based on 6 Sigma	Minimum	Maximum	Assumed Distribution
P, lb	700	70	11.6667	630	770	Bounded normal
L, in	6	0.05	0.0083	5.95	6.05	Bounded normal
a, in	1	0.01	0.0017	0.99	1.01	Bounded normal
b, in	0.5	0.005	0.0008	0.495	0.505	Bounded normal
Yield strength, psi	80000	800	133.3333	79200	80800	Bounded normal

TABLE 11.3 Assumed Variation in Variables

A limit function, G, can be written as

$$G(FS, \sigma_y, \sigma_b) = f(\sigma_y, \sigma_b) - FS_{target} \tag{11.18}$$

$$\begin{aligned} G(FS, \sigma_y, \sigma_b) &= f(\sigma_y, M, c, I) - FS_{target} \\ &= (\sigma_y I)/(Mc) - FS_{target} \end{aligned}$$

where FS_{target} is a desired value for factor of safety.

The probabilistic mathematical statement can be stated as follows:

$$P_s = P(G(\sigma_y, M, c, I, FS_{target}) > \text{or} = 0) \tag{11.19}$$

P_s represents the probability of factor of safety greater than FS_{target}. The likelihood of failure is defined as $P_f = 1 - P_s$

The Monte Carlo simulation has been used to estimate the likelihood of failure (P_f) for FS to attain a desired magnitude. The risk of a certain magnitude of FS obtained through the analysis is shown in Figure 11.6. It provides the influence of statistical variation of stress and yield strength of material. Material properties and applied load are assumed to have statistical variations as described above in Table 11.3. The dimension of the cross section of the beam has manufacturing tolerances. Only guarantee is that the specified dimension stays within the tolerance band after manufacturing. When many beams are manufactured in a similar way, the final dimension will have a statistical dimension, even though the dimension of each beam is expected to be within the tolerance band.

A review of Figure 11.6 demonstrates that FS may range from about 1.51 to 1.67. This means that there is the likelihood of failure

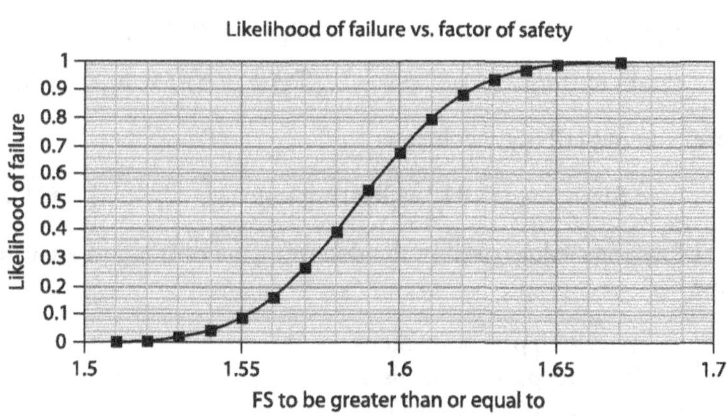

Figure 11.6 Likelihood of failure vs. factor of safety.

of FS about 0.0 below 1.51 and almost 1.0 (100%) above 1.67. The deterministic approach predicts the factor of safety to be 1.59. A factor of safety greater than 1.59 has 0.55 or 55% likelihood of failure and the likelihood of failure of 0.45 (45%) to be less than 1.59. If the assumption of scatter in variables is correct, then design may not be safe under assumed scatter.

11.4.2 Effect on Expander Performance due to Variation in Pressure and Temperature

Another design concern of an expander is its performance (efficiency and available power). This example estimates η_{poly}, η_{ad}, gas horse power (GHP) and shaft horse power (SHP). Details of performance of expanders have been discussed in earlier chapter. For this example, expressions for η_{poly}, η_{ad}, GHP and SHP are given below. Estimations for these parameters are done for deterministically based on average values of parameters as well as probabilistic type of analysis by utilizing assumed scatter in the parameters. Results of probabilistic analysis demonstrate that there is a range of these quantities with different likelihood of not achieving certain values. This type of analysis will help in making decision.

Expression of polytropic efficiency is given below:

$$\eta_{poly} = (k/(k-1))\ln(T_2/T_1)/\ln(P_2/P_1) \tag{11.20}$$

Expression of adiabatic efficiency is given below:

$$\eta_{adia} = \left(1-(P_2/P_1)^{\eta_{poly}((k-1)/k)}\right)/\left(1-(P_2/P_1)^{((k-1)/k)}\right) \tag{11.21}$$

Expression of GHP is given below:

$$\begin{aligned} GHP &= m\,(k/(k-1))RT_1(1-(P_2/P_1)^{((k-1)/k)})\ \text{ft lb/s} \\ &= m\,(k/(k-1))RT_1(1-(P_2/P_1)^{((k-1)/k)})/550\ \text{hp} \end{aligned} \tag{11.22}$$

Expression of SHP is given below:

$$SHP = \eta_{adia}\,GHP - \text{Bearing loss} \tag{11.23}$$

The average values used in the deterministic estimation are listed below.

$$P_1 = 25.8\ \text{psig} = 40.50\ \text{psia}$$
$$P_2 = 0.3\ \text{psig} = 15.0\ \text{psia}$$
$$T_1 = 1300°\text{F} = 1760°\text{R}$$
$$T_2 = 1000°\text{F} = 1460°\text{R}$$

$$k = 1.3$$
$$R = 52.73037543$$
$$m = 300$$
$$\text{Bearing loss} = 120 \text{ hp}$$

Deterministic estimation follows:

$$\text{Pressure ratio} = 15/40.5 = 0.370370$$
$$\text{Temperature ratio} = 1460/1760 = 0.82955$$
$$k/(k-1) = 1.3/(1.3-1) = 4.33333$$
$$(k-1)/k = 0.23077$$

Estimation of polytropic efficiency is given below.

$$\begin{aligned}\eta_{poly} &= (k/(k-1))\ \ln(T_2/T_1)/\ln(P_2/P_1)\\ &= (4.33333)\ \ln(0.82955)/\ln(0.370370)\\ &= 0.8153\end{aligned}$$

Estimation of adiabatic efficiency is given below.

$$\eta_{poly}((k-1)/k) = (0.8153)(\ 0.23077) = 0.18815$$
$$\begin{aligned}\eta_{adia} &= (1 - (P_2/P_1)^{\eta_{poly}((k-1)/k)})/(1 - (P_2/P_1)^{((k-1)/k)})\\ &= (1 - 0.370370^{0.18815})/(1 - 0.370370^{0.23077})\\ &= 0.83213\end{aligned}$$

Estimation of GHP is given below:

$$\begin{aligned}GHP &= m\ (k/(k-1))RT_1(1 - (P_2/P_1)^{((k-1)/k)})\ \text{ft lb/s}\\ &= m\ (k/(k-1))RT_1(1 - (P_2/P_1)^{((k-1)/k)})/550\ \text{hp}\\ &= (300.)(4.33333)\ (52.73037543)(1760)(1 - 0.370370^{0.23077})/550\\ &= 44933.4\ \text{hp}\end{aligned}$$

Estimation of SHP is given below:

$$\begin{aligned}SHP &= \eta_{adia}\ GHP - \text{Bearing loss}\\ &= (0.83213)(44933.4) - 120\\ &= 37270.6\ \text{hp}\end{aligned}$$

Probabilistic Estimation

Limit functions for η_{poly}, η_{ad}, GHP, and SHP are defined below.

Next, estimation of the likelihood of the above quantities is conducted with assumed scatter using the parameters listed in Table 11.4.

	Mean	Total Variation	Std Dev	Minimum	Maximum
P1, psia	40.5	4	0.33333	38.5	42.5
P2, psia	15	0.4	0.03333	14.8	15.2
T1, 0R	1760	50	4.16667	1735	1785
T2, 0R	1460	50	4.16667	1435	1485
m, lb/sec	300	60	5	270	330
Bearing Loss, hp	120	24	2	108	132

TABLE 11.4 Scatter Magnitude of Variables

Analysis for polytropic efficiency is given below.

$$\eta_{poly} = (k/(k-1)) \ln (T_2/T_1)/\ln (P_2/P_1)$$
$$= f(k, T_1, T_2, P_1, P_2)$$

$$G(\eta_{polytarget}, k, T_1, T_2, P_1, P_2) = (k/(k-1)) \ln(T_2/T_1) /\ln(P_2/P_1) - \eta_{polytarget} \quad (11.24)$$

$P_s = P(G(\eta_{polytarget}, k, T_1, T_2, P_1, P_2) > \text{or} = 0)$

P_s represents the probability of η_{poly} greater than $\eta_{polytarget}$.

The likelihood of failure is defined as $P_f = 1 - P_s$

Plotting of the likelihood of failure versus η_{poly} is given in Figure 11.7.

A probabilistic calculation using Monte Carlo simulation is done for the likelihood of failure of polytropic efficiency. It is shown in Figure 11.7. It is evident that after including the influence of variations in parameters, η_{poly} of about 0.73 has about 0 likelihood of failure whereas η_{poly} of 0.83 has about 1.00 likelihood of failure. However, there is 0.85 likelihood of achieving η_{poly} of 0.8153 as estimated by the deterministic approach based on assumed scatter.

Analysis for adiabatic efficiency is given below.

$$\eta_{adia} = (1 - (P_2/P_1)^{\eta_{poly}((k-1)/k)})/(1 - (P_2/P_1)^{((k-1)/k)})$$
$$= f(\eta_{poly}, k, T_1, T_2, P_1, P_2)$$

$$G(\eta_{adtarget}, k, T_1, T_2, P_1, P_2) = (1 - (P_2/P_1)^{\eta_{poly}((k-1)/k)})/ (1 - (P_2/P_1)^{((k-1)/k)}) - \eta_{adtarget} \quad (11.25)$$

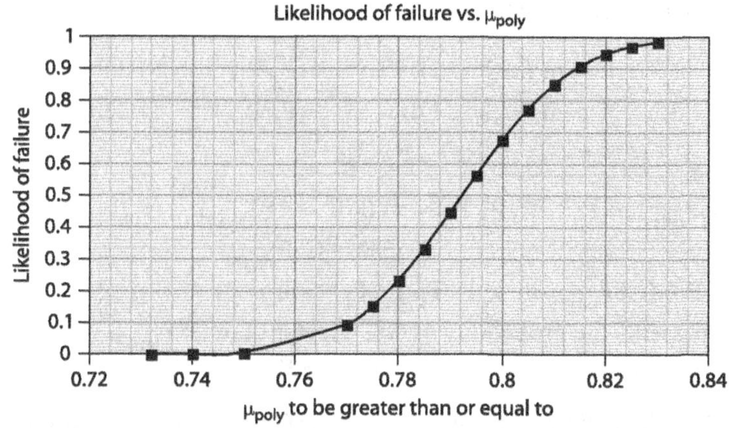

FIGURE 11.7 Likelihood of failure vs. polytropic efficiency.

$P_s = P(G(\eta_{adtarget}, k, T_1, T_2, P_1, P_2) > \text{or} = 0)$
P_s represents the probability of η_{ad} greater than $\eta_{adtarget}$.

The likelihood of failure is defined as $P_f = 1 - P_s$

The plot of likelihood of failure versus η_{adia} is given in Figure 11.8.

A probabilistic calculation using a Monte Carlo simulation is done for the likelihood of failure of adiabatic efficiency. It is shown in Figure 11.8. It is evident that after including the influence of variations in parameters, η_{adia} of about 0.77 has about 0 likelihood of failure whereas η_{adia} of 0.88 has about 1.00 likelihood of failure. However, there is 0.95 likelihood of achieving η_{adia} of 0.832 as estimated by the deterministic approach based on assumed scatter.

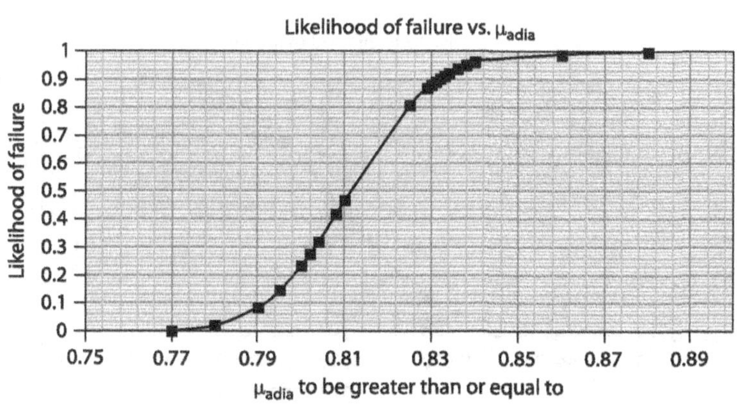

FIGURE 11.8 Plot of likelihood of failure vs. adiabatic efficiency.

Analysis for GHP is given below.

$$\begin{aligned} GHP &= m\,(k/(k-1))RT_1(1-(P_2/P_1)^{((k-1)/k)})\ \text{ft lb/s} \\ &= m\,(k/(k-1))RT_1(1-(P_2/P_1)^{((k-1)/k)})/550\ \text{hp} \\ &= f(k, m, R, T_1, P_1, P_2) \end{aligned}$$

$$G(GSP_{target}, m, k, R, T_1, P_1, P_2) = m\,(k/(k-1))RT_1(1-(P_2/P_1)^{((k-1)/k)})/550 - GSP_{target} \quad (11.26)$$

$P_s = P(G(GSP_{target}, m, k, R, T_1, P_1, P_2) > \text{or} = 0)$

P_s represents the probability of GSP greater than GSP_{target}.

The likelihood of failure is defined as $P_f = 1 - P_s$.

The plot of likelihood of failure versus GHP is given in Figure 11.9.

A probabilistic calculation using a Monte Carlo simulation is done for the likelihood of failure of GHP. It is shown in Figure 11.9. It is evident that after including the influence of variations in parameters, GHP of about 44,000 hp has about 0 likelihood of failure while GHP of 49,000 hp has about 1.00 likelihood of failure. However, there is only 0.08 likelihood of failure for a creep rupture life of 44,933 hp as estimated by the deterministic approach based on assumed scatter.

Analysis for SHP is given below.

$$\begin{aligned} SHP &= \eta_{adia}\,GHP - \text{Bearing loss} \\ &= f(\eta_{adia}, k, m, R, T_1, P_1, P_2, \text{Bearing loss}) \end{aligned}$$

$$G(SHP_{target}, \eta_{adia}, GHP, \text{Bearing loss}) = \eta_{adia}\,GHP - \text{Bearing loss} - SHP_{target} \quad (11.27)$$

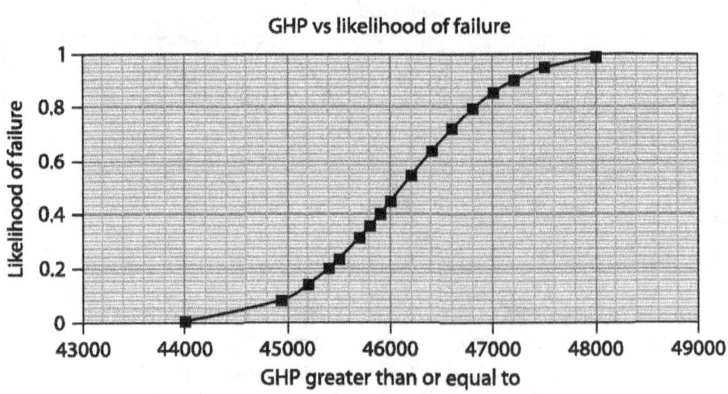

FIGURE 11.9 Plot of likelihood of failure vs. GHP.

$P_s = P(G(SHP_{target}, \eta_{adia}, GHP, \text{Bearing loss}) > \text{or} = 0)$
P_s represents the probability of SHP greater than SHP_{target}.

The likelihood of failure is defined as $P_f = 1 - P_s$.

The plot of likelihood of failure versus SHP is given in Figure 11.10.

A probabilistic calculation using Monte Carlo simulation is done for the likelihood of failure of SHP. It is shown in Figure 11.10. It is evident that after including the influence of variations in parameters, SHP of about 35,000 hp has about 0 likelihood of failure whereas SHP of 40,000 hp has about 1.00 likelihood of failure. However, there is only 0.68 likelihood of failure of achieving a creep rupture life of 37,270 hp as estimated by the deterministic approach based on assumed scatter.

11.4.3 Probabilistic Analysis of Natural Frequency of a Blade Modeled as a Beam

The next level of analysis for blade vibration is to model the blade as a beam. A previous example dealt with a simple spring-mass-damper system. This example considers blade as a beam to demonstrate the process. Singh (1985) demonstrated the implications of geometrical variations on the randomness of blades' natural frequency. He showed that it is instructive to use the transformation of random variables to demonstrate the process. For simplicity, he assumed that the geometrical variation is reflected through the cross-sectional property of the blade (e.g., the moment of inertia of the cross section). Based on the basic equation of the first mode of a cantilever beam, he developed the equations for pdf's of frequency and magnifier. The calculated results for a case were reported.

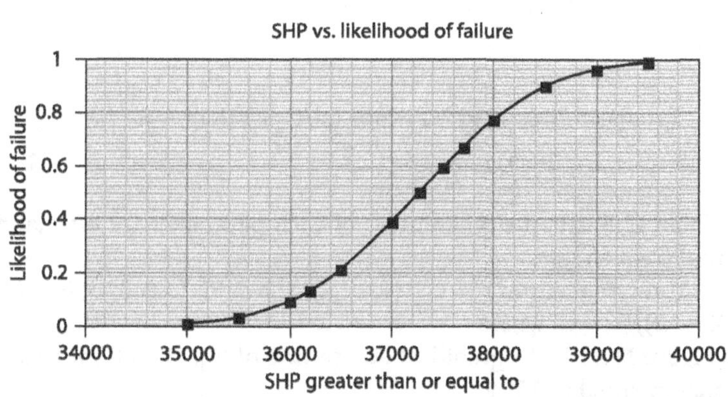

FIGURE 11.10 Plot of likelihood of failure vs. SHP.

The expression for the natural frequency of a cantilever beam is given below:

$$\omega_n = C_n\,((Ei)/(\rho l))^{1/2} \tag{11.28}$$

where E = modulus of elasticity
i = moment of inertia
ρ = mass
l = length of the beam
C_n = constant for a mode (3.52 for mode 1)

Singh (1985) provided the following equations leading to the expression of pdf for ω_n after using the method of transformation of random variables:

$$di/d\omega_n = (2\rho l\,\omega_n)/(C_n^2 E)$$
$$f_\Omega(\omega_n) = f_I\,(I)\,|(2\rho l\,\omega_n)/(C_n^2 E)|$$

For simplicity, the pdf for the moment of inertia is taken as normal distribution and the expression is as follows:

$$f_I\,(I) = (1/(\sigma_I(2\pi)^{1/2}))\,\exp(-(i - i_0)^2/(2\sigma_I^2))$$

i_0 = Mean value of the moment of inertia
σ_I = Standard deviation of the moment of inertia

$$f_\Omega(\omega_n) = (1/(\sigma_I(2\pi)^{1/2})\,\exp(-(i - i_0)^2/(2\sigma_I^2))|(2\rho l\,\omega_n)/(C_n^2 E)|$$
$$i = \omega_n^2 \rho l/C_n^2 E$$
$$i - i_0 = \omega_n^2 \rho l/C_n^2 E - \omega_{n0}^2\,\rho l/C_n^2 E$$
$$(i - i_0)^2 = (\rho l/C_n^2 E)^2\,(\omega_n^2 - \omega_{n0}^2)^2$$

Finally,

$$f_\Omega(\omega_n) = (1/(\sigma_I(2\pi)^{1/2})\,\exp(-(\rho l/C_n^2 E)^2\,(\omega_n^2 - \omega_{n0}^2)^2|/(2\sigma_I^2))\,(2\rho l\,\omega_n)/|(C_n^2 E) \tag{11.29}$$

Table 11.5 lists the magnitudes of each parameter for estimating natural frequency.

Natural frequency, $\omega_n = C_n\,((Ei)/(\rho l))^{1/2} = 3.52((3.0e07 \times 0.0021)/(0.05 \times 2))^{1/2} = 2794$ Hz

Figure 11.11 is the graphical description of Eq. (11.33) for natural frequency equal to 2794.

Modulus, psi	3.00E+07
Density, lb/in	0.05
Damping ratio	0.003
Moment of inertia, in	0.0021
Standard dev of moment of inertia	0.000021
Length, in	2
Constant, C	3.52

TABLE 11.5 List of Magnitudes of Each Parameter for Estimation of Natural Frequency

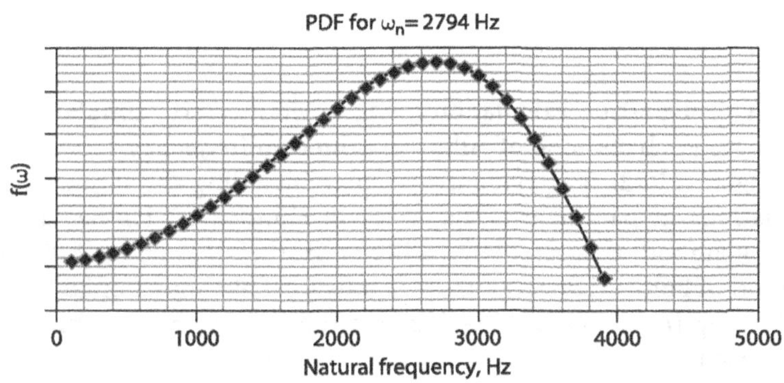

FIGURE 11.11 Probability density function.

The probabilistic analysis for natural frequency is given below.

A basic frequency, Eq. (28), of a beam is used to define a limit function:

$$\omega_n = C_n\,((EI)/(\rho l))^{1/2} = f(C_n, E, I, \rho, l)$$

$$G(\omega_{ntarget}, C_n, E, I, \rho, l) = C_n\,((EI)/(\rho l))^{1/2} - \omega_{ntarget} \tag{11.30}$$

$P_s = P(G(\omega_{ntarget}, C_n, E, I, \rho, l) >$ or $= 0)$

P_s represents the probability of natural frequency greater than $\omega_{ntarget}$. The likelihood of failure is defined as $P_f = 1 - P_s$.

Next the estimation of the likelihood of failure is conducted with assumed scatter of the parameters as listed in Table 11.6.

The plot of likelihood of failure versus frequency is given in Figure 11.12.

	Total % Deviation	Mean	Standard Dev	Minimum	Maximum
Moment of inertia, in^4	12	0.0021	0.000021	0.001974	0.002226
Modulus, psi	15	3.00E+07	375000	27750000	32250000
Density, lb/in	20	5.00E−02	0.00083333	0.045	0.055
Length, in	2	2	0.00333333	1.98	2.02
Damping ratio	15	0.003	0.0000375	0.002775	0.003225
Constant, C		3,52			

Table 11.6 Variation in Geometric and Material Properties

A probabilistic calculation using a Monte Carlo simulation is done for the likelihood of failure of achieving frequency. It is shown in Figure 11.12. It is evident that after including the influence of variations in parameters, frequency of about 2700 Hz has about 0.0 likelihood of failure while a life of 2850 Hz has about 1.00 like lihood of failure. However, there is only 0.50 likelihood of failure of achieving frequency of 2794 Hz as estimated by the deterministic approach.

Singh and Ewins (1988) developed equations for receptances of a spring mass damper system that closely resembles a bladed disc system. They used the model presented by Johnson and Bishop (1956).

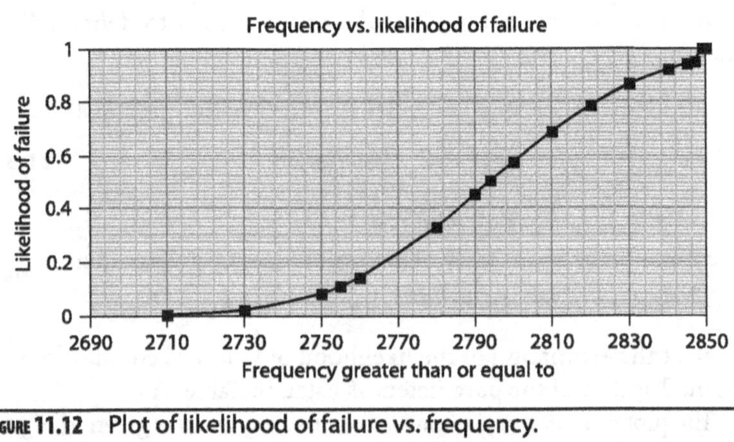

Figure 11.12 Plot of likelihood of failure vs. frequency.

They presented equations and results for random mistuned bladed discs, also for a single blade in both pdf and CDF forms. Singh (1992) used the concept to use the cross receptances to develop pdf and CDF for bladed disc system.

11.4.4 Probabilistic Aspect of Resonant Response of a Blade during Vibration

High-cycle fatigue failure is attributed to resonance of natural frequency and exciting frequency. This is an important condition that is checked for safe operation of a blade. Conditions of resonance for a blade and bladed disc were discussed in detail in an earlier chapter. Resonance of blades may increase the gas stress many times over. For reliability evaluation, however, stress needs to be determined. The basic logic has been that the bending stress caused by the gas loading on the blade may increase many folds depending on the dynamic characteristics of the rotating blade. This was seen through the simple example of the spring-mass-damper example. The analysis of a bladed disc system is complicated. However, designers have used a simple formula to design blades.

A simple Eq. (11.31) to estimate blade vibratory stress was given and is stated below:

$$\sigma_{vib} = S\,(\pi/\delta)\,(RRF)\,\sigma_{hp} \tag{11.31}$$

where σ_{vib} = calculated blade vibratory stress

σ_{hp} = steady gas bending stress in the blade

S = stimulus factor; the percent of the gas force that is unsteady (ranges from 0.1 to 0.3)

π/δ = resonant amplification factor. If a critical damping ratio of 0.003 is used, then the log decrement damping (δ) is equal to $.003(2\pi) = .01885$ and the resonant amplification factor is equal to $\pi/.01835 = 166.66$

RRF = resonant response factor depending on the mode of vibration and other factors (ranges from 0 to 1). Weaver and Prohl (1956) defined the resonant response factors (RRF) that range from 0.1 to 0.9. RRF depends on the mode shape of the blade vibration and phasing of the exciting force, which is consistent with the concept of the Safe diagram. In the caof mode shape and shape of the forcing being identical, RRF will equal 1.0. Any value of RRF less than 1.0 signifies a forced response.

Example: A 5-in-long blade is designed to carry a gas load of 1000.0 lb. The other parameters required to estimate vibratory stress are listed below:

$$
\begin{aligned}
P &= 1000.0 \text{ lb} \\
L &= 5.0 \text{ in} \\
c &= 1.0 \text{ in (distance of outer fiber)} \\
I &= 1.50 \text{ in}^4 \text{ (moment of inertia of cross section)} \\
S &= 0.2 \\
\xi &= 0.0038 \\
\delta &= 2 \times 3.14 \times 0.0038 = 0.023876 \\
RRF &= 0.5
\end{aligned}
$$

Static and vibratory stresses are estimated below:

$$\sigma_{hp} = PLc/I = (1000.0)(5.0)(1.0)/1.50 = 3333.33 \text{ psi}$$

Deterministic estimation of σ_{vib}:

$$
\begin{aligned}
\sigma_{vib} &= S\,(\pi/\delta)(RRF)(PLc/I) \\
&= (0.2)(3.14/0.023876)(0.5)(3333.33) \\
&= 43859.63 \text{ psi}
\end{aligned}
$$

Probabilistic analysis of σ_{vib}:

$$
\begin{aligned}
\sigma_{vib} &= S\,(\pi/\delta)\,(RRF)\,\sigma_{hp} = S\,(\pi/\delta)\,(RRF)\,(PLc/I) \\
&= f(S, \delta, RRF, P, L, c, I)
\end{aligned}
$$

$$
\begin{aligned}
G(\sigma_{vibTarget}, S, \delta, RRF, P, L, c, I) &= S\,(\pi/\delta)\,(RRF)\,(PLc/I) - \sigma_{vibTarget} \\
P_s &= P(G(\sigma_{vibTarget}, S, \delta, RRF, P, L, c, I) > \text{ or } = 0)
\end{aligned}
\tag{11.32}
$$

P_s represents the probability of vibratory stress greater than $\sigma_{vibTarget}$. The likelihood of failure is defined as $P_f = 1 - P_s$.

Next, the likelihood of failure estimation of the above quantities is conducted with scatter of the parameters as listed in Table 11.7.

The plot of likelihood of failure versus resonant stress is given in Figure 11.13.

A probabilistic calculation using Monte Carlo simulation is done for the likelihood of failure of resonant stress. It is shown in Figure 11.13. It is evident that after including the influence of variations in parameters, resonant stress of about 40,000 psi has about 0.0 likelihood of failure whereas a stress of 48,000 psi has about 1.00 likelihood of failure. However, there is only 0.50 likelihood of failure of

	% Total Variation	Mean	SD	Minimum	Maximum
Moment, M	30	5000	125	4250	5750
Length, L	2	5	0.008333	4.95	5.05
Distance of outer fiber, c	2	1	0.001667	0.99	1.01
Moment of inertia, I	10	1.5	0.0125	1.425	1.575
Stimulus, S	5	0.2	0.000833	0.195	0.205
Damping ratio, ξ	25	0.0038	7.92E−05	0.00333	0.004275
Resonance response factor, RRF	20	0.5	0.008333	0.45	0.55

TABLE 11.7 Scatter of the Parameters

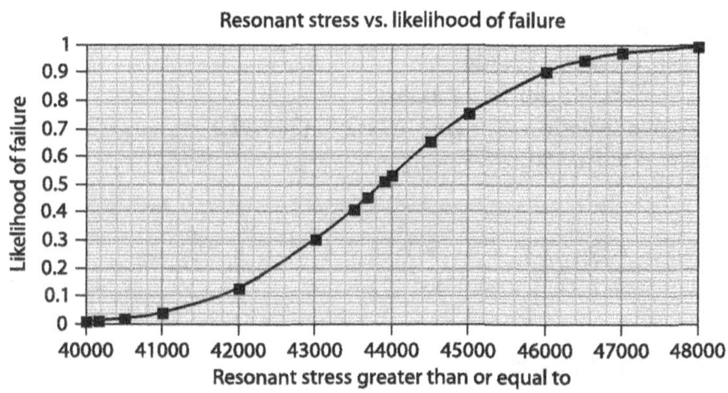

FIGURE 11.13 Plot of likelihood of failure vs. resonant stress.

achieving a resonant stress of 43,859 psi as estimated by the deterministic approach.

11.4.5 Effect of Blockage due to Catalyst on Blade's Stresses

Blockage of inlet passages of expanders owing to catalyst seems to be a common phenomenon. Its effect on the performance of an expander has been noted by designers and operators. This phenomenon also affects vibration characteristics of blades. Effect of blockage on the blade stress was discussed in earlier chapter. Due to change in the forcing pattern, blade dynamic stresses may increase many folds. This increase in dynamic stresses will influence the resulting fatigue life of blade. A method outlined by Kroon (1940) to calculate a magnifier is in case of partial admission for steam turbines. Dynamic stresses are estimated by multiplying the magnifier to the static stresses. This method was discussed in the previous chapter. A method can also be applied to estimate dynamic stresses due to blockage in expander. A simplified equation of a magnification factor was obtained by analyzing the behavior of a blade by using a single spring-mass-damper model.

The equation for magnifier due to blockage is given in Eq. (11.33). Where n is the number of cycles of vibration, ω is the natural frequency, N is the rotor speed, n_1 is the cycle of vibration in the unloaded portion, n_2 is the loaded portion, y_{osc} is the amplitude of vibration, y_{st} is the amplitude owing to a steady load and δ is the logarithmic damping:

$$n = 60\omega/N$$
$$N_1 = \text{number of blades}$$
$$N_2 = \text{number blocked passage}$$
$$\text{Arc of admission, } \theta = (N_1 - N_2) \times (360/N_1)$$
$$n_2 = \theta \times (n/360)$$
$$n_1 = n - n_2$$

$$\mu_1 = y_{osc} / y_{st} = (0.5 + (2 + e^{-n1\delta} + e^{-n2\delta}) / (2(1 - e^{-n\delta})\, e^{\delta/2})) \tag{11.33}$$

Deterministic Analysis

To demonstrate the method, the required inputs (e.g., number of blades, natural frequency of blade, percent blockage, and speed of the blade) are listed in Table 11.8.

Natural frequency, hz	1100
Rotor speed, RPM	4500
Number of blades	60
# of Blocked passage	3
log decrement, δ	0.01885

TABLE 11.8 Input Variables for Kroon's Method

$$n = 60\omega/N = (60)(1100)/4500 = 14.67$$
$$\theta = (N_1 - N_2) \times (360/N_1) = (60 - 3)/(360/60) = 342 \text{ degrees}$$
$$n_2 = \theta \times (n/360) = (342)(14.67/360) = 13.93$$
$$n_1 = n - n_2 = 14.67 - 13.93 = 0.74$$
$$\mu_1 = (0.5 + (2 + e^{-n1\delta} + e^{-n2\delta})/(2(1 - e^{-n\delta})\, e^{\delta/2}))$$
$$= (0.5 + (2 + e^{-(0.74)(0.01885)} + e^{-(13.93)(0.01885)})$$
$$/(2(1 - e^{-(14.67)0.01885)})\, e^{-(0.01885/2)}))$$
$$= 8.20$$

The response is expected to get magnified 8.2-fold.

Probabilistic Analysis

$$\mu_1 = (0.5 + (2 + e^{-n1\delta} + e^{-n2\delta})/(2(1 - e^{-n\delta})\, e^{\delta/2}))$$

$$G(\mu_{1target}, \omega, N, N_1, N_2, \delta) = (0.5 + (2 + e^{-n1\delta} + e^{-n2\delta}) /(2(1 - e^{-n\delta})\, e^{\delta/2})) - \mu_{1target} \quad (11.34)$$

$P_s = P(G(\mu_{1target}, \omega, N, N_1, N_2, \delta) > \text{or} = 0)$

P_s represents the probability of magnifier greater than $\mu_{1target}$.
The likelihood of failure is defined as $P_f = 1 - P_s$.

Next, the likelihood of failure estimation of the above quantities is conducted with scatter of the parameters as listed in Table 11.9.

The plot of likelihood of failure versus magnifier is given in Figure 11.14.

A probabilistic calculation using a Monte Carlo simulation is done for the likelihood of failure of achieving magnifier. It is shown in Figure 11.14. It is evident that after including the influence of variations in parameters, a magnifier of about 7.5 has about 0.0 likelihood of failure while a magnifier of 8.88 has about 1.00 likelihood of failure. However, there is only 0.50 likelihood of failure of achieving magnifier of 8.2 as estimated by the deterministic approach.

11.4.6 Effect of Blade Tip Rubbing due to Deposit of Catalyst

It was discussed in an earlier chapter that during an expander's operation, catalyst may deposit on the inner surface of the sealing shroud of the stage. After sufficient buildup of the deposit, the gap between tip of the blade and the sealing surface may close. As buildup continues the blade tip starts to rub with the hardened deposit. This rubbing action also results in the loss of blade material at the tip. The result

	% Total Variation	Mean	Standard Dev	Minimum	Maximum
Rotor speed, N	10	4500	37.5	4275	4725
Blade freq, ω	20	1100	18.33333	990	1210
Number of blades, N_1	0	60	0	60	60
Number of blocked Passages, N_2	66.67	3	0.166667	2	4
log decrement, δ	20	0.01885	0.000314	0.016965	0.020735

Table 11.9 Scatter of the Parameters

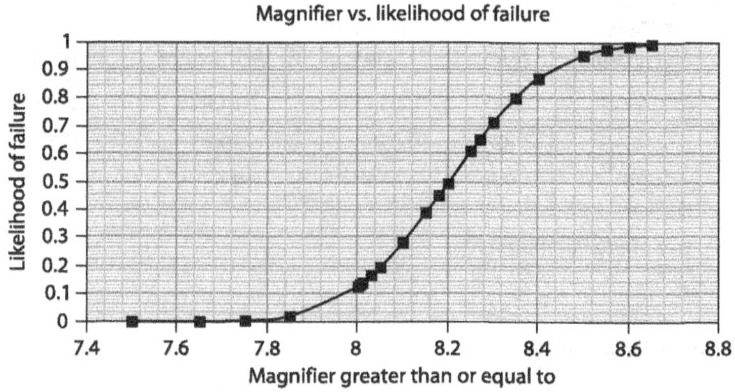

FIGURE 11.14 Plot of likelihood of failure vs. magnifier.

is that the blade gets shorter in each revolution. The effect of a shortened length of the blade affects stresses (centrifugal and bending) and natural frequency of the blade. It should be realized that this may result in changing life due to fatigue. This section explores the effects of any scatter in the assumed loss of length on a blade. Depending on the airfoil shape, the analysis becomes involved. It is usually done by measuring the loss. However, to demonstrate the effect a simple beam formula has been used.

If the reduction in length after rubbing is ΔL, then the reduced length and final pitch diameter are given by

$$L_1 = L - \Delta L$$
$$D_1 = D - \Delta L$$

Let us assume L equals 15 in and D equals 30 in.
For $\Delta L = 0.5$ in

Effect on Centrifugal Stress

The expected reduction in centrifugal stress is given by

$$\text{Ratio} = ((L - \Delta L)(D - \Delta L))/(L\,D) \tag{11.35}$$

% decrease in centrifugal stress $= 100 \times (1 - \text{Ratio}) = 100 \times (1 - ((L - \Delta L)(D - \Delta L))/(L\,D))$

Deterministic Estimation

Percent decrease in stress $= 100 \times (1 - ((15 - 0.5) \times (30 - 0.5))/(15 \times 30) = 4.94\%$

	% Total Variation	Mean	Standard Dev	Minimum	Maximum
L, In	1	15	0.0125	14.925	15.075
D, in	1	30	0.025	29.85	30.15
ΔL, in	25	0.5	0.010417	0.4375	0.5625

TABLE 11.10 Scatter of the Parameters

Probabilistic Analysis

$$\text{Ratio} = ((L - \Delta L)(D - \Delta L))/(L\,D)$$

$$\text{Percent decrease in stress, cfstress} = 100 \times (1 - \text{Ratio}) = 100 \times (1 - ((L - \Delta L) \times (D - \Delta L))/(L\,D))$$

$$G(\text{cfstress}_{target}, L, \Delta L, D) = 100 \times (1 - ((L - \Delta L)(D - \Delta L))/(L\,D)) - \text{cfstress}_{target} \quad (11.36)$$

$P_s = P(G(\text{cfstress}_{target}, L, \Delta L, D) > \text{or} = 0)$

P_s represents the probability of % decrease in centrifugal stress greater than cfstress_{target}.

The likelihood of failure is defined as $P_f = 1 - P_s$.

Next, the likelihood of failure estimation of the above quantities is conducted with scatter of the parameters as listed in Table 11.10.

The plot of likelihood of failure versus σ_{cent} is given in Figure 11.15.

A probabilistic calculation using a Monte Carlo simulation is done for the likelihood of failure of % decrease in centrifugal stress. It is shown in Figure 11.15. It is evident that after including the influence

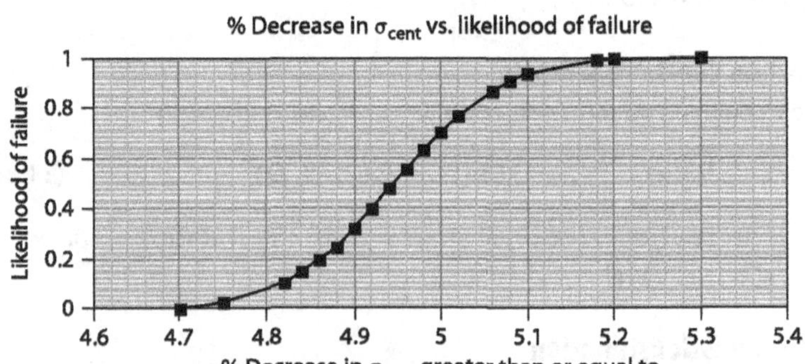

FIGURE 11.15 Plot of likelihood of failure vs. centrifugal stress.

of variations in parameters, % decrease in σ_{cent} of about 4.7% has about 0.0 likelihood of failure while % decrease in σ_{cent} of 5.3% has about 1.00 likelihood of failure. However, there is only 0.50 likelihood of failure of achieving % decrease in σ_{cent} of 4.94% as estimated by the deterministic approach.

Effect on Bending Stress

The stress arising from the gas load is bending stress on the blade. This stress is normally low in magnitude. The dynamic stress may become many folds under fluctuating exciting force. This change must be considered for high-cycle fatigue damage.

Expression of bending stress of a beam (where M is the moment of force, c is the distance of extreme fiber from neutral axis, and I is the moment of inertia of the cross section) is given below:

$$\sigma = Mc/I$$

$$M = PL/2$$

$$\sigma = PLc/2I$$

After rubbing the length is $L_1 = L - \Delta L$

$$\sigma_1 = PL_1c/2I$$

$$\text{Ratio} = L_1/L = (L - \Delta L)/L = 1 - \Delta L/L \qquad (11.37)$$

% decrease in stress $= 100 \times (1 - \text{Ratio}) = 100 \times (\Delta L/L)$

Deterministic Estimation

$$\text{Percent decrease in stress} = 100 \times (\Delta L/L) = 100 \times (0.5/15) = 3.333\%$$

For $\Delta L = 0.5$ in and $L = 15$ in, the % decrease in bending stress is 3.333%.

Probabilistic Analysis

$$\text{Ratio} = L1/L = (L - \Delta L)/L = 1 - \Delta L/L$$

$$\text{Percent decrease in bending, bnstress} = 100 \times (1 - \text{Ratio}) = 100 \times (\Delta L/L)$$

$$G(\text{bnstress}_{target}, L, \Delta L, D) = 100 \times (L, \Delta L) - \text{bnstress}_{target} \qquad (11.38)$$

$P_s = P(G(\text{bnstress}_{target}, L, \Delta L) > \text{or} = 0)$

P_s represents the probability of % decrease in bending stress greater than bnstress_{target}.

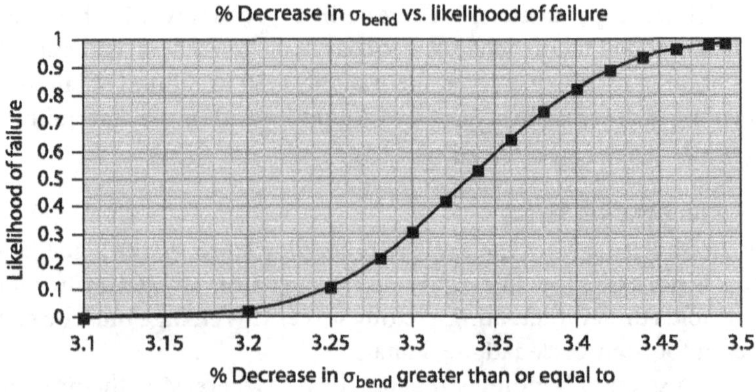

FIGURE 11.16 Plot of likelihood of failure vs. bending stress.

The likelihood of failure is defined as $P_f = 1 - P_s$.

The plot of likelihood of failure versus σ_{bend} is given in Figure 11.16.

A probabilistic calculation using a Monte Carlo simulation is done for the likelihood of failure of % decrease in bending stress. It is shown in Figure 11.16. It is evident that after including the influence of variations in parameters, % decrease in σ_{bend} of about 3.1% has about 0.0 likelihood of failure while % decrease in σ_{bend} of 3.5% has about 1.00 likelihood of failure. However, there is only 0.50 likelihood of failure of achieving % decrease in sbend of 3.33% as estimated by the deterministic approach.

Effect on Natural Frequency

The change in length also affects natural frequency of blade. This may become important for resonance condition for high-cycle fatigue. Natural frequency of a beam is given by the following expression:

$$\text{Change in length} = \Delta L$$
$$\text{Final length} = L - \Delta L$$

Ratio of Frequency

$$\omega'/\omega = (L/(L - \Delta L))^{5/2} = 1/(1 - \Delta L/L)^{5/2} \tag{11.39}$$

$$\%\text{ increase in frequency} = 100 \times (\text{ratio} - 1) = 100 \times (1/(1 - \Delta L/L)^{5/2} - 1)$$

Deterministic Analysis

For $\Delta L = 0.5$ in and $L = 15$ in, the % increase in natural frequency is 8.85%.

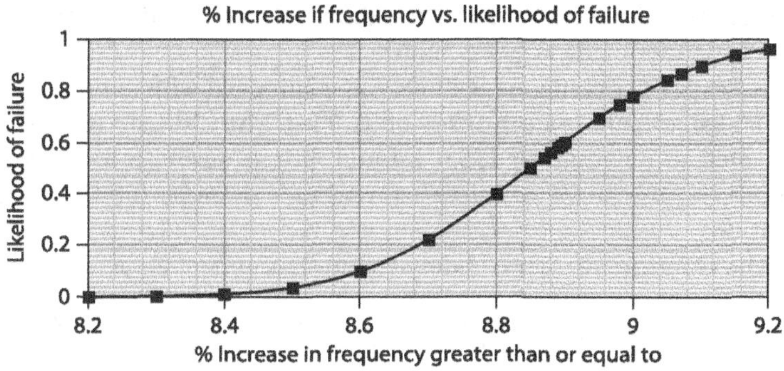

FIGURE 11.17 Plot of likelihood of failure vs. frequency.

Probabilistic Analysis

Percent increase in frequency, freq $= 100 \times (\text{Ratio} - 1)$
$= 100 \times (1/(1 - \Delta L/L)^{5/2} - 1)$

$$G\,(\text{freq}_{\text{target}}, L, \Delta L) = 100 \times (1/(1 - \Delta L/L)^{5/2} - 1) - \text{freq}_{\text{target}} \tag{11.40}$$

$P_s = P\,(G\,(\text{freq}_{\text{target}}, L, \Delta L) > \text{or} = 0)$

P_s represents the probability of % increase in frequency greater than $\text{freq}_{\text{target}}$.

The likelihood of failure is defined as $P_f = 1 - P_s$.

The plot of likelihood of failure versus frequency is given in Figure 11.17.

A probabilistic calculation using a Monte Carlo simulation is done for the likelihood of failure of achieving frequency. It is shown in Figure 11.17. It is evident that after including the influence of variations in parameters, frequency of about 2700 Hz has about 0.0 likelihood of failure while a life of 2850 Hz has about 1.00 likelihood of failure. However, there is only 0.50 likelihood of failure of achieving frequency of 2794 Hz as estimated by the deterministic approach.

11.4.7 Effect of Scatter in Stress and Material Property on Factor of Safety

Risk or reliability estimation for hcf loading is traditionally done through a Goodman equation. This method was explained in an earlier chapter. An expander blade is used as an example; however, the technique can be applied to any mechanical component for hcf loading. In the presence of scatter in the parameters, the factor of safety estimation will have a range with a different likelihood. An example of an expander blade is given below.

An expander blade made of IN 738 is imposed with a mean stress of 20.0 ksi and alternating stress of 32.0 ksi. The ultimate strength is estimated to be equal to 135.0 ksi and fatigue strength to be equal to 60.0 ksi. The factor of safety estimation is done by the Goodman equation.

The Goodman factor of safety equation is given below:

$$FS = (\sigma_a/\sigma_e + \sigma_m/\sigma_{ult})^{-1} \tag{11.41}$$

Deterministic Analysis

$$FS = (\sigma_a/\sigma_e + \sigma_m/\sigma_{ult})^{-1} = (32/60 + 20/135)^{-1} = 1.467$$

Probabilistic Analysis

Figure 11.18 describes graphically the influence of scatter in stresses and material properties on the factor of safety.

The limit function G for this case can be written as

$$G\,(FS_T, \sigma_a, \sigma_e, \sigma_m, \sigma_{ult}) = (\sigma_a/\sigma_e + \sigma_m/\sigma_{ult})^{-1} - FS_T \tag{11.42}$$

where FS_T is a desired value for FS.

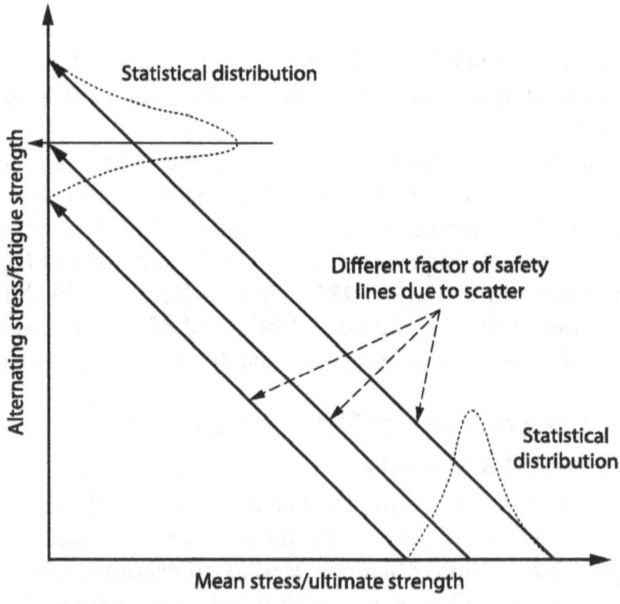

Figure 11.18 A probabilistic aspect of the Goodman diagram.

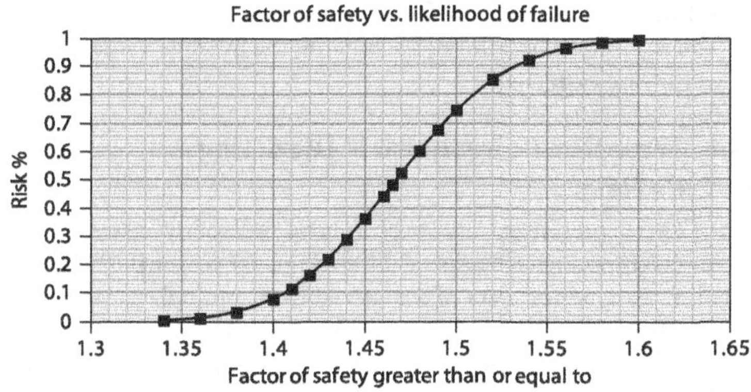

Figure 11.19 Plot of likelihood of failure vs. FS.

The probabilistic mathematical statement is stated as follows:

$$P_s = P\,(G\,(FS_T, \sigma_a, \sigma_e, \sigma_m, \sigma_{ult}) > \text{or} = 0) \tag{11.43}$$

P_s represents the probability of occurrence for a desired value of FS_T.

The likelihood of failure is defined as $P_f = 1 - P_s$.

Eq. (11.43) has been used to conduct a Monte Carlo–type analysis.

The magnitude of the likelihood of failure for each value of FS_T has been estimated and is shown in Figure 11.19.

Next, the likelihood of failure estimation of the above quantities is conducted with scatter of the parameters as listed in Table 11.11.

This simple example demonstrates that there is approximately 0.0 likelihood of failure of achieving a factor of safety equal to 1.34 while

	Total Variations, % of Mean	Mean	Standard Dev	Minimum	Maximum
Ultimate strength, ksi	30	135	3.375	114.75	155.25
Fatigue strength, ksi	40	60	1.5	51	69
Mean stress, ksi	30	20	0.5	17	23
Alternating stress, ksi	40	32	1.066666667	25.6	38.4

Table 11.11 Scatter of the Parameters

the likelihood of failure of a factor of safety of 1.6 is approximately 1.00. The likelihood of failure of achieving the average value of 1.46 is 0.50, as may be expected.

11.4.8 Probabilistic Treatment of Life during High-Cycle Fatigue

Singh (2001) presented the probabilistic reliability assessment of mechanical components. Figure 11.20 explains the reasoning behind this evaluation. Fatigue test data of a material from multiple samples when plotted individually as an S-N diagram shows a band of curves. The band represents the scatter of the test data—either multiple possible lives for a stress level or for the same number of cycles, there might be many stress levels representing variability in material. This variation results from the influence of variations in operating loads and geometrical tolerances, and also includes the inaccuracy in the estimation procedure.

The expression of life (N) was discussed and provided by Singh (2001). It is restated below in Eq. (11.48). Uncertainty in the stress

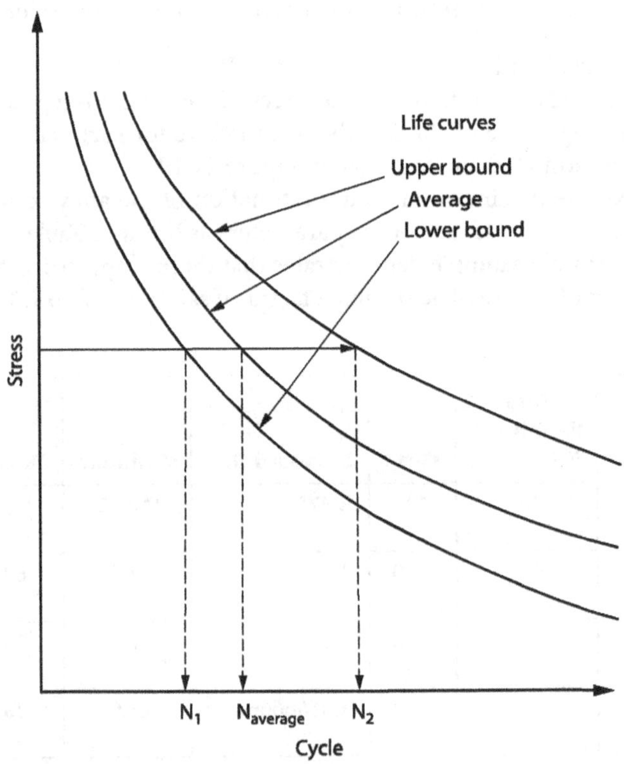

FIGURE 11.20 Depicting scatter of data in S-N curve.

due to variations in loading and the observed scatter in the material properties can be described by statistical distributions. After quantifying this uncertainty or scatter in the variables one can estimate the probability of hcf life having a prescribed "limit."

$$N = (\sigma_a/(\sigma_f' - \sigma_m))^{(1./b)} \text{ cycles} \tag{11.44}$$

Eq. (11.44) has been used to estimate the deterministic hcf life for the expander blade. Estimated value of σf′ equal to 185.0 ksi has been used.

Deterministic Estimation

$$N = (\sigma_a/(\sigma_f' - \sigma_m))^{(1./b)} = (32.0/(185.0 - 20.0))^{-1/0.062}$$
$$= 2.61 \times 10^{11} \text{ cycles}$$

Probabilistic Analysis

The limit function G can be written as

$$G(FN_{target}, \sigma_a, \sigma_f', \sigma_m, b) = (\sigma_a/(\sigma_f' - \sigma_m))^{(1./b)} - FN_{target} \tag{11.45}$$

The probabilistic mathematical statement can be stated as follows:

$$P_s = P\,(G\,(FN_{target}, \sigma_a, \sigma_f', \sigma_m, b) > \text{or} = 0) \tag{11.46}$$

P_s represents the probability of hcf life greater than FN_{target}.

The likelihood of failure is defined as $P_f = 1 - P_s$.

Since the stress magnitude and material properties are not single valued, a probabilistic-type analysis was performed to account for the variability. Table 11.12 lists the assumed variations of these parameters to be used in the probabilistic analysis.

	Total Variations, % of Mean	Mean	Standard Dev	Minimum	Maximum
Mean stress, ksi	30	20	0.5	17	23
Alternating stress, ksi	40	32	1.066666667	25.6	38.4
σf'	40	183	4.575	155.55	210.45
b	40	−0.062	−0.005166667	−0.031	−0.093

TABLE 11.12 Assumed Variation in Properties (Typical of IN 738)

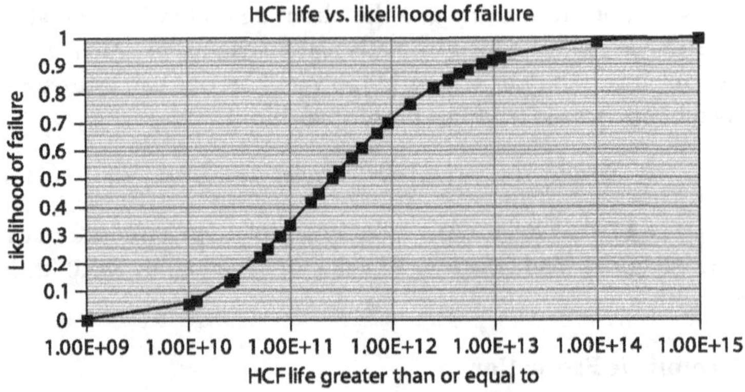

FIGURE 11.21 Likelihood of failure assessment of hcf life for blade.

Next, the likelihood of failure estimation of the above quantities is conducted with scatter of the parameters as listed in Table 11.12.

The plot of likelihood of failure versus HCF life is given in Figure 11.21.

The blade design was reanalyzed using a probabilistic analysis that incorporated the variations in properties identified in Table 11.12. The plot in Figure 11.21 shows the risk versus life cycle for design evaluation; this, in turn, can be compared with the deterministic result. The risk analysis gives a range of life from 1E+09 cycles to 1E+15 cycles. The estimated deterministic life of about 2.61E11 has about a 0.50 likelihood of failure.

11.4.9 Probabilistic Analysis of Life during Low-Cycle Fatigue

Low cycle is a phenomenon when imposed stresses are high enough to yield the material. In any mechanical structure, stresses may be very high in some localized area. Due to the cycling of applied load a crack may starts at these locations. Basic deterministic treatment was discussed in a previous chapter. A probabilistic type of analysis is presented here that will assign a likelihood of initiation of a crack after a certain number of cycles.

To account for the variability in material properties, it is customary to test multiple samples. Figure 11.22 shows the results depicting variations in Dε versus Nf data for the material. It shows that different samples might fail at different numbers of applied cycles even when the applied strain range is identical. Figure 11.22 depicts the possible variation resulting in statistical distribution of life for each strain level. Due to loading variation strain is expected to have variation that

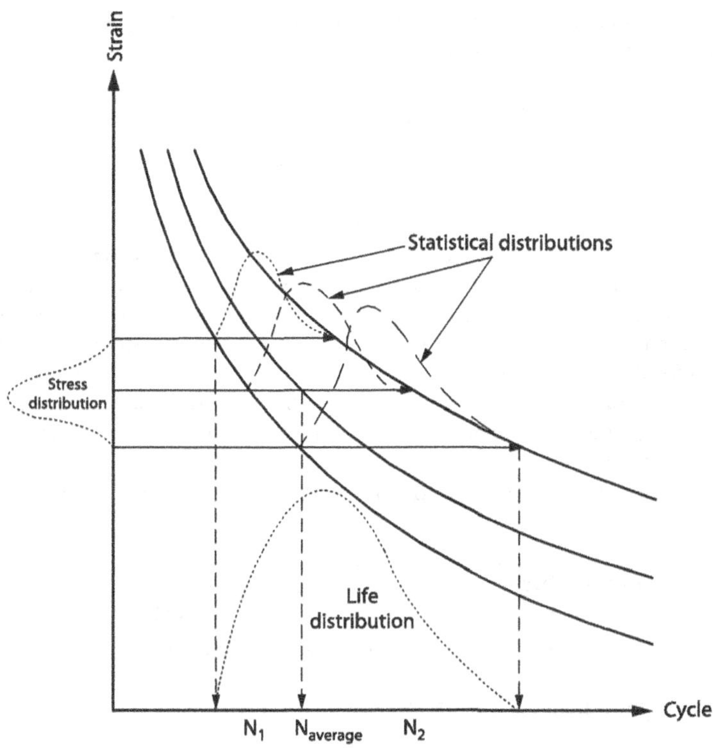

Figure 11.22 Depicting scatter of data in S-N curve.

is shown on the vertical axis. Due to these variations, the resulting life will have scatter and a possible statistical distribution, which is shown at horizontal axis.

The strain versus life relationship was discussed in an earlier chapter and is stated below.

$$\Delta\varepsilon_{Total}/2.0 \quad \text{or strain amplitude, } \varepsilon_{Total} = A\,N_f^b + C\,N_f^d \qquad (11.47)$$

The first term of the equation relates to the damage done due to a high-cycle fatigue and the second term is controlled by plastic strains that contribute to damage due to low-cycle fatigue. A second term will be used to estimate the lcf cycles.

$$\varepsilon_{plastic} = C\,N_f^d \qquad (11.48)$$

$$N_f = (\varepsilon_{plastic}/C)^{1/d} \qquad (11.49)$$

The limit function is given by the following:

$$G(N_{target}, \varepsilon_{plastic}, C, d) = (\varepsilon_{plastic}/C)^{1/d} - N_{target} \qquad (11.50)$$

The equation to estimate the probability of lcf life greater than a target life, N_{target}, is given below.

$$P_s = P(G(N_{target}, \varepsilon_{plastic}, C, d) > \text{or} = 0)$$

The likelihood of failure is defined as $P_f = 1 - P_s$.

Example: Singh et al. (2005) presented a case of assessing lcf life of an intake assembly for an expander. Due to localized high stress, there was a need to estimate the number of start-up and shut-down cycles beyond which there is a possibility of lcf crack initiation. A finite element model of the assembly was created, and analysis was performed to estimate stresses and strains. The FEA model for the component is shown in Figure 11.23. Figure 11.24 is the contour plot of stresses. At the location of high stress the inelastic or plastic strain was estimated. This analysis demonstrated a method to estimate lcf life for high load at high temperature.

The normal operating conditions data for the example case are as follows:

Inlet temperature	1400°F
Discharge temperature	1150°F
Nosecone flange inner ring temperature	800°F

Following loading conditions were considered:

- Mechanical loads—pressure, gravity, and piping
- Thermal loads with temperature gradient 1400°F/1100°F at intake flange and 1400°F/800°F at nose cone flange

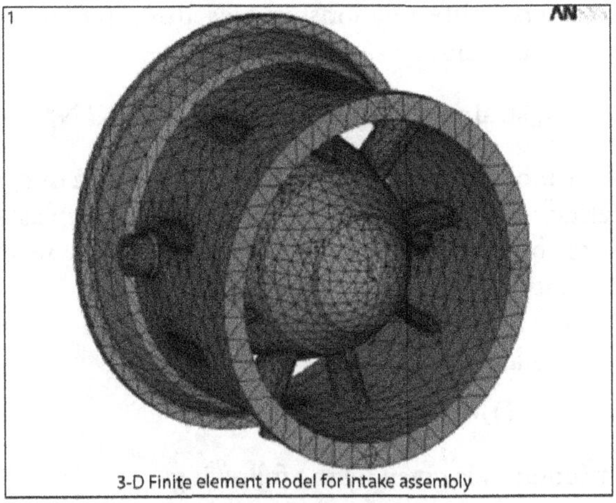

FIGURE 11.23 FEA model for the example intake assembly.

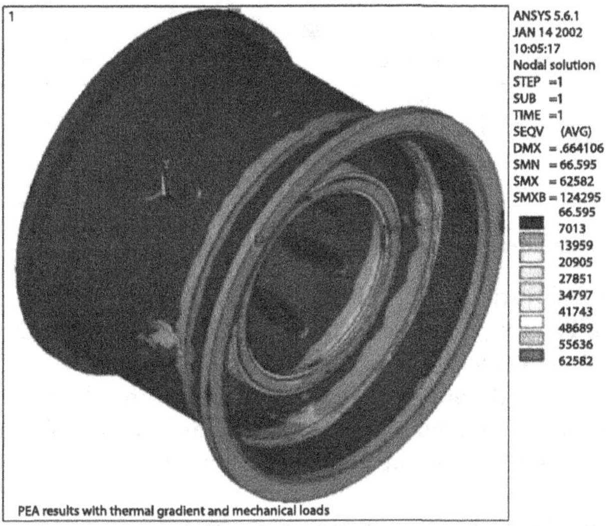

Figure 11.24 Equivalent stress under mechanical and thermal loading.

The estimated plastic strain from analysis was 0.6 in/in.

$$\text{Constant, } C = 36.19$$
$$d = -0.5946$$
$$N_f = (\varepsilon_{plastic}/C)^{1/d}$$
$$N_f = (0.6/36.19)^{-1/0.5946}$$
$$= 987.0 \text{ cycle}$$

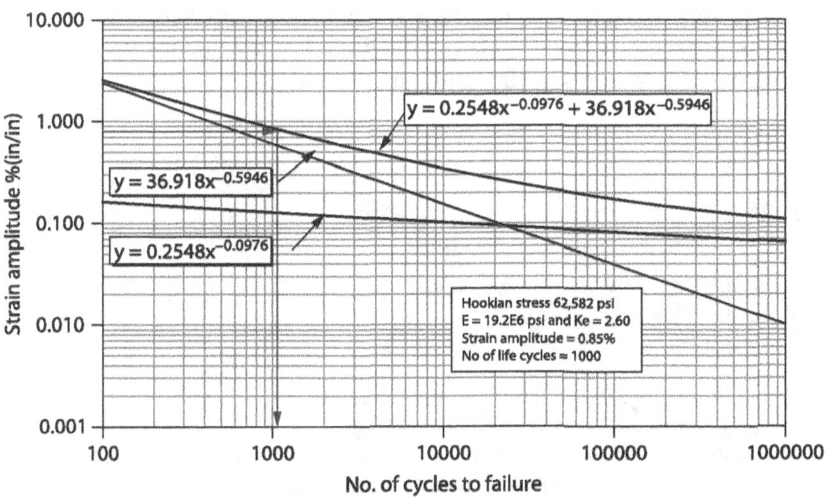

Figure 11.25 Life cycle chart for SS347/348 (0.004 in/s) (Singh, 2005).

	Mean	Total % Variation	Standrad Dev	Minimum	Maximum
% Plastic 8	0.6	20	0.010	0.54	0.66
Constant	36.19	30	0.905	30.7615	41.6185
Power	−0.5946	30	−0.015	−0.50541	−0.68379

TABLE 11.13 Scatter of the Parameters

A probabilistic LCF life evaluation is done using the following variation. For simplicity, estimation was done using only the plastic portion of the strain-life relation.

Next, the likelihood of failure estimation of the above quantities is conducted with scatter of the parameters as listed in Table 11.13.

The plot of likelihood of failure versus LCF cycle is given in Figure 11.26.

The intake casing design was reanalyzed using a probabilistic analysis that incorporated the variations in properties identified in Table 11.13. The plot in Figure 11.26 shows the risk versus lcf life cycle for that design evaluation, which can be compared with the deterministic result. The risk analysis gives a range of life from 650 cycles to 1650 cycles. The estimated deterministic life of about 987 has about 0.50 likelihood of failure.

11.4.10 Probabilistic Evaluation of Creep Rupture Life

Creep of material may occur at a certain combination of applied stress and temperature at the damage location. Treatment of creep was presented in the previous chapter. This section considers variation or scatter in parameters to estimate the likelihood of creep damage. We demonstrate the process in the following example.

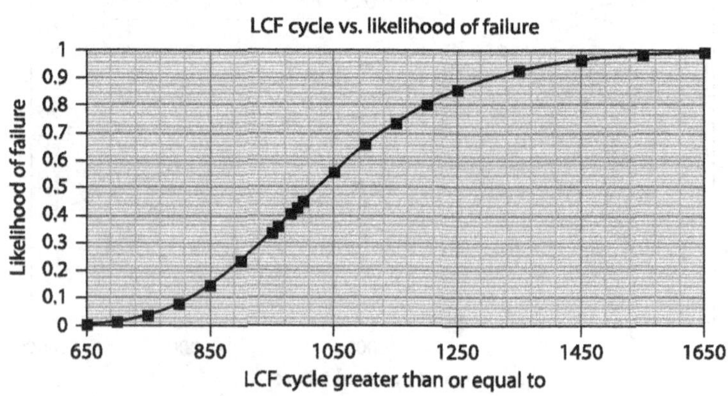

FIGURE 11.26 Likelihood of failure vs. operational cycle.

Example: An expander blade made of IN 738 is subjected to creep under simple tension at a stress of 39 ksi at 1268°F and the estimated rupture life is 323 years. Estimate creep rupture life is if the blade is continuously operated at a temperature of 1450°F.

Larson-Miller Method

$$t_{r(1268)} = 323 \times 365 \times 24\,\text{h} = 2{,}829{,}000\,\text{h}$$
$$\text{LMP} = (T + 460)\,(\log_{10} t_r + C)$$
$$= (1268 + 460)\,(\log_{10}(2{,}829{,}000) + 20) = 45{,}708$$

$$\log_{10} t_{r(1450)} = \text{LMP}/(T + 460) - C$$
$$= 45{,}708/(1450 + 460) - 20 = 3.93$$
$$t_{r(1450)} = 10^{3.93}\,\text{h} = 8529\,\text{h} = 0.97\ \text{years}$$

There is a significant reduction in creep rupture life for an increase of temperature from 1268°F to 1450°F.

It reduces from 323 years to 0.97 years.

To further explore the creep analysis, another example is provided next. An expander blade is subjected to a stress equal to 39.0 ksi at a temperature equal to 1269°F. Figure 11.27 shows a typical Larson-Miller curve for the material.

The relationship between LMP and stress is given below.

$$\text{LMP} = (0.0009\,\sigma^2 - 0.243\sigma + 54.3) \times 1000 \qquad (11.51)$$

and

$$\text{LMP} = (T + 460)\,(\log_{10} t_r + C) \qquad (11.52)$$

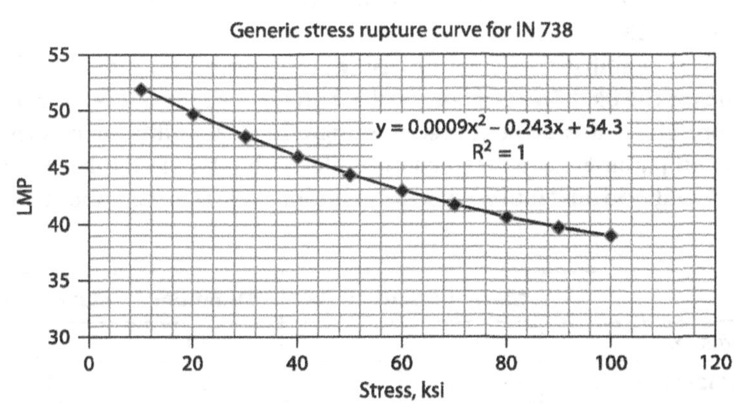

FIGURE 11.27 Typical Larson-Miller curves for IN 738.

Estimation of creep rupture life in hours (h).

$$(\log_{10} t_r + C) = LMP/(T + 460) = (1000.0)\,(0.0009\,\sigma^2 - 0.243\sigma + 54.3)/(T + 460)$$

$$\log_{10} t_r = (1000.0)\,(0.0009\,\sigma^2 - 0.243\sigma + 54.3)/(T + 460) - C$$

$$t_r = 10^{((1000)\,(0.0009\,\sigma 2 - 0.243\sigma + 54.3)/(T + 460)) - C} \quad (11.53)$$

Deterministic creep rupture life:

$$\begin{aligned} t_r &= 10^{((1000)\,(0.0009\,\sigma 2 - 0.243\sigma + 54.3)/(T + 460)) - C} \\ &= 10^{((1000)\,(0.0009(39)2 - 0.243(39) + 54.3)\,(1269.0 + 460)) - 20} \\ &= 10^{6.7} \\ &= 5.2 \times 10^6\,\text{h} \end{aligned}$$

Probabilistic Analysis

The limit function G can be written as:

$$G(t_{rtarget}, \sigma, T, C) = 10^{((1000)(0.0009\,\sigma 2 - 0.243\sigma + 54.3)/(T + 460)) - C} - t_{rtarget} \quad (11.54)$$

The probabilistic mathematical statement can be stated as follows:

$$P_s = P\,(G\,(t_{rtarget}, \sigma, T, C) > \text{or} = 0)$$

P_s represents the probability of creep rupture life greater than $t_{rtarget}$.

The likelihood of failure is defined as $P_f = 1 - P_s$.

Next the likelihood of failure estimation of the above quantities is conducted with scatter of the parameters as listed in Table 11.14.

The plot of likelihood of failure versus creep life is given in Figure 11.28.

A probabilistic calculation using a Monte Carlo simulation is done for the likelihood of failure of achieving creep life. It is shown in Figure 11.28. It is evident that after including the influence of variations in parameters, a creep rupture life of about 1.0E06 h has about 0.0 likelihood of failure whereas a life of 3.0E07 h has about 1.00

	Mean	Standard Dev	Minimum	Maximum
Stress, ksi	39	2	30	45
Temperature, °F	1269	12	1197	1341

TABLE 11.14 Scatter of the Parameters

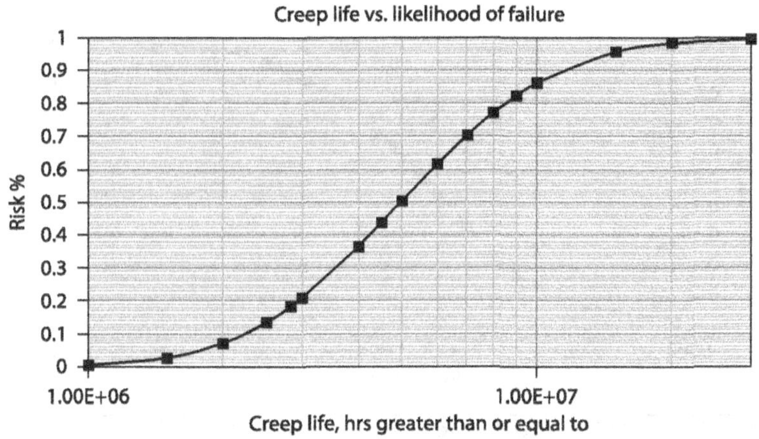

FIGURE 11.28 Creep rupture life (years) vs. likelihood of failure.

likelihood of failure. However, there is only 0.50 likelihood of failure of achieving a creep rupture life of 5.2E06 h as estimated by the deterministic approach.

11.5 Estimation of Business Risk

Method of estimation of likelihood of failure and its implication in design decision have been presented. It becomes important to note that in the presence of inherent scatter in parameters; there is always a possibility of failure, no matter how small it may be. In a business environment, this information is crucial to make business and operational decisions. Whether decision needs to be made during the purchase of new equipment or during an overhaul or during possible repair and replacement. Many times decisions need to be made on cost and life of equipment depending on how long it is expected to be operated.

The risk to business is dependent on the technical risk (likelihood of failure). Business has to deal with the consequences of any unscheduled failure. Many different scenario of consequence may exist depending on equipment type and operational method. Therefore, possible consequence of failure may depend on many factors including mentioned earlier. It is expected that any enterprise should be able to define it.

To explain this process, implication on the net income influenced by possibility of a blade failure due to hcf of an expander is explored. Analysis presented may be tailored to any mechanical component and different operational requirements.

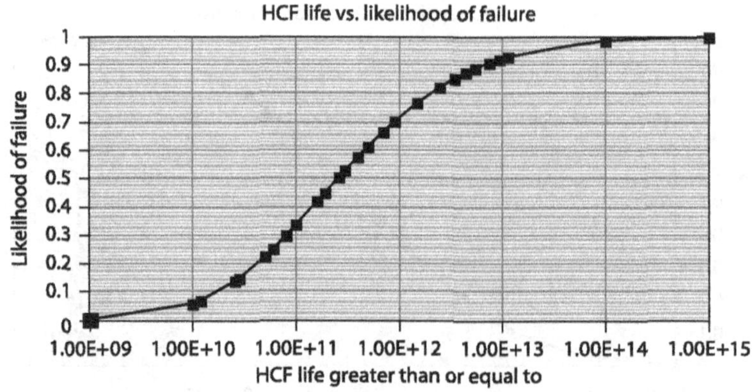

FIGURE 11.29 Likelihood of failure assessment of hcf life for blade.

In section 11.4.8, a result of estimation of likelihood of failure for hcf life of an expander blade has been presented in Figure 11.21. For this discussion it is presented in Figure 11.29.

Dynamic stress arising from the vibration of blade is the primary cause of high-cycle fatigue damage. The dynamic stresses may be due to resonance and/or forced response. Life in the number of days depends on the natural frequency of blade. Previous probabilistic life equations are adjusted to include the natural frequency that converts the number of cycles into year. Eq. (11.55) is given below.

If the natural frequency under consideration of a blade is ω (Hz), then life in days is given as follows:

$$N = (((\sigma_a/(\sigma_f' - \sigma_m))^{(1./b)})/(\omega \times 3600 \times 24) \text{ days}$$

The limit function G can be written as

$$G(N_{target}, \sigma_a, \sigma_f', \sigma_m, b, \omega) = (((\sigma_a/(\sigma_f' - \sigma_m))^{(1./b)}) /(\omega x3600x24) - N_{target} \tag{11.55}$$

The probabilistic mathematical statement can be stated as follows:

$$P_f = P\,(G\,(N_{target}, \sigma_a, \sigma_f', \sigma_m, b, \omega\,)) > \text{or} = 0)$$

P_f represents the probability of hcf life greater than N_{target}.

The likelihood of failure is defined as $P_f = 1 - P_s$

$$\omega = 3000 \text{ cycles/s}$$

Figure 11.30 is the plot of likelihood of failure versus days of operation.

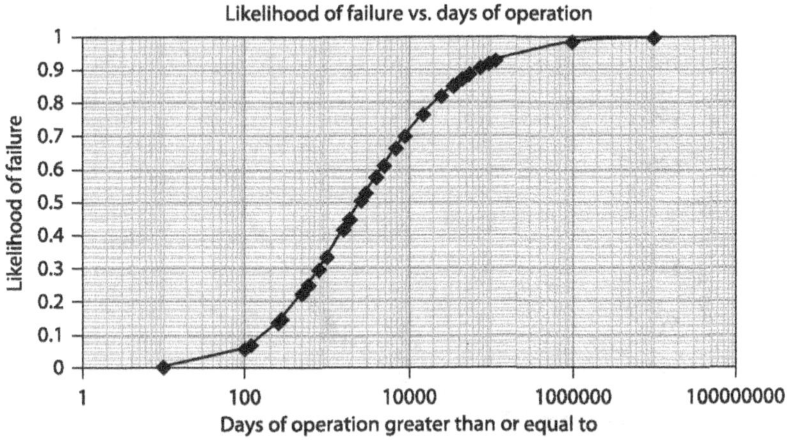

Figure 11.30 Likelihood of failure vs. days of operation.

$$\text{Business risk} = \text{The likelihood of failure} \times \text{Consequences of failure} - \text{Fixed cost of repair} \tag{11.56}$$

For this example, consequences of failure is taken as the cost of production loss per operational day and the fixed cost of repair includes labor cost plus cost of parts. The likelihood of failure is not multiplied to the fixed cost because it has to incur no matter when the blade fails. Traditionally, business risk is estimated at any duration of operation. This may be estimated after 1 year of service or 5 years of service. It should be noted that the likelihood of failure is a function of duration of operation and the accumulation of production loss.

Equation for business risk is modified and provided as Eq. (11.57).

$$\begin{aligned}\text{Business risk} = {} & (\text{The likelihood of failure as a function days of operation}) \times (\text{Days of operation} \\ & \times \text{production loss per day}) \\ & - \text{Fixed cost of repair}\end{aligned} \tag{11.57}$$

Results of the analysis is presented below. Through this for a given duration the expected income may be estimated (Figure 11.31).

For example, let us assume that the cost of production loss is \$1.5 million per day. Fixed cost is equal to \$5.0 million.

Therefore,

$$\begin{aligned}&\text{Business risk (function of days of operation)} \\ &\quad = (\text{The likelihood of failure as a function days of operation, Figure 11.30}) \\ &\quad\quad \times (\text{Days of operation} \times 1.5\text{ M}) - 5.0\text{M}\end{aligned}$$

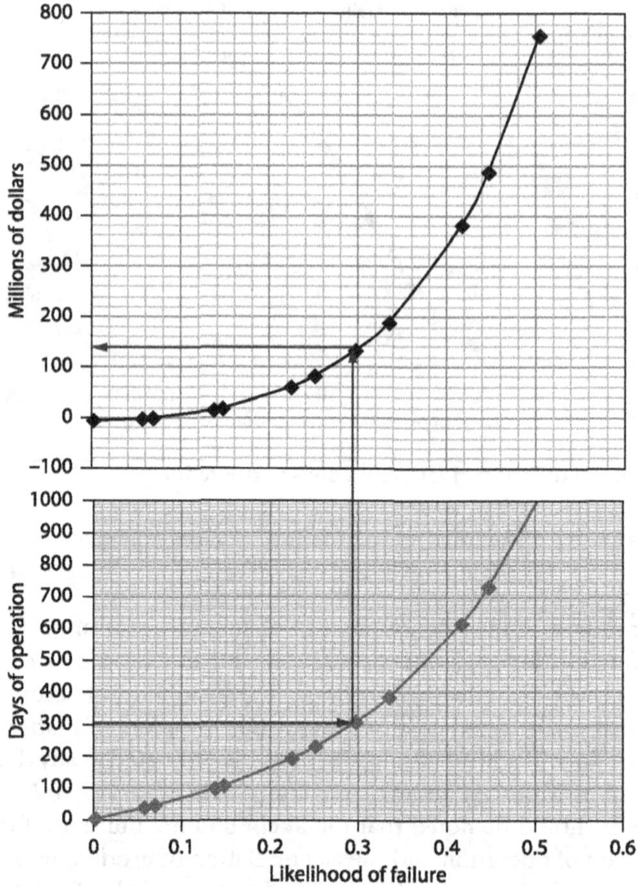

Figure 11.31 Estimation of business risk.

Information provided in Figure 11.30 is used to estimate business risk using Eq. (11.57). Depending on the duration of operational days, business risk will vary due to varying failure likelihood with time. Figure 11.31 provides the result of this analysis. This type of chart helps in making proper decisions based on what business risk an enterprise is willing to take.

Results of this demonstration example are provided as a simple nomograph, as shown in Figure 11.31. It is simple to use. For example, if one needs to estimate the business risk after 300 days of operation, then follow the lines from the lower portion of the graph with an arrow to find net possible income on the top portion.

11.6 Summary

The basic probabilistic theory has been explained to consider how a "true" analysis must incorporate the uncertainty inherent in any calculation. It is possible to quantify the likelihood of failure associated with a design. Thus, design decisions do not need to incorporate the factors of safety (factors of ignorance) to accommodate the uncertainty. Rather, the likelihood of failure may be quantified and design decisions made on the acceptable level of risk.

Risk can be managed when likelihood of failure is quantified and enhancements in analytical tools and capabilities make that determination a reality. The probabilistic analysis tools are available and should be used. They are real but require more application to grow and flourish. The time is right, and the needs are there. New frontiers of design are being sought from the basis of size, loads, speeds, pressures, materials, costs, and delivery times.

It is believed that this represents a significant change in demands on our design approach. The time seems to have arrived to integrate the probabilistic analysis into our design tool kit.

Bibliography

Balfoort, John P. "Power Recovery and Hot Gas Expanders." Proceedings of the Third Turbomachinery Symposium, 1974.

Bennett, E. " Predicting Turbomachinery Erosion Rates", *ANSYS Advantage* 5, no. 2 (2011).

Bennett, E., and A. Ivashchenko. " Predicting Erosion with CFD," *Turbomachinery International*, May/June 2011.

Betke, W. C. "FCCU Expanders: What's the Payout?" *Hydrocarbon Processing*, April 1980.

Betke, William C., Jr. Evaluation of Turbomachinery for the F.C.C. Process, Elliott Hot Gas Expander Conference, Elliott Company, 1967.

Binder, A., and R. Romey. "Secondary Flow Effects and Mixing of the Wake Behind a Turbine Stator," *Transactions of ASME* 105 (1983).

Cameron, Joseph. "Material Selection for Power Recovery Turbines," Elliott Hot Gas Expander Conference, Elliott Company, 1967.

Carbonetto, B., and G. Hoch. "Advances in Erosion Prediction of Axial Flow Expanders," Proceedings of the 28th Turbomachinery Symposium, 1999.

Castien, R. T. P. "Controlling FCCU Power Recovery Trains During a Breaker Trip," Hydrocarbon Processing, 1999.

Chen, Wen Jeng, and Edgar Gunter, Edgar. *Introduction to Dynamics of Rotor-bearing Systems* (Bloomington, IN: Trafford Publishing, 2005).

Childs, Dara W. *Turbomachinery Rotordynamics with Case Studies* (Minter: Spring, 2013).

Childs, Dara. *Turbomachinery Rotordynamic Phenomena, Modeling, and Analysis* (New York: Wiley Interscience, 1993).

Couch, K., L. Bell, and J. Yarborough. "Concepts for an Overall Refinery Energy Solution Through Novel Integration of FCC Flue Gas Power Recovery," NPRA Annual Meeting, March 2006.

Dowson, P., D. M Rishel, and N. S. Bornstein. "Factors and Preventive Measures Relative to the High Temperature Corrosion of Blade/Disk Components in FCC Power Recovery Turbines." Proceedings of the 24th Turbomachinery Symposium, 1995.

Geary, Carl. "Elliott Power Recovery Design Philosophy," Elliott Hot Gas Expander Conference, Elliott Company, 1967.

Hamed, A. "Influence of Secondary Flow on Turbine Erosion," ASME Paper 88-GT-264 (1988).

Hamed, A., W. Tabakoff, and R. Wenglarz. "Erosion and Deposition in Turbomachinery," *Journal of Propulsion and Power* 22, no. 2 (March-April 2006).

Ishii, J., and S. Honami. "A Three-Dimensional Turbulent Detached Flow with a Horsehoe Vortex," *Journal of Engineering for Gas Turbines and Power,* 1986, pp. 125–130.

Jeske, H. O., and H. Sandstede. "Turbine Blade Behavior in Gas Streams Containing High Dust Loads," Forschung im Ingenieurwesen Bs. 51

Kosters, Charles H. "Solid Particle Erosion," Elliott Hot Gas Expander Conference, Elliott Company, 1967.

Linden, D. H. "Catalyst Deposition in FCCU Power Recovery Systems," 7th Annual Fluid Cat Cracking Symposium, Venice, Italy, 1986.

Linden, D. H. "Catalyst Deposits in FCCU Power Recovery Systems Can Be Controlled," *Oil and Gas Journal*, December 15, 1986.

Linden, D. H. "Expander Blade Erosion Prediction," 3rd Expanders Users Conference, Houston, TX, 1995.

Linden, D. H. "Expander Field Remnant Life Analysis and Assessment," Proceedings of the Twenty-Fifth Turbomachinery Symposium, 1995.

Linden, D. H. "FCC Expander Erosion/Deposition Updates," 2nd Expanders Users Conference, Houston, TX, 1993.

Linden, D. H. "FCC Expander Blade Corrosion Testing," 3rd Expanders Users Conference, Houston, TX, 1995.

Linden, D. H. "FCC Expander Corrosion," 1st Expanders Users Conference, Houston, TX, 1991.

Linden, D. H. "Photographic Techniques for Monitoring Turbo Expanders," 2nd Expanders Users Conference, Houston, TX, 1993.

Linden, D. H., and C. F. Conroy. 1985, "Successful Application of Stand Alone FCCU Expander/Generator Sets," Ingersoll Rand, 1985.

Linden, D. H., and W. Mathers. "Operating Experience with a 13,000 SHP, Two-Stage FCC Power Recovery Expander," ASME Paper 84-GT -206 (1984).

Meyer, J. B. "Operating Experience with Multi-Stage Process-Gas Turbo Expander in FCC Plants," ASME IGT Symposium. Beijing, China, 1985.

Miller, H. "Blade Erosion in FCCU Power Recovery Expanders," Turbomachinery Maintenance Congress, Singapore, 1988.

Rao, J. S. *Rotor Dynamics* (New Age International, 1996).

Renard, Daniel J., "Application of Turbomachinery to F.C.C. Process," Elliott Hot Gas Expander Conference, Elliott Company, 1967.

Rucklye, I. J., "Power Recovery by a Flue Gas Turbo Expander on a Fluidized Catalytic Cracking Unit," C44/84, ImechE, 1984.

Samurin, Norman. "TH Testing at Elliott Company," Elliott Hot Gas Expander Conference, Elliott Company, 1967.

Schowalter, D., M. Agrawal, A. Chilka, and R. Russell. "Avoid Equipment Failure with Erosion Prediction Tools," Hydrocarbon Processing, 2006.

Schuster, C. "FCC Computer Simulation", Elliott Company, 1985.

Seib, D. C. "Stress Rupture Behavior of Waspaloy and IN-738LC at 600°C in Low-Oxygen Gaseous Environments Containing Sulfur," Superalloys 2000, TMS, 2000.

Seib, D. C., and F. S. Pettit. "Simulation Testing of Alloys and Coatings for Gas Expander Applications," Dresser Rand Technology Conference, 1997.

Stettenbenz, L. M. "Minimizing Erosion and Afterburn in Power Recovery Gas Expander," ASME Petroleum Mechanical Engineering and Pressure Vessels and Piping Conference, Denver, Colorado, 1970.

Sturley, R. A., H. F. Keller, and J. A. Cameron. "The Ranking of Power Recovery Turbine Blade Materials and Coatings for Resistance to Erosion," Elliott Co., 1983.

Vance, John H. *Rotordynamics of Turbomachinery* (New York: Wiley Interscience, 1988).

References

Allen, R. C. "Steam-Turbine Blading," *Transactions of ASME*, November 1940, pp. 689–710.

API Recommended Practice 684. API Standard Paragraphs Rotordynamic Tutorial: Lateral Critical Speeds, Unbalance Response, Stability, Train Torsionals, and Rotor Balancing, 2nd ed., American Petroleum Institute, August 2005.

API Standard 617. Axial and Centrifugal Compressors and Expander-Compressors for Petroleum, Chemical, and Gas Industry Services, 7th ed., American Petroleum Institute, July 2002.

Calandra, Leo. "FCC EXpander History," CONMEC, Inc., State of the Art Roundtable, 1998.

Carbonetto, B., and G. L. Hoch. "Advances in Erosion Prediction of Axial Flow Expanders." Proceedings of the 28th Turbomachinery Symposium, Turbomachinery Laboratory, Texas A&M University, College Station, TX, 1999, pp. 1–7.

Cencula, J., Wu, R., Ho, J., Singh, M., Cahoon, B., Higgins, E., 2006, "Turbo Expander Wheel Damage Analysis - A Case Study," Proceedings of the Turbomachinery Symposium, Turbomachinery Laboratory, Texas A&M University, College Station, Texas.

Coffin, L. F. Jr. "The Effect of Frequency on High-Temperature, Low-Cycle Fatigue." Proceedings of the Air Force Conference on Fatigue and Fracture of Aircraft Structures and Materials, AFFDL-TR-70-144, 1971, pp. 301–309.

Coffin, L. F. "A Study of the Effects of Cyclic Thermal Stresses on a Ductile Metal," *Transactions of ASME* 76 (1954), pp. 931–950.

Conway, J. B, R. H. Stentz, and J. T. Berling. "Fatigue, Tensile, and Relaxation Behavior of Stainless Steels," published by Technical Information Center, Office of Information Services, U.S. Atomic Energy Commission, 1975.

Davidson, T., Salamone, D., Gunter, E.J., 1992, "Rotor Bearing and Shaft Dynamics Redesign of a Double-Overhung Turboexpander for Reliability Improvement," Proceedings of the Twenty-First Turbomachinery Symposium, Turbomachinery Symposium, Turbomachinery Laboratory, Texas A&M University, College Station, Texas, pp. 35–56.

Dickert, E. J. "Process Overview and History of Power Recovery," Elliott Hot Gas Expander Conference, Elliott Company, 1967.

Drosjack, M. J., J. F. Felton, G. Seamon, and T. R. Griffin "What to Do when 'Classic' RCA Doesn't Give the Answer—Apply Detailed Engineering Analysis," Proceedings of the 35th Turbomachinery Symposium, 2006.

Drosjack, M., M. Singh, and C. Rutan. "Determining Risk Associated Inspection and Overhaul of Turbomachinery: Yesterday, Today, and Tomorrow," NPRA, Reliability and Maintenance Conference and Exhibition, New Orleans, LA, May 24–27, 2005.

Ewins, D. J. "A Study of Resonance Coincidence in Bladed Discs," *Journal of Mechanical Engineering Science* 12, no. 5 (1970).

Ewins, D. J. "Studies to Gain Insight into the Complexities of Blade Vibration Phenomena," Conference on Vibration in Rotating Machinery, ImechE, C184/76 (1976), pp. 165–172.

Ewins, D. J. "The Effects of Detuning upon the Forced Vibrations of Bladed Disk," *Journal of Sound and Vibration* 9, no. 1. (1969), pp. 65–79.

Ewins, D. J. "Vibrational Characteristics of Bladed Disk Assemblies," *Journal of Mechanical Engineering Science* 15, no. 3 (1973), pp. 165–185.

Geary, Carl H. "Power Recovery and the Eddy Current Brake," Elliott Hot Gas Expander Conference, Elliott Company, 1967.

Graham, J. A., K. J. Marsh, and F. J. Appl, eds. *Fatigue Design Handbook*, SAE, Pub. No. AE-4, Society of Automotive Engineers, Warrendale, PA, 1968.

Halford, G. R., M. H. Hirschberg, and S. S. Manson. "Temperature Effects on the Strain Range Partitioning Approach for Creep-Fatigue Analysis," *Fatigue at Elevated Temperatures,* ASTM STP-520, 1973, pp. 658–667.

Haugen, E. B. *Probabilistic Mechanical Design* (New York: John Wiley & Sons, 1980).

Johnson, D. C., and R. E. D. Bishop. "The Modes of Vibration of a Certain System Having a Number of Equal Frequencies," *Journal of Applied Mechanics, Transactions of ASME*, September 1956, pp 379–384.

Jumonville, J, "Tutorial on Cryogenic Turboexpanders," 2010, Proceedings of the Thirty-Ninth Turbomachinery Symposium, Turbomachinery Laboratory, Texas A&M University, College Station, Texas, pp. 147–154.

Kroon, R. P. "Turbine-Blade Vibration Due to Partial Admission," *Journal of Applied Mechanics,* 1940, p. A161.

Larson, F. R., and J. Miller. "A Time-Temperature Relationship for Rupture and Creep Stresses," *Transactions of ASME* 74, 1952, pp. 765–775.

Lipson, C., and N. J. Sheth. *Statistical Design and Analysis of Engineering Experiments* (New York: McGraw-Hill, 1973).

Manson, S. S., and G. R. Halford. "A Method for Estimating High-Temperature Low-Cycle Fatigue Behavior of Materials." Proceedings of International Conference on Thermal and High-Strain Fatigue, the Metals and Metallurgy Trust (London), 1967, pp. 54–70.

Manson, S. S., and G. R. Halford. *Fatigue and Durability of Structural Materials* (Materials Park, OH: ASM International, 2006).

Manson, S. S. "NASA TN2933," 1953.

Manson, S. S. "The Challenge to Unify Treatment of High-Temperature Metal Fatigue—A Partisan Proposal Based on Strain Range Partitioning," *Fatigue at Elevated Temperatures,* ASTM STP 520, 1973, pp. 744–782.

Manson, S. S., and G. R. Halford. "Practical Implementation of the Double Linear Damage Rule and Damage Curve Approach for Treating Cumulative Fatigue Damage," *International Journal of Fracture* 17 (1981), pp. 169–192.

Manson, S. S. *Experimental Mechanics* 5, no.7, (1965), p. 193.

Manson, S. S., G. R. Halford, and D. A. Spera. ,"The Role of Creep in High-Temperature Low-Cycle Fatigue," in *Advances in Creep Design*, ed. A. I. Smith and A. M. Nicholson (New York: Halsted Press, 1971), pp. 229–249.

Manson, S. S., G. R. Halford, and M. H. Hirschberg. "Creep-Fatigue Analysis by Strain Range Partitioning," in *Design for Elevated Temperature Environment*, ed. S. Y. Zamrik, (ASME, 1971), pp. 12–24, disc. pp. 25–28.

Miner, M. A. "Cumulative Damage in Fatigue," *Journal of Applied Mechanics* 12 (1945).

Mitchell, M. R. "Fundamentals of Modern Fatigue Analysis for Design," in *Fatigue and Microstructure* (Materials Park, OH: ASM International, 1979). pp. 385–437.

Ong, C., Cencula, J., Wu, R., Bielecki, M., Lelanno, M., 2011, "Turbo-Expander Compressor Active Magnetic Bearing Trips Reduction – A Case Study," Proceedings of the Fortieth Turbomachinery Symposium, Turbomachinery Laboratory, Texas A&M University, College Station, Texas, pp. 450–471.

Pigott, R. "Turbine Blade Vibration Due to Partial Admission," *International Journal of Mechanical Sciences* 22 (1980), pp 247–264.

Prohl, M. "A Method of Calculating Vibration Frequency and Stress of a Banded Group of Turbine Buckets," *Transactions of ASME*, Paper No. 56-A-116 (1956).

Sadeghbeigi, Reza. *Fluid Catalytic Cracking Handbook* (Houston: TX: Gulf Professional Publishing, 2000).

Siddall, J. N. *Probabilistic Engineering Design* (New York: Marcel Dekker, 1983).

Singh, M. P. "Predicting the Randomness of Forced Vibration Response of a Bladed Disc," ImechE Conference, Bath, UK, C432/128, 1992.

Singh, M. P., and D. J. Ewins. "A Probabilistic Analysis of a Mistuned Bladed Turbine Disc," ImechE Conference, Edinburgh, UK, C229/88, 1988.

Singh, M. P., and J. J. Vargo. "Reliability Evaluation of Shrouded Blading Using the SAFE Interference Diagram," *Journal of Engineering for Gas Turbine and Power* 111 (1989), pp. 601–609.

Singh, M. P., and M. Drosjack, M. 2008, "Emerging Advanced Technologies to Assess Reliability of Industrial Steam Turbine Blade Design." Proceedings of the 37th Turbomachinery Symposium, Turbomachinery Laboratory, Texas A&M University, College Station, TX, 2008, pp. 169–188.

Singh, M. P. "Probabilistic HCF Life Estimation of a Mechanical Component." Proceedings of ASME International Mechanical Engineering Congress and Exposition, New York, November 11–16, 2001.

Singh, M. P. 1985, "Turbine Blade Dynamics: A Probabilistic Approach," in *Vibrations of Blades and Bladed Disk Assemblies*, ASME Book No. H000335, 1985, pp. 41–48.

Singh, M. P., B. K. Thakur, and W. E. Sullivan. "Assessing Useful Life of Turbomachinery Components." Proceedings of the 34th Turbomachinery Symposium, Turbomachinery Laboratory, Texas A&M University, College Station, TX, 2005, pp. 177–192.

Singh, M. P., D. H. Donald, W. E. Sullivan, and J. Hudson. "Probabilistic Life Assessment of Impeller with Discontinuities." Proceedings of the 33th Turbomachinery Symposium, Turbomachinery Laboratory, Texas A&M University, College Station, TX, 2004, pp. 15–23.

Singh, M. P., J. E. Cencula, R. Wu, J. Ho, and E. Higgins. "Successful Solution to a Turboexpander Wheel Failure, Operating in an Erosive Environment of Geothermal Steam." Proceedings of MEMEC, Bahrain, November 4–7, 2007. Paper 07-438.

Singh, M. P., J. J. Vargo, D. M. Schiffer, and J. D. Dello. "Safe Diagram–A Design Reliability Tool for Turbine Blading." Proceedings of the 17th Turbomachinery Symposium, Turbomachinery Laboratory, Texas A & M University, College Station, TX, 1988, pp. 93–101.

Singh, M. P., T. Matthews, and C. Ramsey. "Fatigue Damage of Steam Turbine Blade Caused by Frequency Shift Due to Solid Buildup: A Case Study." Proceedings of the 23rd Turbomachinery Symposium, Turbomachinery Laboratory, Texas A&M University, College Station, TX, 1994, pp. 107–114

Singh, M., *Recent Advances in Solids and Structures*, ed. H. H. Chung and Y. W. Kwon, ASME PVP, vol. 381, Book No. G01095, 1998.

Singh, M., 2013, "History of Evolution, Progress and Application of Safe Diagram for Tuned and Mistuned Systems." Proceedings of the 42nd Turbomachinery Symposium, Turbomachinery Laboratory, Texas A&M University, College Station, TX, 2013.

Singh, M., and D. Schiffer. "Vibrational Characteristic of Packeted Bladed Disc." Presented at ASME Design Engineering Technical Conference, Washington, DC, September 1982, 82-DET-137.

Singh, M., and G. Lucas. *Blade Design and Analysis* (New York: McGraw-Hill, 2001).

Swanson, Erik, Maslen, Eric, Li, Guoxin, Cloud, C. Hunter, 2008, Rotordynamic Design Audits of AMB Supported Machinery, Proveedings of the Thirty Seventh Turbomachinery Symposium, pp. 133–158.

Tabakoff, W. "Turbomachinery Alloys Affected by Solid Particles," ASME Paper 88-GT-295 (1988).

Weaver, F., and M. Prohl. "High Frequency Vibration of Steam Turbine Buckets, ASME Paper No. 56-A-119 (1956).

Index

* Page numbers with an italic *f* indicate figures; page numbers with an italic *t* indicate tables.

D

E

F

T

U

V

W

Z